High Speed Digital Design

Design of High Speed Interconnects and Signaling

High Speed Digital Design

Design of High Speed Interconnects and Signaling

Hanqiao Zhang

Steven Krooswyk

Jeff Ou

AMSTERDAM • BOSTON • HEIDELBERG • LONDON
NEW YORK • OXFORD • PARIS • SAN DIEGO
SAN FRANCISCO • SINGAPORE • SYDNEY • TOKYO
Morgan Kaufmann is an imprint of Elsevier

Acquiring Editor: Todd Green
Editorial Project Manager: Lindsay Lawrence
Project Manager: Mohanambal Natarajan
Designer: Alan Studholme

Morgan Kaufmann is an imprint of Elsevier
225 Wyman Street, Waltham, MA 02451 USA

ISBN: 978-0-12-418663-7

British Library Cataloguing-in-Publication Data
A catalogue record for this book is available from the British Library.

Library of Congress Cataloging-in-Publication Data
Library of Congress Control Number: 2015939631

For information on all MK publications
visit our website at www.mkp.com

Contents

About the Authors/Contributors

ABOUT THE AUTHORS

Hanqiao Zhang is an Analog Engineer at Intel and holds a PhD degree in Electromagnetics and Microwave Engineering from Clemson University. Hanqiao joined Intel Xeon product electrical validation team in 2011, where he worked on generations of Intel high-speed digital systems. He developed methodologies for validating high-speed interfaces, such as PCI Express and Quick Path Interface (QPI). Hanqiao is now a signal integrity engineer with Intel Data Center Group. He is involved in mission-critical high-performance servers signal integrity design, bring up, validation and debug. Hanqiao is a regular contributor to journals including IEEE Transactions, Journal of Applied Physics and Applied Physics Letters. Hanqiao has 10 years of experience on novel passive RF/microwave component design.

Steven Krooswyk has been at Intel since 2003 when we joined as a signal integrity engineer for EPSD server development. In 2009, Steve transitioned into the signal integrity lead for PCI Express in Intel's Enterprise Platform Technology Division (EPTD). In addition to server products, his experience includes involvement in the PCI Express 3.0 and 4.0 specifications. He holds a B.S. and M.S. in electrical engineering from the University of South Carolina.

Jeff Ou joined Intel in 1999 as an analog design engineer in CMOS RF transceiver design. In 2006, Jeff transitioned to Xeon processor product design team in Server Development Group (SDG) developing a serial I/O module configurable for PCI Express and Quick Path Interface (QPI). Since then Jeff has been involved in several generations of Xeon products from design to post silicon validation. In 2012, Jeff was recognized as a tech lead in SDG, and continued to develop the cutting-edge high speed serial I/O modules for server products. Jeff holds a PhD degree in EECS from UC Berkeley and is a member of IEEE.

ABOUT THE CONTRIBUTORS

David Blankenbeckler is a Principal Engineer at Intel, where he has worked since 1995. He has spent the last 14 years doing I/O and memory electrical validation for Intel chipsets and CPUs. David graduated from Clemson University with a BS in Electrical Engineering and then obtained a MS in Management in Science and Technology at Oregon Graduate Institute. Today he lives in South Carolina where he continues to pursue his passion for family, technology, and his love-hate relationship with extreme endurance sports.

Tommy Cheung joined Intel in 1999 and is presently an Engineering Manager, with 15 years of experience in signal integrity and electrical validation of high speed interfaces. Tommy graduated from the University of Washington with a BS in Electrical Engineering, and also has an MBA from Babson College. Today he resides in Portland, Oregon, and outside of engineering, he spends his time on family, photography, and the culinary arts.

Pelle Fornberg is a Senior Staff Engineer at Intel Corporation, where he focuses on high-speed differential I/O validation for Intel client products. He joined Intel in 2002 after receiving the B.S. and M.S. degrees in Electrical Engineering from the University of Colorado at Boulder. Pelle currently lives in Portland, Oregon where he enjoys the many outdoor activities the area has to offer including skiing, snowboarding, mountain biking and sailing.

Kai Xiao is a technical lead in the Data Center Group at Intel Corporation, responsible for the signal integrity of the serial I/O interfaces on server platforms. He joined Intel at DuPont, WA, in 2005, after he received his Ph.D. degree from the EMC Laboratory of the Missouri University of Science & Technology (a.k.a. UMR). He received his BS and MS degrees, both in Electrical Engineering, from Tsinghua University in Beijing, in 1997 and 2000, respectively. His areas of interests include signal integrity, power integrity, and computational electromagnetics. He currently serves as the secretary of Technical Committee of Computational Electromagnetics (TC-9) of the IEEE EMC society.

Henry Peng is a Engineering Manager at Intel Corporation. His focus is in high speed off die interconnect for the high performance computing (HPC) segment. He joined Intel in 1997 after receiving a BS degree in Electrical Engineering at University of Washington (Seattle, Washington), and M.S. degrees in Electrical Engineering from the University of Cincinnati (Cincinnati, Ohio).

Jiangqi He is a principal engineer at Intel Corporation, data center group. His focus area includes power delivery, voltage regulator, and integrated voltage regulator at silicon, signal integrity and EMI/EMC. He obtained his B.S. in physics from Xiamen University, China and Ph.D. in electrical engineer from Duke university, 1992 and 2000 respectively. He owns more than 80 USA and international patents and published more than 50 peer reviewed papers.

CHAPTER

Transmission line fundamentals

1

All the mathematical sciences are founded on relations between physical laws and laws of numbers, so that the aim of exact science is to reduce the problems of nature to the determination of quantities by operations with numbers.
— James Clerk Maxwell

The chapter introduces electromagnetics and presents the origin physics of Maxwell's equations. Electromagnetic wave propagation equations in both free space and conductive media are derived. Transmission line theory laying the foundation for signal integrity analysis and interconnect design is discussed. Commonly used transmission lines in today's high-speed systems and new development trends for them are presented in the last section of the chapter.

BASIC ELECTROMAGNETICS

Starting with the introduction of integral and derivative forms of Maxwell's equations, four physical laws composing Maxwell's equations are explained. Four fundamental electromagnetic field vectors, building blocks of electromagnetics, are also presented. Lastly the propagation of electromagnetic waves is covered.

ELECTROMAGNETICS FIELD THEORY

Maxwell's equations

By introducing the concept of displacement current, Maxwell's summarized the famous equation set describing the electromagnetic phenomenon that electric field can induce magnetic field and vice versa. By combining the equations representing four electromagnetic physics laws, the integral form of Maxwell's equations are written as below:

$$\begin{cases} \oint_l \boldsymbol{H}\cdot d\boldsymbol{l} = \int_s \left(\boldsymbol{J}_c + \dfrac{\partial \boldsymbol{D}}{\partial t}\right)\cdot d\boldsymbol{S} \\ \oint_l \boldsymbol{E}\cdot d\boldsymbol{l} = -\int_s \dfrac{\partial \boldsymbol{B}}{\partial t}\cdot d\boldsymbol{S} \\ \oint_S \boldsymbol{D}\cdot d\boldsymbol{S} = \int_v \rho dv \\ \oint_S \boldsymbol{B}\cdot d\boldsymbol{S} = 0 \end{cases} \tag{1.1}$$

Symbols used are defined as follows: electric field intensity $\boldsymbol{E}$ (V/m), electric flux density $\boldsymbol{D}$ (C/m^2), magnetic field intensity $\boldsymbol{H}$ (A/m), and magnetic flux intensity $\boldsymbol{B}$ (T). ***Jc*** is the conducting current density (A/m^2), σ is the media conductivity (S/m). $\boldsymbol{E}$ and $\boldsymbol{D}$ and $\boldsymbol{B}$ and $\boldsymbol{H}$ are dependent. Relationships between $\boldsymbol{D}$ and $\boldsymbol{E}$, $\boldsymbol{J_c}$ and $\boldsymbol{E}$, and $\boldsymbol{B}$ and $\boldsymbol{H}$ in isotropic media are:

$$\boldsymbol{D} = \varepsilon \boldsymbol{E},\ \boldsymbol{B} = \mu \boldsymbol{H},\ \text{and}\ \boldsymbol{J}_c = \sigma \boldsymbol{E} \tag{1.2}$$

Electric field intensity $\boldsymbol{E}$ is defined as the electric force experienced by a unit positive charge in an electric field:

$$\boldsymbol{E} = \frac{\boldsymbol{F}}{q} \tag{1.3}$$

Symbol q in the equation is the quantity of charge on the test charge experiencing the force. $\boldsymbol{F}$ is the force experienced by the test charge.

Electric flux density $\boldsymbol{D}$ is used to define the electric field in dielectric materials where the dielectric can be polarized by an applied electric field. The induced dielectric polarization density is defined as $\boldsymbol{P}$:

$$\boldsymbol{D} = \varepsilon_0 \boldsymbol{E} + \boldsymbol{P} \tag{1.4}$$

where ε_0 is the electric permittivity of free space, $\varepsilon_0 = 8.85 \times 10^{-12}$ F/m.

In a linear and isotropic media, $\boldsymbol{P}$ is defined as:

$$\boldsymbol{P} = \chi_e \varepsilon_0 \boldsymbol{E} \tag{1.5}$$

where χ_e is the electric susceptibility of the dielectric material. It is a measure of how easily it polarizes in response to an electric field. The electric flux density $\boldsymbol{D}$ is further written by:

$$\boldsymbol{D} = \varepsilon_0 \varepsilon_r \boldsymbol{E} = \varepsilon \boldsymbol{E} \tag{1.6}$$

where $\varepsilon = \varepsilon_0(1 + \chi_e)$, the electric permittivity of the dielectric material, and $\varepsilon_r = 1 + \chi_e$ is the relative permittivity of the dielectric.

Magnetic field strength or magnetic flux density $\boldsymbol{B}$ is a vector used to describe magnetic field. It relates the magnetic force experienced by a particle carrying a charge of q coulomb to the magnetic field the charge is passing through at a speed of v,

$$\boldsymbol{F} = qv \times \boldsymbol{B} \tag{1.7}$$

As dielectric being polarized by an applied electric field, magnetic media can be magnetized by an applied magnetic field. The magnetization is defined as $\boldsymbol{M}$. In a linear and isotropic media, $\boldsymbol{M}$ is defined as:

$$\boldsymbol{M} = \chi_m \boldsymbol{H} \tag{1.8}$$

where χ_m is the magnetic susceptibility.

The magnetic flux density can be written as:

$$\boldsymbol{B} = \mu_0 \boldsymbol{H} + \mu_0 \boldsymbol{M} \tag{1.9}$$

where μ_0 is the permeability of free space, $\mu_0 = 4\pi \times 10^{-7}$ H/m. And,

$$\boldsymbol{B} = \mu_0 \mu_r \boldsymbol{H} = \mu \boldsymbol{H} \tag{1.10}$$

where $\mu = \mu_0(1 + \chi_m)$ is the permeability of the magnetic media. $\mu_r = 1 + \chi_m$ is the relative permeability of the media.

Magnetic field is given by:

$$\boldsymbol{H} = \frac{\boldsymbol{B}}{\mu_0} - \boldsymbol{M} \tag{1.11}$$

Maxwell's equations formulate the interactions of vector fields $\boldsymbol{D}$, $\boldsymbol{E}$, $\boldsymbol{B}$, and $\boldsymbol{H}$. They are the cornerstones for the study of electromagnetic field and electromagnetic waves. The integral form of Maxwell's equation describes the relationship between different vector fields in certain regions—for example, a symmetric distribution of charges and currents. However, in practical applications less symmetric situations and/or vector fields at a certain location in the region are usually desired. In these cases the differential form of Maxwell's equations are more often used. The differential forms of Maxwell's equations are summarized below:

$$\begin{cases} \nabla \times \boldsymbol{H} = \boldsymbol{J}_c + \dfrac{\partial \boldsymbol{D}}{\partial t} \\ \nabla \times \boldsymbol{E} = -\dfrac{\partial \boldsymbol{B}}{\partial t} \\ \nabla \cdot \boldsymbol{D} = \rho \\ \nabla \cdot \boldsymbol{B} = 0 \end{cases} \tag{1.12}$$

Each equation of Maxwell's equations represents a physics law observed in experiments. Their details are discussed in following sections.

Ampere's law

Starting with Ampere's law with displacement current correction, the origin of Ampere's law states that magnetic fields can be generated by electrical current:

$$\oint_l \boldsymbol{H} \cdot d\boldsymbol{l} = \mathrm{I} \tag{1.13}$$

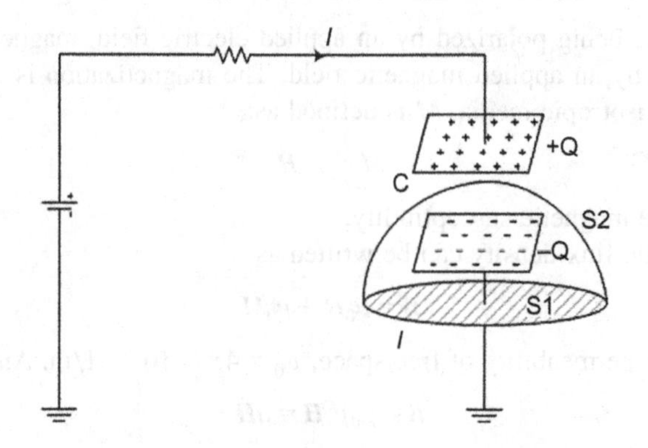

FIGURE 1.1

Current going through different surfaces around a capacitor being charged.

The equation shows that the circulation or the line integral of the magnetic field is equal to the sum of the current inside the curl. However, this form of Ampere's law does not apply to a non-continuous conducting current. Consider the situation in Figure 1.1: when a capacitor is being charged there will be a continuous conducting current outside of the capacitor. However, inside the capacitor there will be no conducting current. The magnetic circulation has different values depending on which surface surrounded is selected. When the current inside the surface S1 is ***I***, and in contrast the current inside the surface S2 is 0, both the surfaces are bounded by path *l*. Maxwell's correction by introducing displacement current filled this gap in the original Ampere's law. The correction shows that not only does a continuous conducting current induce a magnetic field, but also a changing electric field induces a magnetic field. Ampere's law can be rewritten as:

$$\oint_l \boldsymbol{H} \cdot dl = I_c + I_d$$

where I_c is the conducting current and I_d is the displacement current.

Since,

$$I = \int_s \boldsymbol{J} \cdot d\mathbf{S}$$

$$\oint_l \boldsymbol{H} \cdot dl = \int_s (\boldsymbol{J}_c + \boldsymbol{J}_d) \cdot d\mathbf{S} = \int_s \left(\boldsymbol{J}_c + \frac{\partial \boldsymbol{D}}{\partial t}\right) \cdot d\mathbf{S} \tag{1.14}$$

where $\boldsymbol{J}_c$ and $\boldsymbol{J}_d$ are conducting current density and displacement current density, respectively.

Faraday's law

Faraday's law summarizes that a voltage or electromotive force (EMF) can be produced by the altering magnetic flux in an electric circuit. The induced EMF (V) is equal to the negative change rate of magnetic flux Φ_B:

$$\varepsilon = -\frac{d\Phi_B}{dt} \tag{1.15}$$

where ε is EMF. Φ_B is the magnetic flux (W_b). Φ_B is defined as:

$$\Phi_B = \int_s \boldsymbol{B} \cdot d\boldsymbol{S} \tag{1.16}$$

$\boldsymbol{S}$ is the surface bounded by a contour.

When magnetic flux changes the induced EMF will generate electric field in the electric circuit, usually a wire loop, which in turn produces electric current. EMF is the integral of the electric field along a closed loop:

$$\varepsilon = \oint_l \boldsymbol{E} \cdot d\boldsymbol{l} = -\frac{d\Phi_B}{dt} \tag{1.17}$$

where $\boldsymbol{S}$ is the surface bounded by a contour. From equations 1.16 and 1.17 we can get:

$$\oint_l \boldsymbol{E} \cdot d\boldsymbol{l} = -\int_s \frac{\partial \boldsymbol{B}}{\partial t} \cdot d\boldsymbol{S} \tag{1.18}$$

Gauss's law

Gauss's law states that the total electric flux going out of a closed surface is the ratio of the total charge within the surface and the permittivity of the media. It relates the distribution of electric charge to the resulting electric field:

$$\Phi_E = \frac{Q}{\varepsilon} \tag{1.19}$$

where Φ_E is the electric flux flowing through a closed surface, and Q is the total charge enclosed by the surface.

Since:

$$\Phi_E = \oint_S \boldsymbol{E} \cdot d\boldsymbol{S}$$

$$\oint_S \varepsilon \boldsymbol{E} \cdot d\boldsymbol{S} = \oint_S \boldsymbol{D} \cdot d\boldsymbol{S} = \int_v \rho dv \tag{1.20}$$

where v is the volume enclosed by the closed surface S, and ρ is the charge density within the surface.

Gauss's law for magnetism

Gauss's law for magnetism is the fourth fundamental equation to summarize the electromagnetic phenomenon. It states that the net magnetic flux out of any closed surface is zero:

$$\oint_S \boldsymbol{B} \cdot d\boldsymbol{S} = 0 \tag{1.21}$$

For any given closed surface the magnetic flux inward the surface will equal the flux outwards the surface. This is because the source of a magnetic field is always a current or a magnetic dipole, and there is no known magnetic monopole exists.

PROPAGATION OF PLANE WAVES

Electromagnetic waves, e.g., high-speed signal on a transmission line, propagating through space or media are time-varying electromagnetic fields in nature. This section derives electromagnetic wave equations from Maxwell's equations. It starts with discussions on uniform plane wave in an infinite ideal media and moves on to discuss the propagation of plane wave in conductive media.

Uniform plane wave

In a linear, homogeneous, nonconductive, isotropic and source-free media Maxwell's first two equations can be written as:

$$\begin{cases} \nabla \times \boldsymbol{H} = \dfrac{\partial \boldsymbol{D}}{\partial t} \\ \nabla \times \boldsymbol{E} = -\dfrac{\partial \boldsymbol{B}}{\partial t} \end{cases} \tag{1.22}$$

Taking the curl of the first curl equation and using the second curl equation:

$$\nabla \times \nabla \times \boldsymbol{H} = \varepsilon \nabla \times \frac{\partial \boldsymbol{E}}{\partial t} = \varepsilon \frac{\partial}{\partial t} \nabla \times \boldsymbol{E} = -\varepsilon\mu \frac{\partial^2 \boldsymbol{H}}{\partial t^2}$$

By using vector identity:

$$\nabla \times \nabla \times \boldsymbol{H} = \nabla(\nabla \cdot \boldsymbol{H}) - \nabla^2 \boldsymbol{H}$$

in free media:

$$\nabla \cdot \boldsymbol{H} = 0$$

we get the wave equation for $\boldsymbol{H}$:

$$\nabla^2 \boldsymbol{H} - \varepsilon\mu \frac{\partial^2 \boldsymbol{H}}{\partial t^2} = 0 \tag{1.23}$$

In the same manner, the wave equation for $\boldsymbol{E}$ is given as:

$$\nabla^2 \boldsymbol{E} - \varepsilon\mu \frac{\partial^2 \boldsymbol{E}}{\partial t^2} = 0 \tag{1.24}$$

Assuming the plane wave has an electric field in the x direction and a magnetic field in the y direction. It is propagating in the z direction. By definition $\boldsymbol{E}$ and $\boldsymbol{H}$ are uniform on the wavefront:

$$\frac{\partial}{\partial x} = \frac{\partial}{\partial y} = 0$$

Equations 1.23 and 1.24 can be simplified to:

$$\begin{cases} \dfrac{\partial^2 H_y}{\partial z^2} - \varepsilon\mu \dfrac{\partial^2 H_y}{\partial t^2} = 0 \\ \dfrac{\partial^2 E_x}{\partial z^2} - \varepsilon\mu \dfrac{\partial^2 E_x}{\partial t^2} = 0 \end{cases} \tag{1.25}$$

The above equations in phasor form are:

$$\begin{cases} \dfrac{\partial^2 H_y}{\partial z^2} + \omega^2 \varepsilon\mu H_y = 0 \\ \dfrac{\partial^2 E_x}{\partial z^2} + \omega^2 \varepsilon\mu E_x = 0 \end{cases} \tag{1.26}$$

A constant $k = \omega\sqrt{\varepsilon\mu}$ is defined as the phase constant or wave number. It is a spatial frequency of a wave that equals to cycles per unit distance. Equation 1.26 reduces to:

$$\begin{cases} \dfrac{\partial^2 H_y}{\partial z^2} + k^2 H_y = 0 \\ \dfrac{\partial^2 E_x}{\partial z^2} + k^2 E_x = 0 \end{cases} \tag{1.27}$$

One solution of equation 1.27 is:

$$\begin{cases} H_y(z) = H^+ e^{-jkz} \\ E_x(z) = E^+ e^{-jkz} \end{cases} \tag{1.28}$$

The solution in time domain is:

$$\begin{cases} H_y(z,t) = H^+ \cos(\omega t - kz) \\ E_x(z,t) = E^+ \cos(\omega t - kz) \end{cases} \tag{1.29}$$

H^+ and E^+ are wave amplitudes. The wave propagates in the $+z$ direction.

Some other important wave propagation parameters are deduced from 1.29. Phase velocity is how fast the wavefront with the same phase moves in space. Mathematical expression of this statement is:

$$\omega t - kz = \text{constant}$$

Phase velocity then can be calculated as:

$$v = \frac{dz}{dt} = \frac{\omega}{k} = \frac{1}{\sqrt{\varepsilon\mu}} = \frac{v_0}{\sqrt{\varepsilon_r \mu_r}} \tag{1.30}$$

where v_0 is the speed of light and ε_r and μ_r are the relative permeability and permittivity of the medium.

Wavelength in the media is given as:

$$\lambda = vt = \frac{v}{f} = \frac{v_0}{f\sqrt{\varepsilon_r \mu_r}} = \frac{\lambda_0}{\sqrt{\varepsilon_r \mu_r}} = \frac{2\pi}{k} \tag{1.31}$$

where λ_0 is the wavelength in free space.

When one component of the wave is known, the other field vector can be calculated. By using the electric field curve equation of Maxwell's equations:

$$\nabla \times \boldsymbol{E} = -\mu \frac{\partial \boldsymbol{H}}{\partial t} \tag{1.32}$$

For the plane wave traveling in the $+z$ direction, whose electric and magnetic fields are in x and y directions, the above equation becomes:

$$\frac{\partial E_x}{\partial z} = -j\omega\mu H_y$$

$$H_y = \frac{j}{\omega\mu}\frac{\partial E_x}{\partial z} = \frac{j}{\omega\mu}\frac{\partial (E^+ e^{-jkz})}{\partial z} = \frac{k}{\omega\mu}E_x = \frac{E_x}{\eta} \tag{1.33}$$

where $\eta = \omega\mu/k = \sqrt{\mu/\varepsilon}$ is the wave impedance of the medium. In free space:

$$\eta_0 = \sqrt{\frac{\mu_0}{\varepsilon_0}} = 120\pi \tag{1.34}$$

The wave discussed above is a transverse electromagnetic (TEM) wave, whose electric and magnetic fields are orthogonal to each other and orthogonal to the wave propagation direction.

Uniform plane wave in conductive media

In contrast to ideal media, which is free of charges and currents, electromagnetic waves propagating in conductive media generate conductive currents. From equation 1.12:

$$\nabla \times \boldsymbol{H} = \boldsymbol{J}_c + \frac{\partial \boldsymbol{D}}{\partial t} = \sigma \boldsymbol{E} + j\omega\varepsilon \boldsymbol{E} = (\sigma + j\omega\varepsilon)\boldsymbol{E} \tag{1.35}$$

Define a complex dielectric constant as:

$$\tilde{\varepsilon} = \varepsilon - j\sigma/\omega \tag{1.36}$$

By using 1.36:

$$\nabla \times \boldsymbol{H} = j\omega\tilde{\varepsilon}\boldsymbol{E} \tag{1.37}$$

By replacing the dielectric constant with 1.36, wave propagation equations in conductive media become:

$$\begin{cases} \nabla^2 \boldsymbol{H} + \omega^2 \tilde{\varepsilon}\mu \boldsymbol{H} = 0 \\ \nabla^2 \boldsymbol{E} + \omega^2 \tilde{\varepsilon}\mu \boldsymbol{E} = 0 \end{cases} \tag{1.38}$$

The equations can be further simplified for uniform plane waves discussed in the previous section, where the electric field is in the x direction and the magnetic field is in the y direction:

$$\begin{cases} \frac{\partial^2 H_y}{\partial z^2} + \omega^2 \tilde{\varepsilon}\mu H_y = 0 \\ \frac{\partial^2 E_x}{\partial z^2} + \omega^2 \tilde{\varepsilon}\mu E_x = 0 \end{cases} \tag{1.39}$$

A complex wave number is used for the conductive media:

$$\gamma = \omega\sqrt{\tilde{\varepsilon}\mu} = \omega\sqrt{\varepsilon\mu}\sqrt{1 - j\frac{\sigma}{\omega\varepsilon}} \tag{1.40}$$

$$\gamma = a + j\beta \tag{1.41}$$

where a is the attenuation constant and β is the phase constant.

For the positive traveling wave:

$$\begin{cases} H_y = H^+ e^{-\gamma z} = H^+ e^{-az} e^{-j\beta z} \\ E_x = E^+ e^{-\gamma z} = E^+ e^{-az} e^{-j\beta z} \end{cases} \tag{1.42}$$

The wave amplitude decays exponentially as it propagates in the conductive media. Relation between magnetic and electric field can be written as:

$$H_y = \frac{j}{\omega\mu}\frac{\partial E_x}{\partial z} = \frac{j}{\omega\mu}\frac{\partial(E^+ e^{-j\gamma z})}{\partial z} = \frac{\gamma}{\omega\mu}E_x = \frac{E_x}{\tilde{\eta}} \tag{1.43}$$

Where $\tilde{\eta} = \omega\mu/\gamma$ is the complex wave impedance of the conductive media.

A special case is the copper transmission lines used in high-speed systems, where $\sigma \gg \omega\varepsilon$ and the media is considered a good conductor. Complex wave number and wave impedance can be approximated as:

$$\gamma = a + j\beta \approx j\omega\sqrt{\varepsilon\mu}\sqrt{\frac{\sigma}{j\omega\varepsilon}}$$

$$= \sqrt{\frac{\omega\mu\sigma}{2}} + j\sqrt{\frac{\omega\mu\sigma}{2}}$$

$$a \approx \sqrt{\frac{\omega\mu\sigma}{2}} \tag{1.44}$$

$$\beta \approx \sqrt{\frac{\omega\mu\sigma}{2}} \tag{1.45}$$

Wave impedance can be calculated as:

$$\tilde{\eta} \approx \sqrt{\frac{\mu}{-j\frac{\sigma}{\omega}}} = \sqrt{\frac{\omega\mu}{\sigma}}\sqrt{j} = \sqrt{\frac{\omega\mu}{2\sigma}} + j\sqrt{\frac{\omega\mu}{2\sigma}} \tag{1.46}$$

Electromagnetic waves tend to attenuate when propagating in good conductive media. Their attenuation rate is proportional to frequency, media conductivity and media permittivity. In other words, high-frequency electromagnetic waves only exist near the surface of a good conductor or magnetic media. This effect is called skin effect. The complex wave impedance equation indicates that the electric and magnetic fields have a 45 degrees phase difference.

Power flow and the Poynting vector

Electromagnetic waves carry energy while they are propagating. The energy flux density, i.e., the rate of energy transmission per unit area, is called the Poynting vector. In the lossless media, Poynting vector of the positive z traveling wave is defined as:

$$\boldsymbol{S} = \boldsymbol{E} \times \boldsymbol{H}^* = (\hat{x}E^+e^{-jkz}) \times (\hat{y}H^+e^{jkz}) = \hat{z}\frac{E^{+2}}{\eta} \tag{1.47}$$

In conductive media the electric field and magnetic field are written as:

$$\begin{cases} \boldsymbol{H} = \hat{y}H^+e^{-az}e^{-j\beta z} \\ \boldsymbol{E} = \hat{x}E^+e^{-az}e^{-j(\beta z-\varphi)} \end{cases} \tag{1.48}$$

where φ is the phase difference between electric and magnetic field.

In conductive media the Poynting vector is written as:

$$\begin{aligned} \boldsymbol{S} = \boldsymbol{E} \times \boldsymbol{H}^* &= (\hat{x}E^+e^{-az}e^{-j(\beta z-\varphi)}) \times (\hat{y}H^+e^{-az}e^{j\beta z}) \\ &= \hat{z}E^+H^+e^{-2az}e^{j\varphi} \end{aligned}$$

$$p = \frac{1}{2}\text{Re}(\boldsymbol{E} \times \boldsymbol{H}^*) = \frac{1}{2}E^+H^+e^{-2az}\cos\varphi \tag{1.49}$$

TRANSMISSION LINE THEORY

Electromagnetics theories describe the relationships between electric fields and magnetic fields, and how the electromagnetic wave propagates in a given media. Transmission line is the most common manmade medium used in high-speed systems to transmit high-speed signals from chip to chip, chip to add-in card, and system to system. This section discusses the basic theories of transmission lines, especially for the lossless transmission lines.

In high-speed systems, electromagnetic wave wavelength is shorter than or comparable to the length of the transmission line. The transmission line connections between circuit components cannot be ignored anymore. The lumped element model is often used under this condition. The power of the electromagnetic waves transmitted on transmission line in a high-speed system gets attenuated due to transmission line conductor resistance. The resistive loss gets even

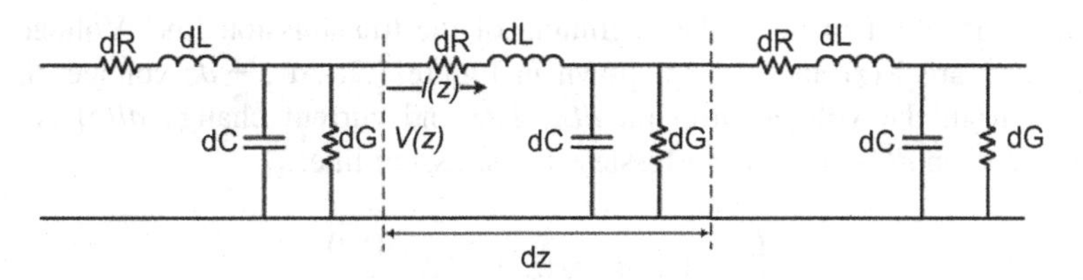

FIGURE 1.2

Lumped element model of transmission line.

more severe at high data rates because of the skin effects. This is called the distributed resistance effect, dR. Dielectric effects also become significant at high data rates. Dielectric loss is due to the imperfect insulating dielectric used in between signal strip and ground, which adds to the resistance loss. The equivalent shunt resistor here is called distributed conductance, dG. Alternating electric field induces a magnetic field around the transmission line and causes an inductance effect, which is called distributed inductance, dL. Lastly, the shunt capacitance between signal and ground is represented by distributed capacitance, dC. Voltage and current on the line are not only time relevant but also location dependent.

With the lumped element model, transmission lines can be segmented and each unit can be modeled with lumped elements as shown in Figure 1.2. The model laid the foundation for deduction of the transmission line theory covered in the following sections of this chapter.

Following sections show the derivation of basic transmission line equations and critical parameters of lossless transmission lines. Lossy transmission lines are not discussed here but can be found in other literature.

WAVE EQUATIONS ON LOSSLESS TRANSMISSION LINES

Transmission lines with $dR = dG = 0$ in their lumped element model are called lossless transmission lines. Under some conditions, e.g., low-frequency and short length buses, the loss of transmission line used in today's high-speed systems is small because good conductors and low loss dielectric materials are often used in manufacturing. As a result, it is a good starting point and practice to analyze lossless transmission lines theoretically. In this section lossless transmission line equations will be derived from the lumped element model and Maxwell's equations. The equations describe the dynamics of voltage and current on a transmission line with respect to both time and distance.

Lossless transmission line

A lossless transmission line unit section is used in the analysis. It is stimulated with a sine wave with frequency ω and is terminated with a load resistor Z_L. The spatial origin is set to be at the beginning of the transmission line. Voltage and current at z are $V(z)$ and $I(z)$ as shown in Figure 1.2. At $z + dz$ voltage change $dV(z)$ is from the voltage drop on $dL = Ldz$ and current change $dI(z)$ is from current drop on $dC = Cdz$, for a lossless transmission line:

$$\begin{cases} v(z+dz,t) = v(z,t) - Ldz\dfrac{\partial i(z,t)}{\partial t} \\ i(z+dz,t) = i(z,t) - Cdz\dfrac{\partial v(z,t)}{\partial t} \end{cases} \tag{1.50}$$

For the sinusoidal steady-state condition:

$$\begin{cases} \dfrac{dV(z)}{dz} = -j\omega LI(z) \\ \dfrac{dI(z)}{dz} = -j\omega CV(z) \end{cases} \tag{1.51}$$

The above equations are transmission line equations. They are also called the telegrapher equations.

Wave propagation on a lossless transmission line

Wave equations for $V(z)$ and $I(z)$ can be derived from equation 1.51:

$$\begin{cases} \dfrac{d^2V(z)}{dz^2} + j\omega L\dfrac{dI(z)}{dz} = 0 \\ \dfrac{d^2I(z)}{dz^2} + j\omega C\dfrac{dV(z)}{dz} = 0 \end{cases} \tag{1.52}$$

Using equation 1.51:

$$\begin{cases} \dfrac{d^2V(z)}{dz^2} + \omega^2 LCV(z) = 0 \\ \dfrac{d^2I(z)}{dz^2} + \omega^2 LCI(z) = 0 \end{cases} \tag{1.53}$$

The above equations are the wave equations for lossless transmission lines. The solutions are:

$$\begin{cases} V(z) = V_0^+ e^{-j\beta z} + V_0^- e^{j\beta z} \\ I(z) = I_0^+ e^{-j\beta z} + I_0^- e^{j\beta z} \end{cases} \tag{1.54}$$

From equations 1.51 and 1.54, we get coefficients for $I(z)$:

$$I_0^+ = \frac{V_0^+}{Z_0}, \; I_0^- = -\frac{V_0^-}{Z_0} \tag{1.55}$$

where:

$$Z_0 = \sqrt{\frac{L}{C}} \tag{1.56}$$

$$\beta = \omega\sqrt{LC} \tag{1.57}$$

Z_0 is the characteristic impedance of the lossless transmission line. Characteristic impedance is used for transmission line impedance characterization; it is the ratio of the incident wave voltage to the incident current, or the negative ratio of the reflected wave voltage to the reflected current:

$$Z_0 = \frac{V_i(z)}{I_i(z)} = -\frac{V_r(z)}{I_r(z)} \tag{1.58}$$

Equation 1.56 shows $Z_0 = \sqrt{L/C}$, which indicates that characteristic impedance of a lossless transmission line is only dependent on the unit inductance and capacitance of the transmission line, which are decided by transmission line dimension and substrate materials. Typical types of transmission line characteristic impedances will be discussed in a later section of this chapter.

Phase constant β is the phase change in a unit length. It is easier to remember equation 1.57 by comparing it to the phase constant or wave number of a plane wave $k = \omega\sqrt{\mu\varepsilon}$. Unit inductance and capacitance of transmission lines are equivalent to the permittivity and permeability of a free media mathematically.

Phase velocity v_p is how fast the constant phase wavefront moves:

$$v = \frac{\omega}{\beta} = \frac{1}{\sqrt{LC}} \tag{1.59}$$

Wavelength λ_p is the distance that creates a 2π phase difference:

$$\lambda = \frac{2\pi}{\beta} \tag{1.60}$$

Wavelength on a transmission line is dependent on the dielectric property of the transmission line substrate and the material on top of it.

Incident waves and reflected waves

Equation 1.54 shows that the voltage and current wave equation solutions contain incident and reflected waves:

$$\begin{cases} V(z) = V_0^+ e^{-j\beta z} + V_0^- e^{j\beta z} = V_i(z) + V_r(z) \\ I(z) = \dfrac{V_0^+}{Z_0} e^{-j\beta z} - \dfrac{V_0^-}{Z_0} e^{j\beta z} = I_i(z) + I_r(z) \end{cases} \tag{1.61}$$

Incident waves and reflective waves exist simultaneously and add together to form the actual voltage and current on the transmission line. $V_i(z, t)$ and $I_i(z, t)$ are incident traveling waves propagating from the source of the transmission line to the load termination. $V_r(z, t)$ and $I_r(z, t)$ are reflective traveling waves propagating in the opposite direction of the incident wave.

IMPEDANCE, REFLECTION COEFFICIENT, AND POWER FLOW ON A LOSSLESS TRANSMISSION LINE

Wave propagation properties—e.g., phase constant, wavelength, and characteristic impedance—on a lossless transmission line are discussed in a previous section. Input impedance, reflection coefficient, and transmission power are discussed in this section.

Input impedance and reflection coefficient

Reflection coefficient is used to define the reflected wave with respect to the incident wave. When a load Z_l is connected to the transmission line as shown in Figure 1.3, and the voltage and current at the end of the transmission line are V_l and I_l, by using equation 1.61 and $z = 0$:

$$Z_l = \frac{V_l}{I_l} = \frac{V_0^+ + V_0^-}{V_0^+ - V_0^-} Z_0$$

the reflection coefficient Γ_l at the load is given as:

$$\Gamma_l = \frac{V_0^-}{V_0^+} = \frac{Z_l - Z_0}{Z_l + Z_0} \tag{1.62}$$

If load reflection coefficient is measured, e.g. by a network analyzer, load impedance can be calculated by:

$$Z_l = Z_0 \frac{1 + \Gamma_l}{1 - \Gamma_l} \tag{1.63}$$

The reflection coefficient at any point is given by:

$$\Gamma(z') = \frac{V_0^- e^{-j\beta z'}}{V_0^+ e^{j\beta z'}} = \Gamma_l e^{-2j\beta z'} \tag{1.64}$$

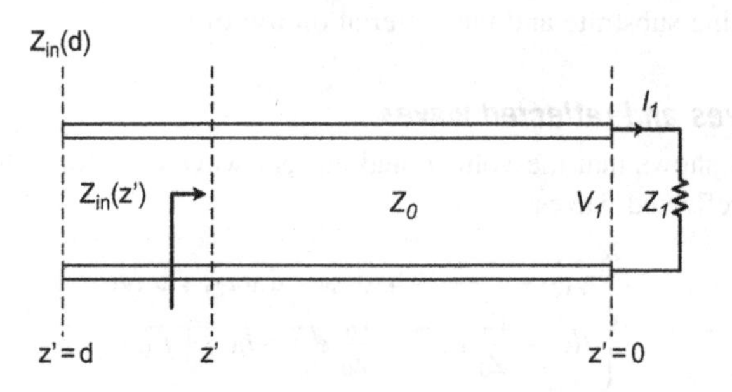

FIGURE 1.3

Illustration of the input impedance locations of a loaded transmission line.

Using equation equation 1.61, reflection coefficients for voltage and current are:

$$\Gamma(z') = \Gamma_V(z') = -\Gamma_I(z') \tag{1.65}$$

Equation 1.64 shows reflection coefficient is a function of location and the reflection coefficient at the load. Equation 1.63 shows the load reflection coefficient is dependent on the load impedance and the transmission line characteristic impedance. When load impedance is equal to the characteristic impedance there is no reflection on the transmission line.

With the defined $\Gamma(z')$, voltage and current at any point on the transmission line can be calculated by using equation 1.61 and 1.65:

$$\begin{cases} V(z') = V_i(z') + V_r(z') = V_i(z')[1 + \Gamma(z')] \\ I(z') = I_i(z') + I_r(z') = I_i(z')[1 - \Gamma(z')] \end{cases} \tag{1.66}$$

Input impedance at a given location on the transmission line is defined as the ratio of the voltage to the current, by using equations 1.78 and 1.82:

$$\begin{aligned} Z_{in}(z') &= \frac{V(z')}{I(z')} = \frac{V_0^+ e^{-j\beta z'}(1 + \Gamma_l e^{-2j\beta z'})}{V_0^+ e^{-j\beta z'}(1 - \Gamma_l e^{-2j\beta z'})} Z_0 \\ &= \frac{1 + \Gamma_l e^{-2j\beta z'}}{1 - \Gamma_l e^{-2j\beta z'}} Z_0 \end{aligned} \tag{1.67}$$

Plugging equation 1.63 back into equation 1.67:

$$Z_{in}(z') = Z_0 \frac{Z_l I_l \cos \beta z' + jZ_0 I_l \sin \beta z'}{Z_0 I_l \cos \beta z' + jZ_l I_l \sin \beta z'} = Z_0 \frac{Z_l + jZ_0 \tan \beta z'}{Z_0 + jZ_l \tan \beta z'} \tag{1.68}$$

Power flow on a lossless transmission line

So far the voltage, current, and impedance on a transmission line have been discussed. However, let's not forget the transmission line is used to conduct high-speed signals from one point of the system to another point. Signals traveling along the transmission line are associated with power.

The electromagnetic power associated with the transmission line is with a form very similar to the Poynting vector. However, electrical and magnetic fields are replaced by voltage and current:

$$\begin{aligned} P(z') &= \frac{1}{2}\mathrm{Re}\left[V(z')I^*(z')\right] = \frac{1}{2}\mathrm{Re}\left\{V_i[1 + \Gamma(z')]I_i^*\left[1 - \Gamma^*(z')\right]\right\} \\ &= \frac{|V_i|^2}{2Z_0}\mathrm{Re}\left[1 - \Gamma^*(z') + \Gamma(z') - |\Gamma(z')|^2\right] \end{aligned}$$

$$P(z') = \frac{|V_i|^2}{2Z_0}(1 - |\Gamma(z')|^2) \tag{1.69}$$

Due to the lossless nature of the transmission line, there is no conductive or dielectric loss along the transmission line. The only power loss at any location is caused by the load impedance mismatch. The power reflection coefficient depends on the load impedance and transmission line characteristic impedance.

TRAVELING AND STANDING WAVES ON A TRANSMISSION LINE

Traveling waves and standing waves are discussed in this section. Voltage, current, and impedance distribution along a transmission line in the two scenarios are discussed to understand how high-frequency signal travels. Reflection is the main factor that determines the type of electromagnetic wave propagating on a transmission line.

Traveling waves

Traveling waves travel free of reflection interference on a transmission line. When load impedance matches transmission line characteristic impedance there is no reflection on a transmission line. Under this condition, the voltage and current on the transmission line can be calculated by using equation 1.69 and by eliminating the reflected terms of the equation:

$$\begin{cases} V(z) = \dfrac{(V_0 + Z_0 I_0)e^{-j\beta z}}{2} = V_0^+ e^{-j\beta z} \\ I(z) = \dfrac{(V_0 + Z_0 I_0)e^{-j\beta z}}{2Z_0} = I_0^+ e^{-j\beta z} \end{cases} \tag{1.70}$$

Traveling waves' voltage and current amplitudes are constant along the line. The input impedance at any location of the transmission line can be calculated by definition:

$$Z(z) = \frac{V_0^+ e^{-j\beta z}}{I_0^+ e^{-j\beta z}} = \frac{V_0^+}{I_0^+} = Z_0 \tag{1.71}$$

The input impedance is a constant at any location on of the transmission line and is equal to the its characteristic impedance. Traveling wave is an idea condition for the operation of the high-speed system.

Standing waves

On the other side of the spectrum, a totally reflected wave, also known as $|\Gamma_l| = 1$, will interfere with the incident wave to form the standing wave. Two special cases are the short and open circuited lines:

For open termination $\Gamma_l = \frac{Z_l - Z_0}{Z_l + Z_0} = 1$, current at the termination is $I_l = 0$ and the reflected and incident waves are:

$$\begin{cases} V_l^- = \Gamma_l V_l^+ = V_l \\ I_l^- = -\Gamma_l I_l^+ = -I_l \end{cases} \tag{1.72}$$

$$V_l = V_l^+ + V_l^- = 2V_l^+$$

Using equation 1.67:

$$\begin{cases} V(z') = 2V_l^+ \cos \beta z' \\ I(z') = \dfrac{2jV_l^+}{Z_0} \sin \beta z' \end{cases} \tag{1.73}$$

The transient expressions for voltage and current are:

$$\begin{cases} v(z', t) = 2V_l^+ \cos \beta z' \cos(\omega t + \varphi_l) \\ i(z', t) = \dfrac{2V_l^+}{Z_0} \sin \beta z' \cos\left(\omega t + \varphi_l + \dfrac{\pi}{2}\right) \end{cases} \tag{1.74}$$

Figure 1.4 illustrates the voltage and current distributions for an open terminated transmission line.

The input impedance is given by:

$$Z_{\text{in}}(z') = -jZ_0 \cot \beta z' \tag{1.75}$$

Input impedance of an open terminated lossless transmission line.

For short termination $\Gamma_l = \frac{Z_l - Z_0}{Z_l + Z_0} = -1$, voltage at the termination is $V_l = 0$ and the reflected and incident waves are:

$$\begin{cases} V_l^- = \Gamma_l V_l^+ = -V_l \\ I_l^- = -\Gamma_l I_l^+ = I_l \end{cases} \tag{1.76}$$

$$I_l = I_l^+ + I_l^- = 2I_l^+$$

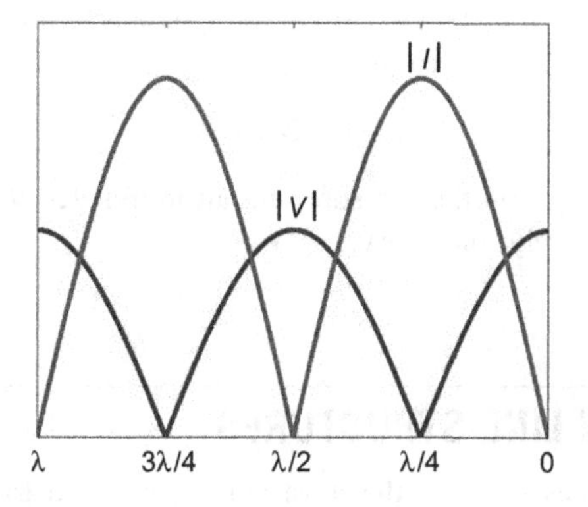

FIGURE 1.4

Transient voltage and current distribution along the open-ended line.

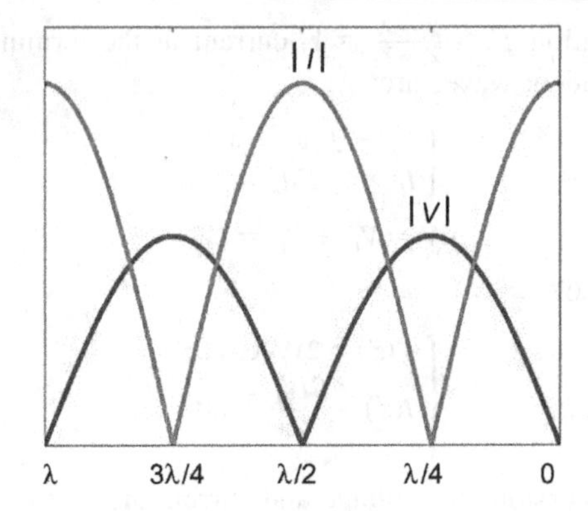

FIGURE 1.5

Transient voltage and current distribution along the shorted line.

Using equation 1.67, voltage and current at z' from the load are:

$$\begin{cases} V(z') = 2jV_l^+ \sin \beta z' \\ I(z') = \dfrac{2jV_l^+}{Z_0} \cos \beta z' \end{cases} \tag{1.77}$$

The transient expressions for voltage and current are:

$$\begin{cases} v(z', t) = 2V_l^+ \sin \beta z' \cos\left(\omega t + \varphi_l + \dfrac{\pi}{2}\right) \\ i(z', t) = \dfrac{2V_l^+}{Z_0} \cos \beta z' \cos(\omega t + \varphi_l) \end{cases} \tag{1.78}$$

Distributions of the voltage and current amplitude are shown in Figure 1.5. Input impedance at z' is given by:

$$Z_{in}(z') = jZ_0 \tan \beta z' \tag{1.79}$$

The input impedance on theof a short circuit terminated loaded lossless transmission line is also purely imaginary.

TRANSMISSION LINE STRUCTURES

This section introduces some of the most commonly used transmission lines for high-speed systems. Their advantages and disadvantages are discussed. Field distributions and propagation characteristics of different types of transmission lines

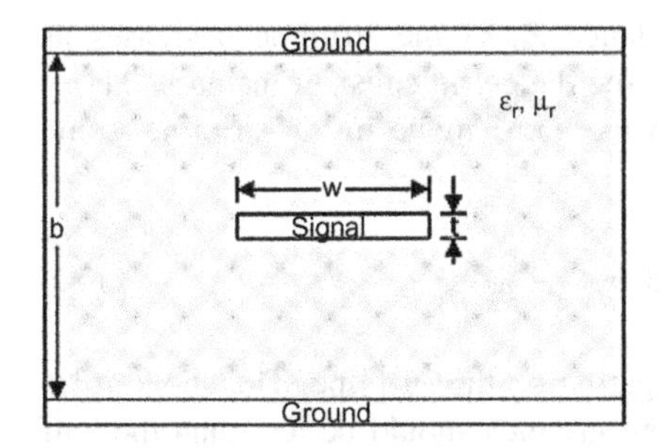

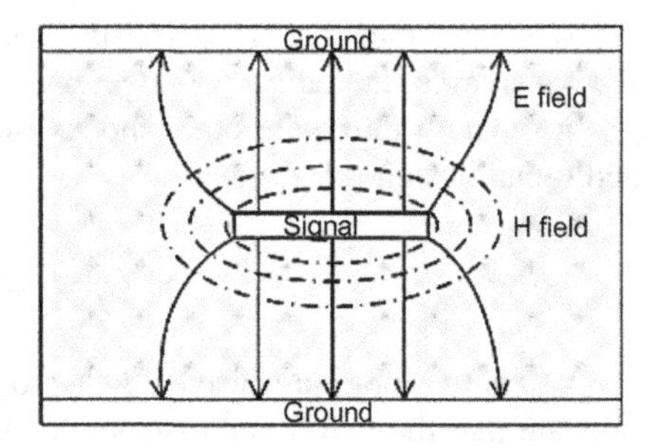

FIGURE 1.6

Cross section of stripline and field distribution.

are bonded to Maxwell's equations and boundary conditions. Important design rules of these transmission lines are introduced.

STRIPLINE

Stripline was invented by Robert M. Barrett of the Air Force Cambridge Research Center in the 1950s [1]. A cross section of a stripline is shown in Figure 1.6.

The idea was to flatten the coaxial cable on both the ground and signal conductors. The flattened center conductor with thickness t is sandwiched dielectrically between a pair of flattened grounds. The distance between the two grounds is b. Space between the grounds is usually filled by one single media that can be either dielectric or air.

The dominant mode of wave propagation on stripline is TEM mode. The uniform dielectric nature of the stripline filling gives stripline some advantages such as being non-dispersive and having no lower cutoff frequency, great isolation, and less radiation. However, stripline requires high fabrication complexity and needs narrower signal traces and greater thickness to get the same characteristic impedance compared to microstrip line [2].

Characteristic impedance of stripline can be calculated by conformal transformation theoretically. However, it is not very practical to solve the elliptic integrals for signal integrity and board design engineers. A simplified equation for stripline design is instead used in practice with good accuracy:

$$Z_0 = \frac{30\pi}{\sqrt{\varepsilon_r}} \frac{b}{W_e + 0.441b} \tag{1.80}$$

where W_e is the effective width of the signal conductor given by:

$$\frac{W_e}{b} = \frac{W}{b} - \begin{cases} 0 & \text{for } \frac{W}{b} > 0.35 \\ \left(0.35 - \frac{W}{b}\right)^2 & \text{for } \frac{W}{b} < 0.35 \end{cases} \tag{1.81}$$

It is seen that the characteristic impedance Z_0 of the stripline decreases as W/b increases. In addition, Z_0 decreases as the dielectric constant increases [1].

In order to eliminate other modes than the TEM mode in the stripline, some additional design rules are:

$$W < \frac{\lambda_{\min}}{2\sqrt{\varepsilon_r}} \text{ and } b < \frac{\lambda_{\min}}{2\sqrt{\varepsilon_r}} \tag{1.82}$$

$\lambda_{\min}$ is the minimum wavelength propagating on the stripline. The above rules indicate that the width and distance between grounds should be less than the half effective wavelength to guarantee TEM mode wave propagation. Another important design rule is to reduce transverse radiation and improve characteristic calculation accuracy with the simplified equation:

$$D > 5W \tag{1.83}$$

where D is the ground plane width.

MICROSTRIP

Microstrip is a modification to the stripline by getting rid of the top half of the dielectric and the top ground plane. A cross section of microstrip is shown in Figure 1.7. It consists of a strip on top of a dielectric layer, usually called substrate, and the bottom ground plane. The width of the conductor strip is W, t is the thickness of the strip, and h is the thickness of the substrate. Microstrip is not only widely used in high-speed computing systems but also the most popular microwave transmission line used by microwave integrated circuits and monolithic microwave integrated circuits (MMICs) because microstrip is easier to fabricate compared to stripline. Since the conductive strip is on top of the printed circuit board, other components of the system can be mounted on top of the board, making the system integration easier [2].

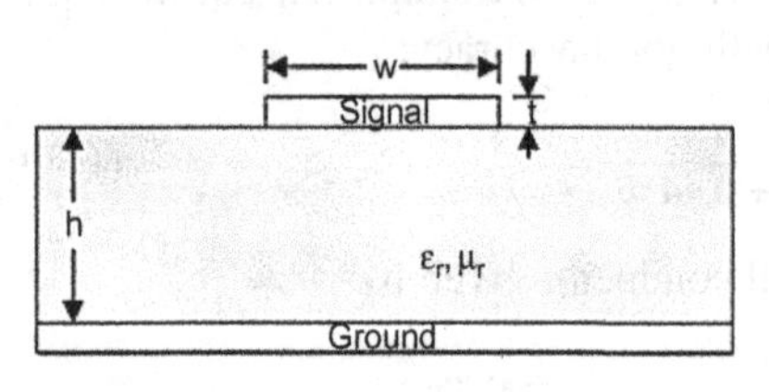

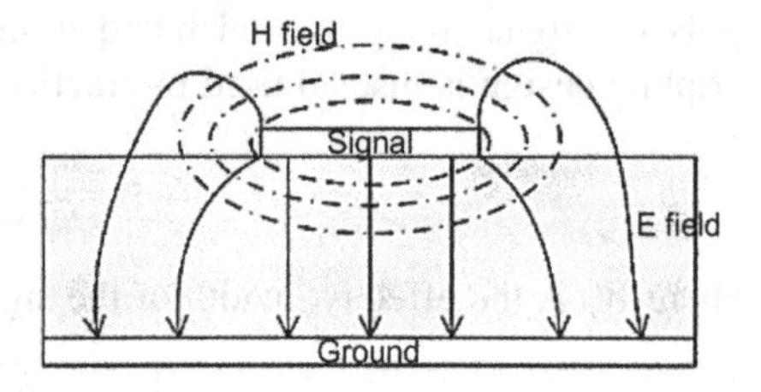

FIGURE 1.7

Cross section of microstrip.

However, the electromagnetic wave propagates in both air and dielectric for microstrip; the dominant mode on microstrip is a hybrid mode of transverse electric (TE) and transverse magnetic (TM) modes. Even though microstrip is dispersive, the velocity difference at the current high-speed frequency range is benign, and the dominant mode on microstrip is categorized as quasi-TEM mode. Another disadvantage is that microstrip does not have as good isolation as stripline. As a result, more crosstalk effect is expected from microstrip. Radiation losses of microstrip also need to be considered in the design.

Electromagnetic waves propagate at the interface of air and dielectric. The effective dielectric constant is defined such that when the strip of the same dimension is uniformly surrounded by the effective dielectric with the effective dielectric constant, its characteristic also keeps the same value. The effective dielectric constant ε_{eff} is in the range [1, ε_r]. It is given by [4]:

$$\begin{cases} \varepsilon_{eff} = \dfrac{\varepsilon_r + 1}{2} + \dfrac{\varepsilon_r - 1}{2}\left[\left(1+12\left(\dfrac{h}{W}\right)\right)^{-1/2} + 0.04\left(1-\left(\dfrac{h}{W}\right)\right)^2\right] & \dfrac{W}{h} < 1 \\ \varepsilon_{eff} = \dfrac{\varepsilon_r + 1}{2} + \dfrac{\varepsilon_r - 1}{2}\left(1+12\left(\dfrac{h}{W}\right)\right)^{-1/2} & \dfrac{W}{h} > 1 \end{cases} \tag{1.84}$$

The caveat here is that the thickness of the strip is not taken into account and the effective dielectric constant is an approximation [2].

With the ε_{eff} being introduced, characteristic impedance of microstrip is given by [4]:

$$\begin{cases} Z_0 = \dfrac{60}{\sqrt{\varepsilon_{eff}}}\ln\left(8\dfrac{h}{W} + 0.25\dfrac{W}{h}\right) & \dfrac{W}{h} < 1 \\ Z_0 = \dfrac{120\pi}{\sqrt{\varepsilon_{eff}}\left[\dfrac{W}{h} + 1.393 + \dfrac{2}{3}\ln\left(\dfrac{W}{h} + 1.444\right)\right]} & \dfrac{W}{h} > 1 \end{cases} \tag{1.85}$$

Characteristic impedance and other import properties, such as phase delay and attenuation of microstrip, can also be analyzed by using commercial software such as Agilent ADS LineCalc.

COPLANAR WAVEGUIDES

As its name indicates, signal strip and ground planes of coplanar waveguide (CPW) are all on the same side of the substrate. A cross section of CPW is shown

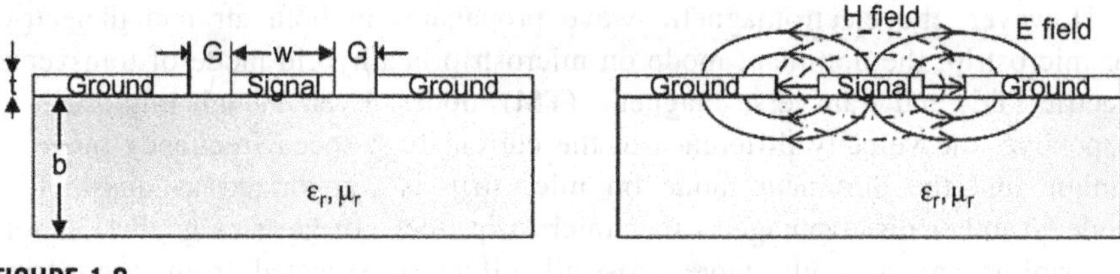

FIGURE 1.8

Cross section of CPW and field distribution.

in Figure 1.8. There are two ground planes on each side of the center strip separated by a gap. The gap is most of the time constant width along the line. The ground plane typically spreads to a large distance.

Field distribution is also shown in Figure 1.8. The electromagnetic wave propagates in both the dielectric and in the air. The inhomogeneous nature of the media in which the transmission line is contained determines that CPW will not support the TEM wave. The effective dielectric constant can be simply estimated to be the average of the dielectric constant of the substrate and the air.

CPW is mostly used in microwave devices and MMICs due to their better isolation than microstrip. Another advantage of CPW for MMIC use is that the conductors are all atop the substrate so that thick substrate can be used and it is not necessary to worry about the substrate etching problem. However, extra steps are required to wire bond the two grounds planes to get them to the same potential; this will keep the CPW mode even in both gaps and prevent unwanted modes from propagating.

Characteristic impedance, phase delay, attenuation, and other properties of the line depend on the dimensions of the gap, center strip, and substrate thickness, as well as conductor and substrate material properties. Closed form equations for CPW characteristic impedance and other properties will not be discussed here. Instead they can be analyzed by using commercial software such as Agilent ADS LineCalc.

NOVEL TRANSMISSION LINES

As discussed in previous sections, each of the transmission lines has pros and cons. Research and development has been conducted to design better performance transmission lines with, for example, lower loss, higher common mode rejection, and less crosstalk. These novel transmission lines either have new signal strip and ground structures [5,6] or take advantage of artificial metamaterials [7]. This section explores two examples of novel transmission lines in detail.

a. Low-crosstalk transmission line using magnetic material

Crosstalk is an important factor to consider in high-speed system design. Undesired interference from active neighbor channels can downgrade the victim link performance. It is caused by mutual capacitive and inductive couplings. Detailed discussion on this topic can be found in Chapter 2. Crosstalk is very common in high-speed system designs, taking place on-chip and in the package, the PCB board, connectors, and on cables, with the strongest crosstalk occurring in packages and connectors [8]. Larger spacing and/or shorter run length and PCB stack up options are explored to resolve the inductance coupling crosstalk. However, this sacrifices design flexibility. Stitching vias to the ground plane can be used to isolate coupling and improve immunity to crosstalk, but it increases fabrication difficulties and board cost. These methods could further deteriorate the data rate capacity of the available routing resources [9]. A transmission line that suppresses mutual inductance and increases the interconnect self-inductance along the link can be achieved by incorporating magnetic materials to the transmission line interconnects.

The proposed magnetic materials need to possess high-saturation magnetization fields and small coercive forces. Their ferromagnetic resonance (FMR) frequency should be beyond 10 gigahertz frequency range so that the magnetic resonant loss does not occur for current high-speed interconnects. Patterned ferromagnetic thin film, Permalloy (Py), used in this specific example, has high permeability and low magnetic and eddy current loss, their FMR frequency having been lifted from hundreds of megahertz up to 20 GHz by Py film geometry patterning. This covers most of today's high-speed interconnect applications.

Fabrication of the Py thin film patterns is compatible with today's complementary metal-oxide-semiconductor (CMOS) micro-fabrication technology. Process flow for Py patterns with 8 and 12 GHz FMR frequency and $\sim$40% self-inductance boost is as follows: Gold transmission lines with 5 μm signal strip width, 0.25 μm thickness, and 1 mm length are fabricated on quartz substrates through a standard liftoff process. A stack of chromium ranges from 5 to 10 nm thick, and patterned 100 nm thick Py thin films are subsequently deposited on top of the transmission line. The chromium layer was used as an adhesion layer. Py patterns with various dimensions and 10 μm length were obtained by e-beam lithography. The space between each Py structure is 80 to 100 nm. Figure 1.9 shows the scanning electron microscope (SEM) picture of a section of the transmission line with Py patterns with different widths.

The crosstalk suppression effect of magnetic material integrated transmission line fabricated using a similar method was reported in reference [9]. Py was deposited on a differential microstrip. The far end crosstalk is compared with a reference line and a Py loaded line. It demonstrates that 200-nm-thick as-deposited Py films provide significant magnetic field shielding. The measurement result shows that a 10 dB crosstalk reduction can be achieved at

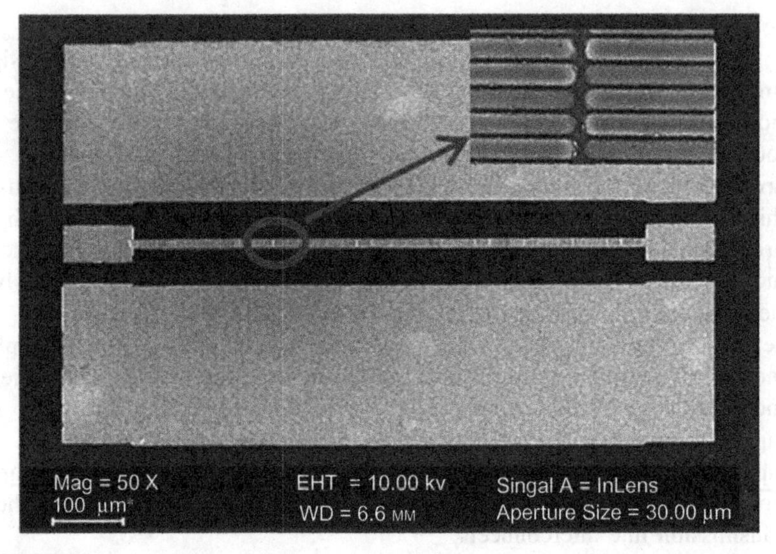

FIGURE 1.9

250 nm and 550 nm wide Py patterns on top of microstrip.

the Nyquist frequency of PCIe Gen3 signal, which translates to a three-time crosstalk voltage reduction. The magnetic shielding effects can be improved by incorporating Py films on the sidewalls and bottoms of the transmission line.

b. Slow wave transmission line

Another novel transmission line is the slow wave transmission line. Slow wave elements (SWE) are promising structures to shrink the size of radio frequency (RF) and mmWave components, which take 80% of the chip and board real estate. SWE have been widely studied to design compact RF passive components. SWE-based transmission lines are also of interest for the design of low loss and low jitter interconnects, as well as delay lines.

As discussed in previous sections, the wavelength, phase velocity, and characteristic impedance of the lossless transmission line are given as:

$$\lambda = \frac{v}{f}$$

$$v = \sqrt{\frac{1}{LC}}$$

$$Z_0 = \sqrt{\frac{L}{C}}$$

The above equations indicate that when the ratio of inductance and capacitance is locked, the wavelength and phase velocity of the

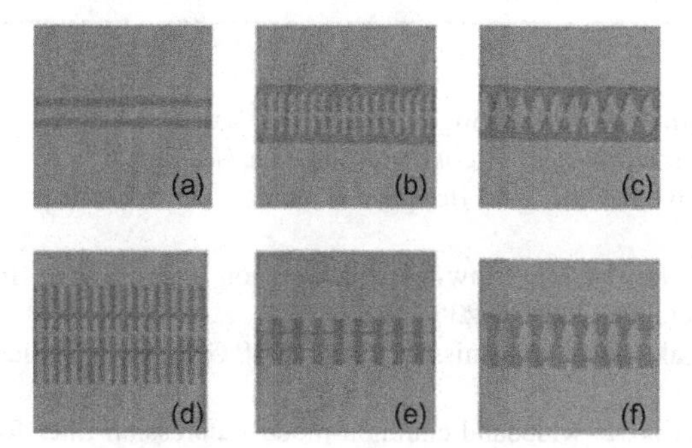

FIGURE 1.10

(a) Regular non-SWS, (b) step-type SWS, (c) zigzag-type SWS, (d) defected ground-type SWS, (e) optimized defected ground with step-type signal structure, and (f) optimized defected ground with zigzag-type signal structure.

electromagnetic wave propagating on the transmission line can be reduced by increasing the L and C values while keeping the characteristic impedance constant. The wavelength reduction at a given frequency is essential because the dimensions of the RF components usually correlate to the wavelength. It is also desirable in transmission line design because the reduced length comes with the benefit of reduced loss.

SWE can be realized by using periodically alternating characteristic impedance transmission line sections. To improve the loss induced by impedance discontinuity, cross-tie periodic structures and inhomogeneously doped semiconductor structures have been explored [10,11]. However, these structures are not fabrication friendly. CPW structure has both ground and signal strips on the same plane, which makes changing L and C relatively easy compared with other structures. As a result, it is a good candidate for SWE design. Numerous studies on different types of CPW SWEs have been reported. New techniques, including the use of defected ground structure and different signal strip shape, have been implemented to achieve higher slow wave effect while maintaining comparative loss [12]. Figure 1.10 shows a series of designed CPW SWEs and a regular CPW for reference.

The results show that more than 42% length reduction can be achieved at the expense of only 0.3 dB insertion loss increase, which translates to a 66% area reduction for the design of a branch line coupler [12].

The two novel transmission line examples can even be combined. Patterned magnetic thin film can be integrated on top of the CPW SWE and boost the slow wave effects [12]. The slow wave effect of the combined design is DC current tunable as applied, thanks to the observation of this phenomenon in reference [13]. This provides a solution in designing RF passive components and high-speed system elements that can work in wide frequency bands.

REFERENCES

[1] Pozar D. Microwave engineering. 4th ed. 2012.
[2] <http://www.microwaves101.com/encyclopedias/stripline>.
[3] Bahl IJ, Trivedi DK. A designer's guide to microstrip line. Microwaves 1977:174–82.
[4] Kim J, et al. Novel CMOS low-loss transmission line structure. In: IEEE radio and wireless conference. 2004. p. 235–38.
[5] Izadian J, et al. Novel transmission line for 40 GHz PCB applications. DesignCon 2011.
[6] Wu S, et al. A novel wideband common-mode suppression filter for gigahertz differential signals using coupled patterned ground structure. IEEE Trans Microw Theory Tech 2009;57(4):848–55.
[7] Sercu S, et al. BER link simulations. DesignCon 2003.
[8] Wang P, et al. Permalloy loaded transmission lines for high-speed interconnect applications. IEEE Trans Electron Devices 2004;51(1):74–82.
[9] Seki S, et al. Cross-tie slow-wave coplanar waveguide on semi-insulating GaAs substrates. Electron Lett 1981;17(25–26):940–1.
[10] Wu K, Vahldieck R. Hybrid-mode analysis of homogeneously and inhomogeneously doped low-loss slow-wave coplanar transmission lines. IEEE Trans Microw Theory Tech 1991;39(8):1348–60.
[11] Rahman BMF, et al. High performance tunable slow wave elements enabled with nano-patterned permalloy thin film for compact radio frequency applications. J Appl Phys 2014;115(17):17A508.
[12] Zhang H, et al. Direct current effects on high-frequency properties of patterned Permalloy thin films. IEEE Trans Magn 2009;45(12):5296–300.

CHAPTER

PCB design for signal integrity

2

There are two kinds of engineers — those who have signal integrity problems, and those who will.
— Eric Bogatin

The printed circuit board (PCB) is one of the greatest bottlenecks in implementing high-speed designs. The relatively inexpensive material for volume board designs is slow to change, compared to the rate at which transistor speeds are increasing. Higher speeds are achieved in transistor I/O while the PCB material and structures see little change, carrying the same bandwidth capabilities used a decade ago and putting a greater burden on the I/O design. The opportunity is at hand to design the best signaling path with today's current technology and leverage the new capabilities that have come into the PCB industry. This chapter discusses important signal integrity analysis techniques, considerations in stack-up design, and common layout practices for successful high-speed design.

DIFFERENTIAL SIGNALING

In order to reduce interference and extend the capabilities of high-speed circuits, many interfaces have adopted a differential signaling architecture over single-ended signaling. Differential signals in PCB design are more resilient against the presence of nearby coupling, allowing for longer lengths and higher speeds in ever denser designs. Differential signaling exchanges noise tolerance for more layout area, using two signal traces. Differential signaling requires strict layout rules that will lead to functional failures if not followed. A sample of such interfaces using differential signaling include USB, Serial ATA, InfiniBand, PCI Express (PCIe), XAUI, and 10-gigabit Ethernet.

Figure 2.1 illustrates the theoretical ideal noise cancellation on a differential signal. Equal and opposite signals are transmitted and absorb a noise source equally during transmission. At the receiver's differentiator, the signals are

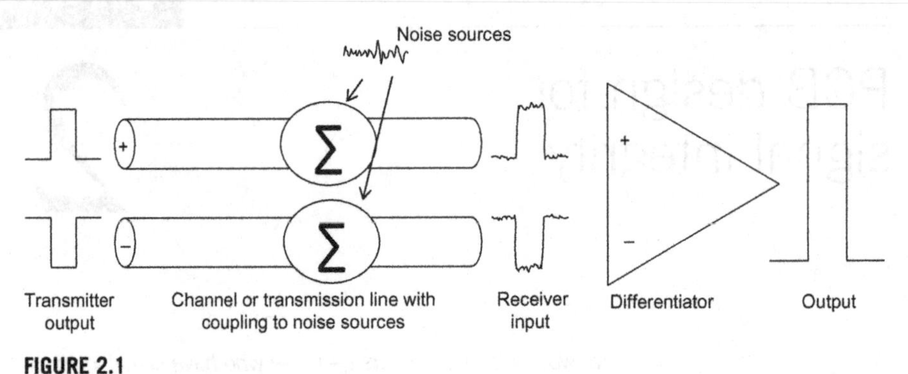

FIGURE 2.1

Differential signaling concept.

subtracted, removing the noise and doubling the signal strength. In reality, the noise cancellation is a function of how tightly the differential signals are coupled. A loosely coupled differential pair receives noise differently on each trace due to the different separations from the aggressing signal, decreasing the effectiveness of differential signaling. Maintaining a tightly coupled pair also reduces the electromagnetic interference (EMI) impact from microstrip traces. The opposing magnetic fields cancel, greatly reducing the field energy.

A tightly coupled differential pair moves return current that otherwise flows on the reference plane onto the opposing signal within the pair. The amount of current returned on the opposite signal depends on the geometry of the trace and the buffer design. When less return current is present on the reference plane, the differential signals are less sensitive to voltage noise appearing on the ground plane and non-idealities in the reference plane. In such cases, it may be possible to route over plane splits or voids and low-noise power planes. Layout guidance for differential signals may be found at the end of this chapter.

IMPEDANCE

Characteristic impedance is the intrinsic and instantaneous property of the geometry cross section and not a function of the traveling length. Impedance is a design variable used to minimize mismatch between transmission lines, vertical components, and transceiver terminations. Impedance terms can be calculated from voltage and current terms; however, for this application we discuss the calculations used to analyze the characteristics of transmission line models by inductance and capacitance before using them in a simulation environment. For the differential pair in Figure 2.2, self- and mutual inductance and capacitance are identified.

The impedance seen by one trace within a differential pair where signals are driven in the same (common) direction is the even mode impedance. The impedance then seen as a collective differential pair is the common-model impedance,

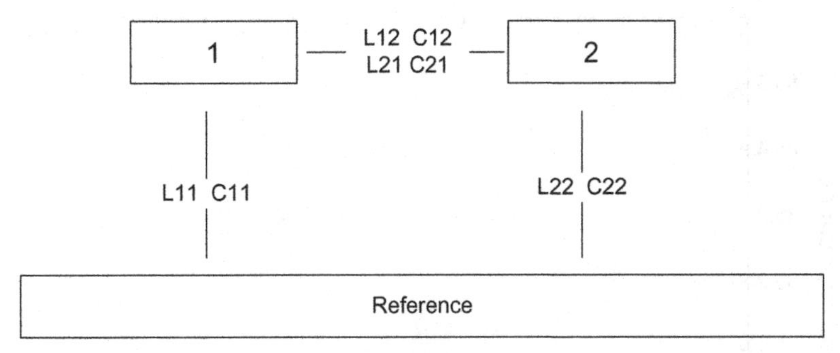

FIGURE 2.2

Inductance and capacitance terms for a differential pair.

useful for understanding the impedance of noise sources common to both signals with a differential pair. Even and common-mode impedances are defined by equations (2.1) and (2.2):

$$Z_{even} = \sqrt{\frac{L_{11} + L_{12}}{C_{11} - C_{12}}} = 2 \times Z_{common} \quad (2.1)$$

$$Z_{common} = 2 \times Z_{even} \quad (2.2)$$

For signals driven in opposite directions within a differential pair, the impedance on one trace is described as the odd mode impedance. The differential pair impedance is twice the odd mode and describes the impedance seen by incident signal of the differential pair. Odd and differential mode impedances are defined by equations (2.3) and (2.4):

$$Z_{odd} = \sqrt{\frac{L_{11} - L_{12}}{C_{11} + C_{12}}} \quad (2.3)$$

$$Z_{diff} = 2 \times Z_{odd} \quad (2.4)$$

Inductive and capacitive terms are related to the stack-up and differential pair design and are most accurately extracted from electric field solvers. A good rule of thumb is to extract L and C terms in the frequency range of test equipment used for impedance qualification in the manufacturing process. Generally, equipment edge rates are in the range of 500 MHz to 1 GHz. The frequency choice is not critical, however, as the slope in impedance over frequency is slow. Figure 2.3 displays the differential impedance over frequency for a stripline differential pair at 85 ohms.

Trace width and dielectric height have the greatest role in the characteristic impedance. A design of experiment (DOE) is used to demonstrate the sensitivity of stack-up terms on differential impedance. The example shown in Figure 2.4 is a differential stripline with parametric sweeps of trace width (*w*), trace space (*s*), core dielectric height (*h1*), core dielectric constant (*er1*), prepreg dielectric height (*h2*), prepreg dielectric constant (*er2*), and trace thickness (*tt*). Stronger slopes of

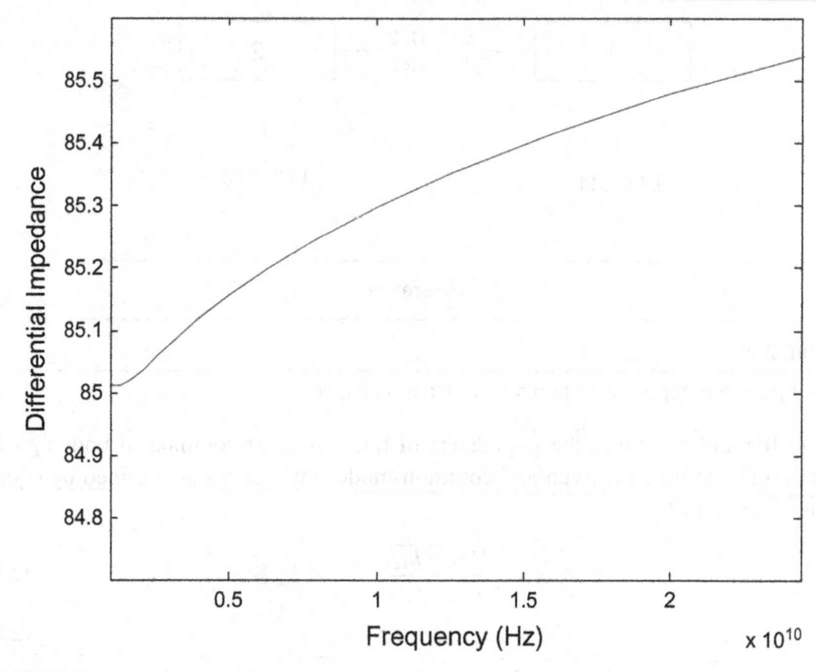

FIGURE 2.3

Differential impedance increase with frequency is negligible.

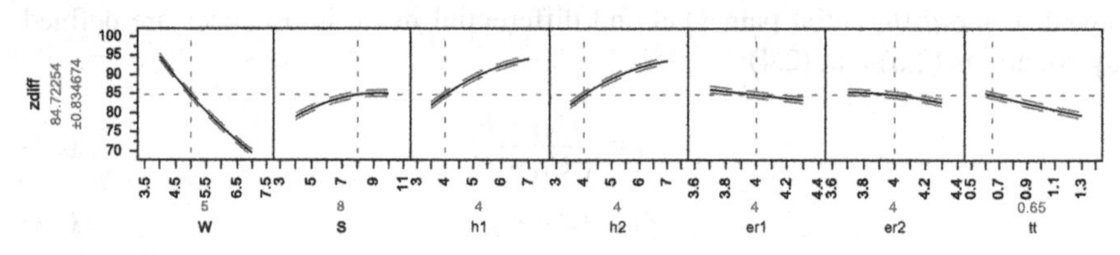

FIGURE 2.4

Parametric sweep sensitivity for stripline differential impedance (mils).

trace width and dielectric heights indicate their significant role in impedance, rooted in their influence on capacitance terms. Increases in width, spacing, and height increase the differential impedance. Increases in dielectric constant and trace thickness decrease the differential impedance.

In order to obtain differential impedance calculations quickly and without complex field solver simulation, equations for microstrip and stripline are offered as approximations for the design phase. Equations for odd mode and differential impedance are offered by Eric Bogatin in *Differential Impedance Finally Made Simple.* Odd mode impedance is offered in equations (2.5), (2.6), and (2.7) for microstrip, symmetrical stripline, and asymmetrical stripline, respectively. Coefficients for the odd mode microstrip impedance calculation are altered from the original equation to improve correlation with field solver results. The

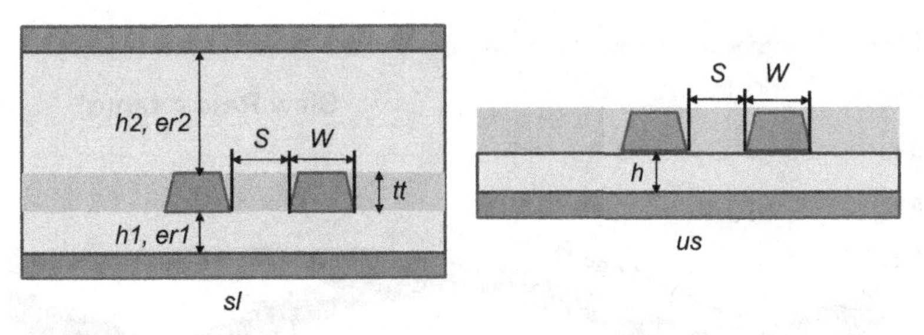

FIGURE 2.5

Stack-up parameters for stripline (left) and microstrip (right).

dielectric constant E_r is an effective dielectric constant and may be approximated as an average between the upper and lower dielectrics. Stack-up parameters for the equations are defined in Figure 2.5:

$$Z_{odd\_us} = \frac{100}{\sqrt{E_r + 1.41}} \ln\left(7.5\frac{h}{w}\right) \tag{2.5}$$

$$Z_{odd\_sym\_sl} = \frac{60}{\sqrt{E_r}} \ln\left(2.35\frac{(h1 + h2 + tt)}{w}\right) \tag{2.6}$$

$$Z_{odd\_asym\_sl} = \frac{80}{\sqrt{E_r}} \ln\left(4.75\frac{h1}{w}\right)\left(1 - \frac{h1}{4 \times h2}\right) \tag{2.7}$$

Equations for differential microstrip and stripline impedance are shown in equations (2.8) and (2.9), respectively:

$$Z_{diff\_us} = 2 \times Z_{odd\_us} \times \left(1 - 0.48e^{\left(-0.96\frac{s}{h}\right)}\right) \tag{2.8}$$

$$Z_{diff\_sl} = 2 \times Z_{odd\_sl}{}^{*}\left(1 - 0.347e^{\left(-2.9\frac{s}{(h1+h2+tt)}\right)}\right) \tag{2.9}$$

TIME DOMAIN ANALYSIS

EYE DIAGRAM

The eye diagram is a general tool commonly used for transmitter quality measurement, receiver stressed eye and jitter injection test calibrations, and system margin tests. The eye diagram is a superposition of a significant population of unit intervals (UI) of a high-speed signal. An eye diagram of a PCIe Gen2 transmitter (TX) signal captured after a long channel on a customer reference board is shown in Figure 2.6. A horizontal span of 1.5 UI is shown. Both amplitude and timing characteristics of the high-speed signal are given by an eye diagram. Several important parameters, such as signal rise/fall time, minimum eye width (EW), minimum eye height (EH), and total jitter (TJ) at a certain bit error rate (BER), can be characterized with the

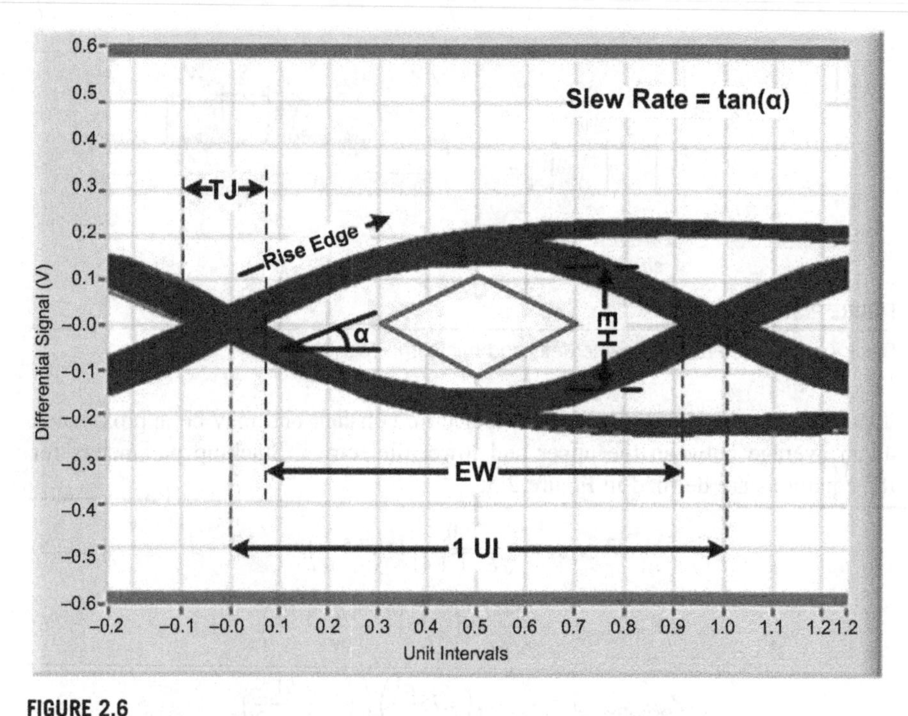

FIGURE 2.6

A PCIe Gen2 TX eye diagram.

eye diagram. Illustrations of these parameters are shown in Figure 2.6. It is clearly seen that the edges of the UIs are spread on the time axis instead of landing at the same ideal position. Timing variation of a high-speed signal bit crossing point from its ideal position is called jitter. Jitter is one critical parameter in high-speed design and will be discussed in detail in the next section. The eye diagram is a direct tool for jitter behavior extraction. When a receiver's sampling point is positioned at the center of an eye, the probability of an error is lower than when the sampling point is close to the edge of the eye. The gap between the edge of the eye and the sampling point is the margin until an error occurs. In Figure 2.6, the eye opening is often compared to an eye mask to determine passing or failing performance, shown by the red diamond at the center of the eye. In order to pass an eye mask of a certain specification, the waveforms of all the accumulated bits must be outside the eye mask area. The purpose of the eye mask is to account for unmeasured (or unsimulated) noise sources from the transmitter or receiver. For example, in the PCIe Gen2 transmitter card electromechanical (TX CEM) spec, the eye mask is 95 ps EW and 225 mV EH, and the PCIe Gen3 TX CEM eye mask is 41.25 ps EW and 46 mV EH for the system transmitter. A broader definition of the eye mask and its application to the system validation and product risk assessment will be discussed in Chapter 7.

JITTER

Jitter describes timing noise impacts on a signal. As shown in Figure 2.6, jitter causes the eye diagram to close and signal edges interfere with the decision-making circuit in the receiver. The consequence is confusing a "1" with a "0" or vice versa by the receiver and generating an error. BER is defined as the ratio of the number of errors received to the number of bits transmitted. BER is an important concept for high-speed system design and TJ of a certain system is defined in terms of BER. Total jitter with respect to BER and its jitter components needs to be determined in order to completely understand the performance of the system. Separating (decomposing) the Tj into different jitter types is a useful diagnosis of the system. For example, only a thorough analysis of the different jitter components makes it possible to identify which element of the problematic system is causing the eye diagram to fail the horizontal eye mask.

Jitter components and budget

Jitter in a high-speed system is a very complicated phenomenon and has drawn broad attention. Mathematical models, e.g., the dual-Dirac model, are often used to understand the jitter behavior of the system. Jitter is usually separated or decomposed to individual components in these models. The values of each component and component coefficients can be tuned to fit the overall system jitter behavior. As a result, a model can be used to extrapolate the jitter performance in certain circumstances, e.g., at ultra-high BER where data capture can be very time consuming. Since each jitter component has its own characteristic and specific physical meaning, jitter separation can be a useful tool to trace back to the root cause for a low system timing margin. Detailed jitter decomposition and separation algorithms (such as dual-Dirac model) are out of the scope of this book.

The most commonly used jitter separation is shown in Figure 2.7. In this chart, the TJ is decomposed into two categories at the first level: random jitter (RJ) and deterministic jitter (DJ). The key differentiation factor for RJ and DJ is whether the component's probability distribution function (PDF) is bounded or unbounded. Theoretically RJ can reach any magnitude. Hence RJ is ubiquitously not quantified by its peak to peak value. Instead RJ is assessed by the standard deviation of its Gaussian distribution, whose tails extend infinitely with decreasing probability. When a model is used to determine the TJ at a BER, proper assessment of RJ's standard deviation is a prerequisite for success. For example, to indirectly calculate TJ to 10^{-12} BER, the RJ contribution is equal to its standard deviation multiplied by 14.1. As a result, a small error in extracting RJ has a significant impact on the accuracy of a TJ estimation at 10^{-12} BER. The PDF for the RJ applied to this signal is also shown in Figure 2.8. Primary sources of RJ are thermal noise, clock/oscillator source, and clock recovery circuits.

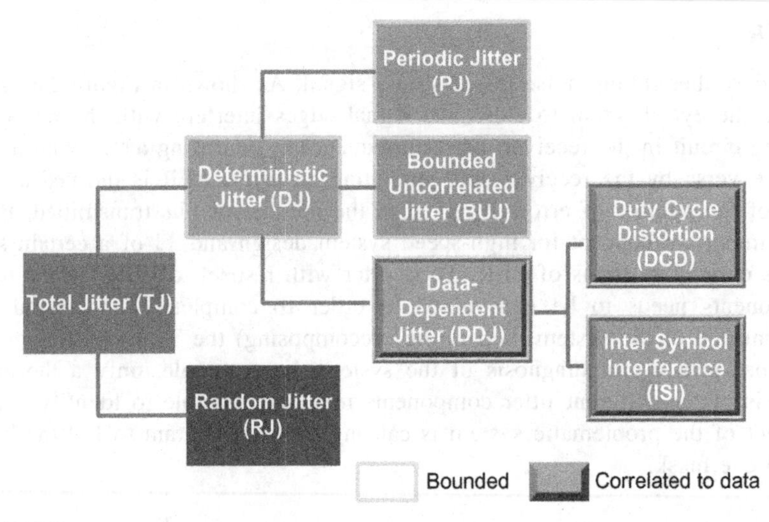

FIGURE 2.7

Jitter components overview.

In contrast, DJ is bounded in nature. The peak to peak value of DJ's PDF is a finite value that is repeatable and predictable. Consequently, DJ PDF peak to peak, aka DJDD, is usually used to assess DJ. DJ can be further decomposed into three categories: periodic jitter (PJ, also named sinusoidal jitter, SJ), bounded uncorrelated jitter (BUJ), and data-dependent jitter (DDJ). PJ and BUJ are not correlated to the data pattern being transferred in the system. An eye diagram for a 10-Gbps clock signal with 0.05 UI of PJ is shown in Figure 2.9. The PDF for the PJ applied to this signal is also shown in Figure 2.9. Sources for PJ are switching power supply noise, strong radio-frequency (RF)/EMI, and unstable clock recovery.

BUJ is bounded in nature too; its PDF is a truncated Gaussian without the infinite tails. As a result, it is evaluated by a peak to peak value. BUJ is uncorrelated to the data transmitting on the link and can occur at low probabilities, making it appear random. BUJ can be caused by crosstalk.

As the name indicates, DDJ correlates to the data stream that is being transmitted on the link. The data pattern itself is the cause of DDJ. Separation of the jitter into DDJ effectively identifies the source of jitter as the channel (DDJ) or the silicon (all others). DDJ usually does not have a distinct statistic distribution as do RJ and PJ. It is usually directly observed on the data bit graph. The signal edges arrive either earlier or later than the ideal time locations. The distribution for the errors or DDJ histogram, negative for edges arrived earlier and positive for the edges arrived later, is obtained after accumulating the specified number of bits. DDJ can be further separated into duty-cycle distortion (DCD) and intersymbol interference (ISI). DCD is the result of two abnormal phenomena in the

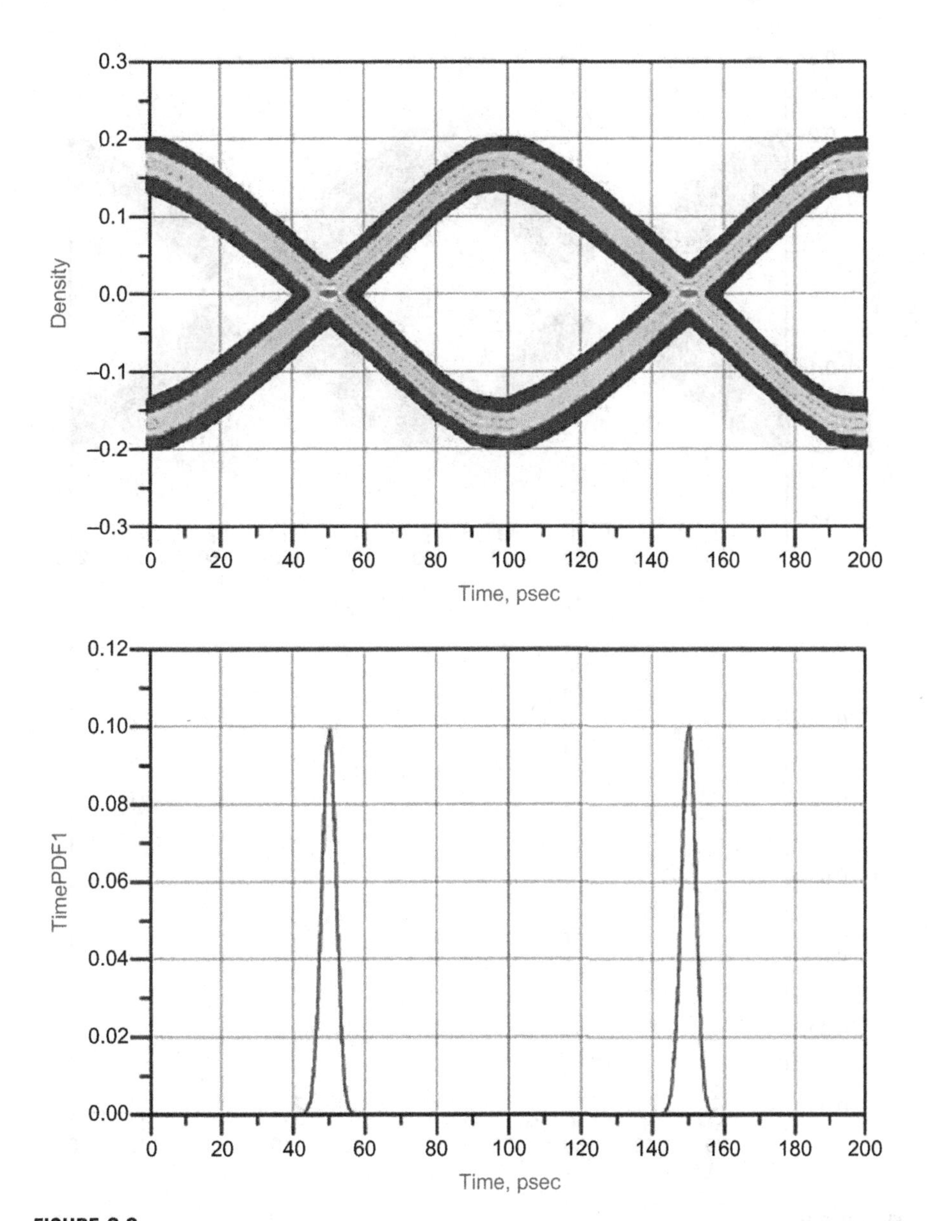

FIGURE 2.8

Random jitter eye diagram and PDF.

waveform—one is the rising and falling edge mismatch due to insufficient voltage supply, e.g., faster rising edge and slower falling edge or vice versa, and the other is the common-mode offset of signal. The result is that the ones in a bit sequence are always having a longer or shorter duration than the zeros. DCD is one of the

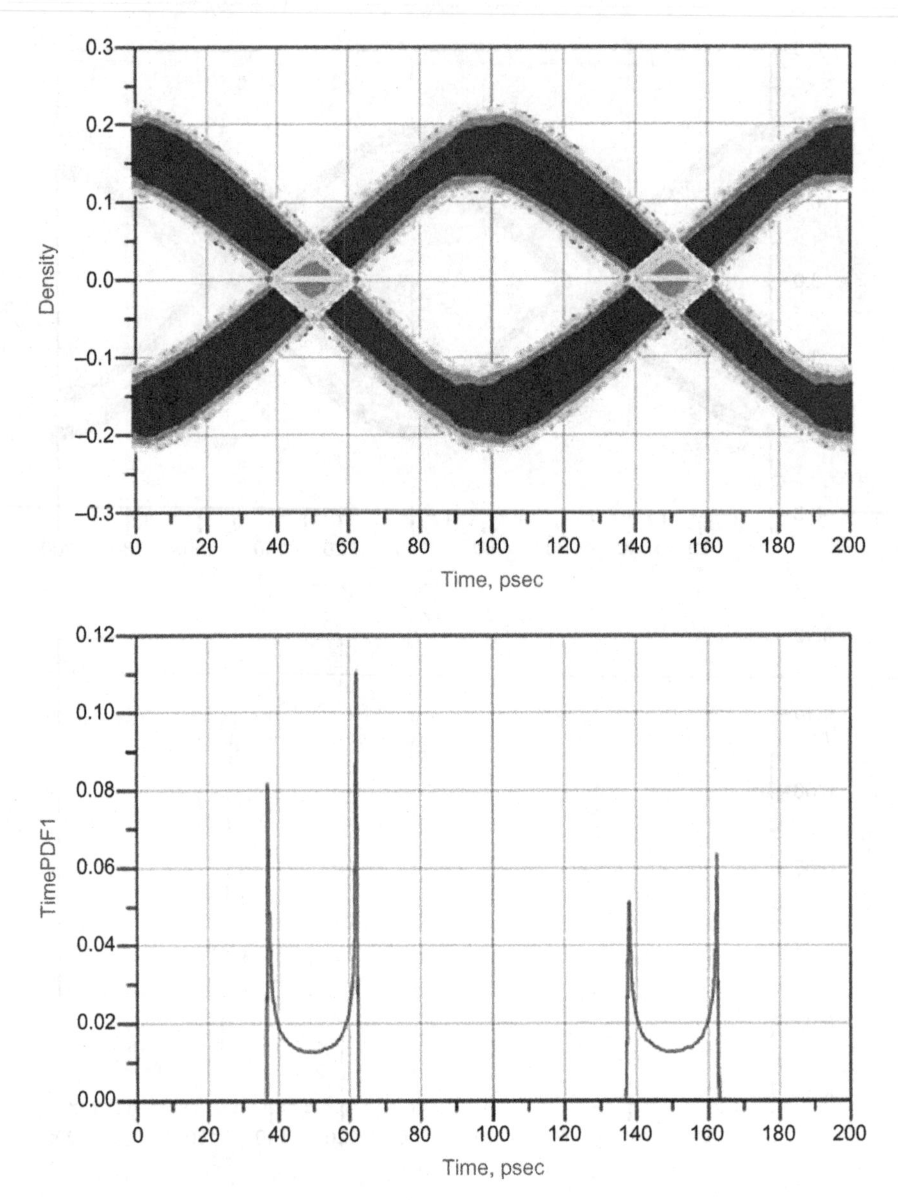

FIGURE 2.9

Periodic jitter eye diagram and PDF.

major contributions to DDJ on a transmitter. ISI, on the other hand, is caused by reflections and bandwidth limitations of the transmission line (called dispersion). Conductor loss and dielectric loss differences at different frequency points cause each single bit in the data stream to deform. The deformed single bit can exceed

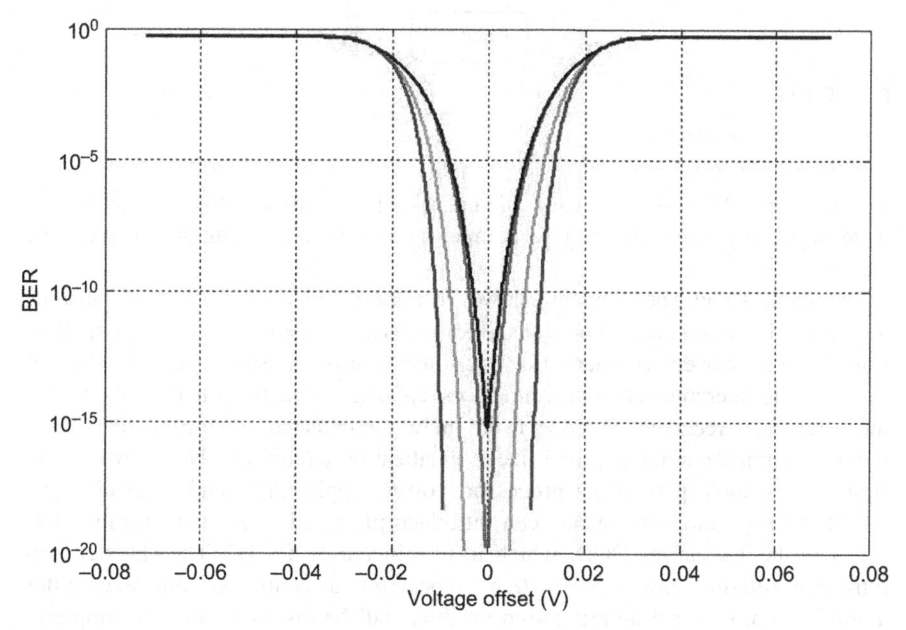

FIGURE 2.10

Bathtub plot.

one UI and can have ripples at time points that are even many UIs away. This single bit deformation in turn will impact the bits that are next to it or a few UIs around the current bit. The ISI can get its worst scenario when a single bit is surrounded by bits of the opposite polarity, and this can ultimately close the eye.

A bathtub curve is a single slice of the 3D intensity or BER eye contour at a specified horizontal or vertical sampling point (ideally, the vertical sampling point is 0 mV). A bathtub curve is shown in Figure 2.10 for a 16-GT/s data rate on a 28-dB (at 8-GHz) channel. A selection of bathtub curves are displayed to show the effect of accumulating noise sources on the margin. The outer curve (brown) is the cumulative density function (CDF) for the channel without any added noise sources. The simulated curve includes the channel loss and reflections after equalization from TX FFE, CTLE, and DFE. At a BER of 1e-12, the width of the bathtub is 22.6 mV of eye height. The next inner curve (teal) adds the simulated crosstalk and decreases the eye height to 15.6 mV. Deterministic jitter sources are added to the next inner curve (magenta). The total deterministic jitter contribution is 6.0 ps out of the UI, for a new eye height of 5.8 mV. In the last curve 0.7 ps RMS of random transmitter jitter is added to the simulation. The final eye height observed at the sample position is 3.4 mV. In this example, receiver uncertainties are not added to the simulation. Inclusion of receiver uncertainties may not be required if they are budgeted into an eye mask. This is common for some connector or pin interoperability specifications such as the PCI Card Electro-Mechanical

FIGURE 2.11

A typical high-speed system link.

Specification. After the inclusion of any sizable eye mask, this channel would have negative margin and require compensation (shorter length, stronger CTLE, etc.).

A simplest high-speed system shown in Figure 2.11 is comprised of transmitter, channel, and receiver. The transmitter consists of pre-driver, driver, equalization circuitry, current compensation loop, termination resisters, etc. The channel may include interconnects, via, connectors, cabling, and active devices like a retimer, etc. The receiver includes even more complicated circuitries like clock recovery circuit, continuous time linear equalization circuit, etc. Not shown in the figure is the logic part of the processor, power supply, PLL, and other chipsets. All of these system components can introduce jitter and noise. For example, DC offset in the TX causes DCD, which in turn increases TX DJ. The channel may introduce filtering, non-linearity (e.g., repeaters), discontinuity, and even jitter amplification. Channel amplification on jitter will be discussed as an example in the next section. As the data stream propagates from the transmitter to the receiver, the signal jitter generally gets worse.

In order to let the receiver successfully determine whether the data stream coming in is a "1" or "0," the received signal's eye must be sufficiently open both horizontally and vertically. This is why in the validation phase of product development an eye mask is often defined to evaluate if the system performance is good enough by comparing the system eye to an eye mask requirement. Most validation coverage schemes will only evaluate a limited set of pre-production parts yet attempt to project the behavior of the total system across high-volume manufacturing (HVM), so the eye mask will need to take this into account. More details on this will be discussed in Chapter 7. All in all, a careful and thorough interface, e.g., PCIe, and design specification will allocate a jitter budget so that each component of the system has known jitter limits and that adequate margin remains when the signal enters the receiver to be recovered. This can be done by a cross-team exercise among circuit designers, specification definition team, signal integrity engineers, and validation architects. A jitter budget example for a PCIe 8.0-GT/s (Gen3) transmitter is shown in Table 2.1.

Jitter amplification example

A band-limited channel having a varying channel response across frequencies and significant insertion loss also contribute to the total jitter. This section gives an example of how a channel amplifies the input jitter. The simulation and analysis are similar to those of F. Rao and S. Hindi in *Frequency Domain Analysis of Jitter Amplification in Clock Channels* in 2012 and may be referred to for more

Table 2.1 8.0-GT/s Specific TX Jitter Parameters

Symbol	Parameter	Value	Units
$T_{TX\text{-}UTJ}$	TX uncorrelated total jitter	31.25	ps PP @ 10^{-12}
$T_{TX\text{-}UDJDD}$	TX uncorrelated deterministic jitter	12	ps PP
$T_{TX\text{-}UPW\text{-}TJ}$	Total uncorrelated PWJ	24	ps PP @ 10^{-12}
$T_{TX\text{-}UPW\text{-}DJDD}$	Deterministic DJDD uncorrelated PWJ	10	ps PP
$T_{TX\text{-}DDJ}$	Data-dependent jitter	18	ps PP

details. In this example, a channel with an insertion loss of 20 dB at 5 GHz and return loss greater than −25 dB is used, similar channel characteristics as in current server platforms using Intel processors. Definitions of insertion loss and return loss will be discussed in later sections of this chapter.

Clock signal represented by a 1010 square wave is transmitted into the channel. The periodic nature of clock pattern cancels out ISI so that jitter at the measured jitter is entirely induced by the transmitter. Periodic jitter and random jitter amplification phenomena are presented in this section. PJ and RJ of Gaussian distribution were applied individually at transitions on the transmitter side. The received signal is computed with the linear superposition method to reduce run time compared to the step response in transient simulation [1, p. 52]. One million UIs were accumulated in each simulation. PJ and RJ PDF are directly extracted at the channel output port.

PJ with amplitude of 5 ps at frequencies from 0.5 GHz to 4.0 GHz were injected to the channels sequentially. The output PJ amplitude of the 0.5-GHz PJ input is used as baseline. PJ amplification factors, i.e., output PJ amplitudes at other frequencies divided by the baseline, as functions of PJ frequency at 10 Gbps data rate are plotted in Figure 2.12. It can be observed that the PJ amplification factor grows exponentially with respect to PJ frequency. Output PJ PDFs at 0-V transitions of three input frequency points, 500 MHz, 2.5 GHz, and 4 GHz at a 10-Gbps data rate, are shown in Figure 2.13. It is clearly seen that peak to peak values of PJ PDFs increased significantly as PJ frequency increases.

Random jitter amplification is also studied. RJ with 1-ps amplitude and Gaussian distribution was injected at the channel input. The standard deviation at the output is measured and compared to the input Gaussian standard deviation. RJ amplification calculation is given in equation (2.10) as defined by Chaudhuri, Anderson, Bryoan, McCall, and Dabral [2, p. 21]:

$$RJ_{Amp} = \frac{Output\ Stdev}{Input\ Stdev} \tag{2.10}$$

Output RJ amplitude at a 6-Gbps data transfer rate is used as baseline. Six-Gbps and higher data rates are of particular interest to align with current high-speed I/O throughput, e.g., Serial-Attached SCSI (Small Computer System Interface), PCIe Gen3, 10GbE (GigaBit Ethernet), and Quick Path Interconnect (QPI).

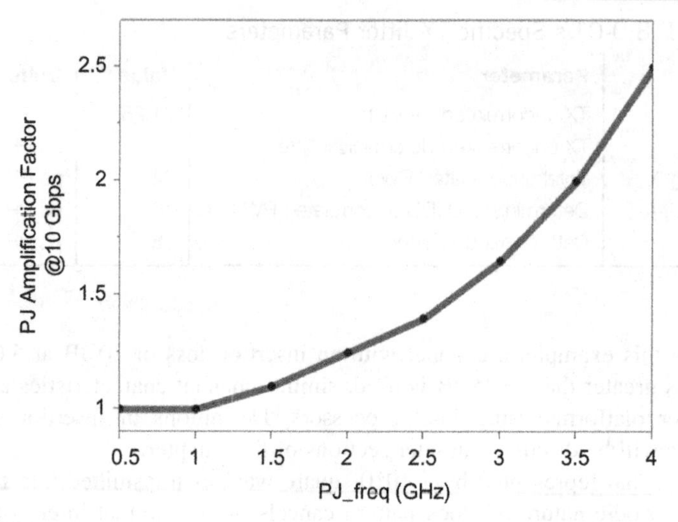

FIGURE 2.12

PJ amplification factor versus PJ frequency.

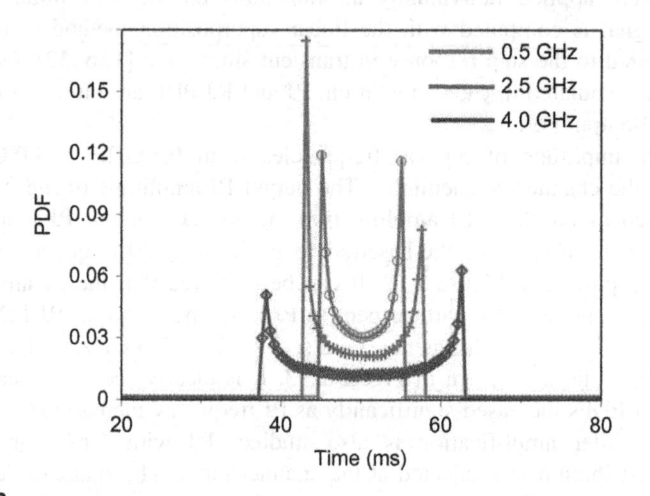

FIGURE 2.13

PJ output PDF at different PJ frequencies.

RJ amplification factors as functions of data rate for different data rates are plotted in Figure 2.14. RJ amplification factor grows exponentially with respect to data rate. Output RJ PDFs at 0-V crossing points at 8-, 12-, and 16-Gbps data rates are overlaid in Figure 2.15. The RJ PDF standard deviation increased from 1.6 ps to 3.2 ps.

Previous research studied the jitter transfer function and amplification factors for PJ, DCD, and RJ in clock channels. Both analytical and measurement results showed

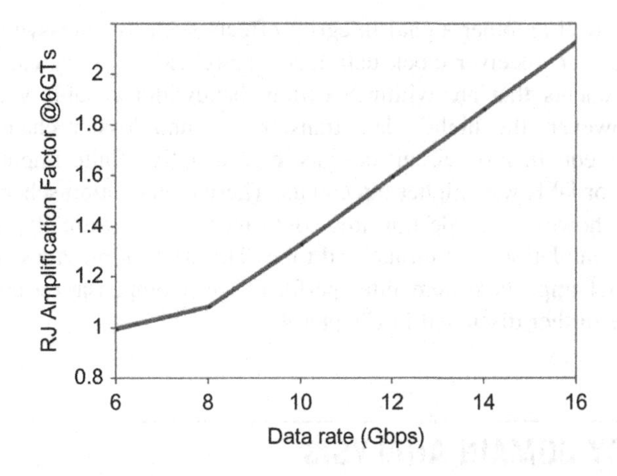

FIGURE 2.14

RJ amplification factor versus data rate.

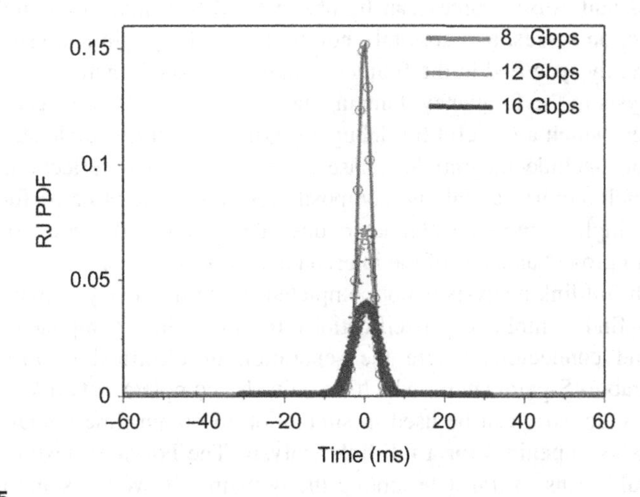

FIGURE 2.15

RJ output PDF at different times for different data rates.

that jitter amplification is the consequence of smaller attenuation at the jitter lower sideband than at the fundamental. Today's high data transfer rate systems, e.g., server platforms, require ever-growing data transfer rates, longer and lossier customized channels, and higher channel densities. All these factors aggravate the jitter amplification in the channel.

Equalization schemes, such as finite-filter equalization in transmitters, continuous time linear equalization (CTLE), and decision feedback equalization (DFE) in receivers, are commonly used to compensate for DDJ due to channel loss dispersion, as well as other signal integrity effects such as inter-symbol interference (ISI). Further, receiver clock data recovery (CDR) circuitry can remove DJ and RJ components that are within a certain bandwidth capability (e.g., below 10 MHz). However, the higher data transfer rate and lossier channels require more complex equalization circuit designs, e.g., adaptive finite impulse response (FIR) designs or DFE with higher tap counts. These equalization schemes are usually effective; however, the design time could increase dramatically, as could the recipe tuning, validation, and turnaround time. This section provides one example of how channel impacts system jitter performance. Compensation circuit design details will be further discussed in Chapter 4.

FREQUENCY DOMAIN ANALYSIS

Channel information represented in the frequency domain is useful in gaining an understanding of the interconnect across frequencies. In the frequency domain, all relevant loss and noise sources can be observed: differential and common-mode losses, reflection noises, and crosstalk noises. Full link margin analysis (pass/fail) is not commonly specified in the frequency domain (though there are some exceptions). Analysis in the frequency domain characterizes the interconnect into multiple capacities, which are useful for debug, design, and intuitive understanding. The measurements include magnitude, phase, transmission, and reflection. Full link analysis, which requires a real and composite response, cannot be performed from one of the single mentioned characteristics. This section discusses quantitative metrics for improved analysis of the interconnect noise sources.

Although full link analysis is not completed in the frequency domain, it is also common to find compliance specifications for individual components, such as packages and connectors, where the separation of electrical characteristics is highly favorable. S-parameters, which contain the complete network response in the frequency domain, can be used in simulation to obtain time domain information (such as eye opening) for a full link analysis. The Fourier transform provides the analytical means to translate among the domains. However, some rules must be followed to ensure adequate translation.

SPECTRAL CONTENT

Information about the spectral content for a given operating frequency can significantly aid in the system design. Power spectral density (PSD) describes the distribution of power over frequency and may be computed with the Fourier transform of a time domain data stream. In this section, the PSD for a random data sequence is considered. The PSD is changed for coded interfaces (where the lower frequency data patterns may not be permitted), and therefore the PSD may be reconsidered when this is the case. As a simplification, the PSD for a lone bit is shown

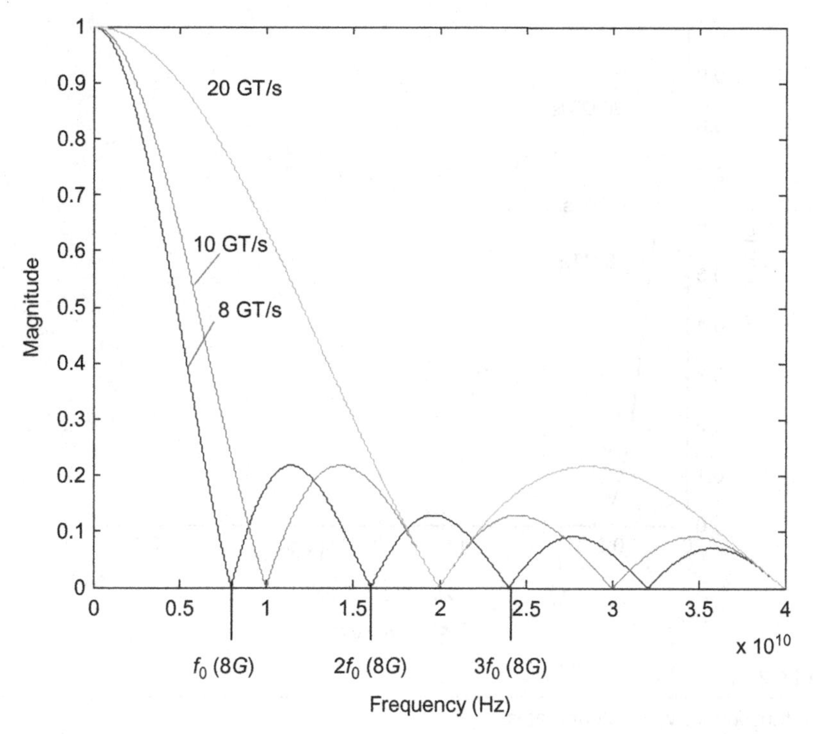

FIGURE 2.16

Sinc function for various data rates.

because it practically represents an infinitely long random data sequence. The PSD of a lone bit that will yield a rectified sync function, defined as a function of the unit interval (*UI*), is described in equation (2.11):

$$PSD_{lone\ bit} = \left|sinc(UI \times \pi \times f)\right| = \left|\frac{\sin(UI \times \pi \times f)}{UI \times \pi \times f}\right| \tag{2.11}$$

Review of the PSD in Figure 2.16 demonstrates where in frequency power will be located for a random data pattern. The magnitude of power quickly decreases with frequency, coming to a null at f_0, where $f_0 = 1/UI$, and every integer multiple thereafter. However, harmonic energy above f_0 appears to warrant concern at a magnitude of approximately 0.2, but this energy may not be transferred through entirely from transmitter to receiver. Understanding where in the frequency spectrum that power is transferred can help designers understand what interconnect features must be optimized.

The calculation for transferred power is further advanced by Healy (2009:4) to include the circuit I/O capability. Healy defines that the transferred power shall include two additional terms than the data pattern spectral content in

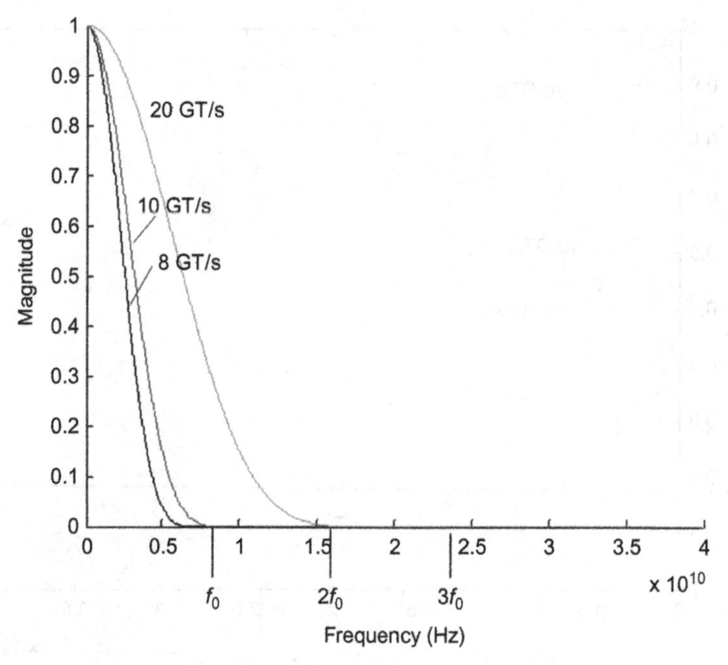

FIGURE 2.17

PWF function for various data rates.

equation (2.12)—the transmitter rise time and receiver bandwidth. The composite response defined by Healy is the power weighting function (PWF), where T_r is the transmitter rise-time and F_{RX} is the receiver bandwidth:

$$PWF(f) = sinc(UI * \pi * f)^2 \left(\frac{1}{1 + \left(f \frac{T_r}{0.2365}\right)^4}\right)\left(\frac{1}{1 + \left(\frac{f}{F_{RX}}\right)^8}\right) \tag{2.12}$$

An example PWF is created in Figure 2.17 for the same data rates in Figure 2.16. A transmitter rise time is chosen as 25 percent of the UI and receiver bandwidth at the Nyquist frequency for each response. The further filtered response of the PWF including the transmitter and receiver has dampened out the importance of higher frequency content. Responses after the first harmonic are not transferred.

INSERTION LOSS

The transferred signal through a medium is characterized in the frequency domain as the insertion loss. Insertion loss is the magnitude of S_{12} in an S-parameter matrix and expressed in dB, where port 1 is input and port 2 is output. Standard nomenclature is to express insertion loss as a negative number for attenuation and positive for gain. Insertion loss is defined in equation (2.13), where V_t is the voltage transmitted and V_r is the voltage received.

$$IL = -20\log_{10}\left|\frac{V_t}{V_r}\right| = -20\log_{10}|S_{12}| \tag{2.13}$$

Insertion loss is a common figure of merit when reviewing the characteristics of a system or single component. For differential signaling, it is standard to review S_{12} of the differential-to-differential matrix, commonly referred to as SDD_{12}, described earlier in this book. Relative comparisons of SDD_{12} in the frequency domain between new designs, design proposals, and known good designs are common. In some cases, industry specifications for insertion loss must be satisfied for components (such as connectors or packages) as well as complete systems. SCD_{12} and SCD_{12} are S-parameter matrices relating differential and common modes. SCD_{12} provides information about the differential symmetry of the link and the amount of differential energy lost as common mode. SCD_{12} describes the susceptibility to convert common-mode noise into differential mode noise. Measured or modeled insertion loss is a critical tool in system and component analysis. Insertion loss may be used in some of the following additional ways:

- Inspect channel DC loss, verifying good measurement contact or properly modeled termination.
- Review for modeling mistakes by comparing against insertion loss expectations.
- Measure the rate of PCB loss per unit length.
- Observe magnitude and frequency of resonances characterizing vias, connectors, or other structures.

In order to understand the relevant frequencies of insertion loss pertaining to the designed interface we must look to the PWF of the interface, as discussed in the previous section.

Two measured channels are shown in Figure 2.18 with a PWF for random data before and after the channel. The throughput of the channel is observed by multiplying the PWF magnitude with $|S_{12}|$. Channel A has a smooth insertion loss and channel B has a significant reflection occurring at 8.0 GHz. This feature may be the result of a significant via stub or other undesirable feature. In this example, the magnitude of the reflection is higher than normal for PCB designs to answer the question: Is the feature detrimental to the point that it must be fixed at 5-GT/s data rates? Figure 2.18 shows that the PWF at the channel output observes a minimal change between the two test channels. The near-zero energy present in the PWF near 8 GHz is the cause for a minimal difference in the transferred power between channel A and channel B. With the knowledge of the PSD of the interface, design engineers can make a first-order assessment of the sensitivity to insertion loss features in the frequency domain without performing channel simulations.

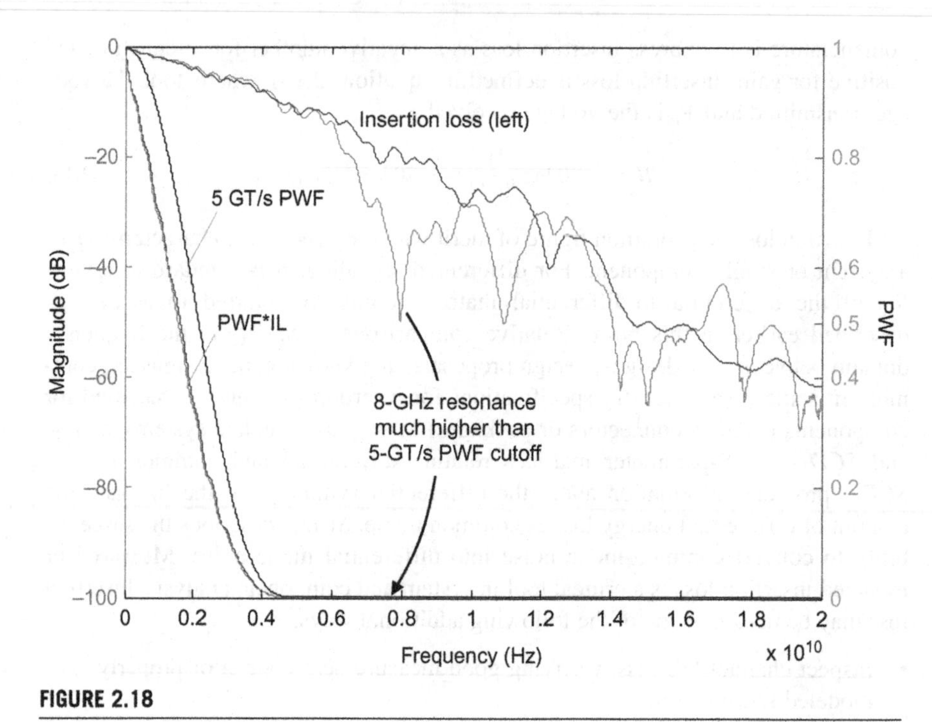

FIGURE 2.18

Insertion loss and 5-GT/s PWF with and without channel.

At higher operating frequencies a greater impact can be expected. An example for the same channel is provided at 20 GT/s in Figure 2.19. The PWF scaled by the insertion loss demonstrates as much as 10 percent signal is present after the channel in the 6- to 7-GHz range, leading to significant differences when this magnitude is reduced to 5 percent or below on channel B. At 20 GT/s the reduction in energy can be expected to have an impact on performance, requiring a change in design.

INTEGRATED INSERTION LOSS NOISE

Insertion loss contains indicators about the discontinuities in a channel. Channel reflections from discontinuities that are not transferred create "valleys" in insertion loss. Discontinuities that are re-reflected and arrive at the receiver create distortions in an otherwise smooth response. These re-reflections will destructively or constructively interfere with the insertion loss profile. The effect these discontinuities have on the insertion loss is to create "ripples" along the insertion loss profile. To illustrate this, a 10-inch transmission line and a 10-inch transmission line with large via stubs are plotted in Figure 2.20. The profile for the transmission line only is smooth, because the transmission line is homogeneous and lacks

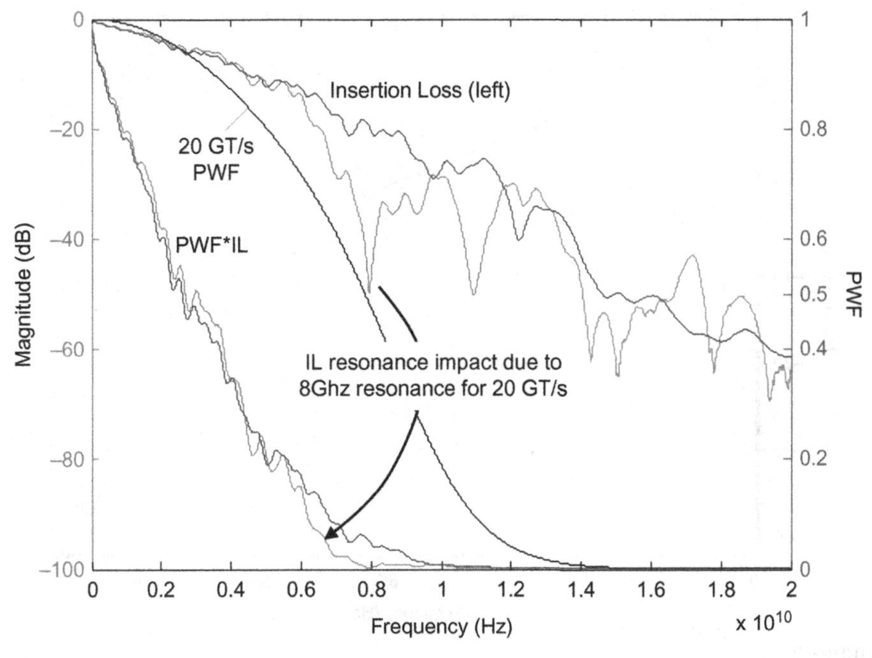

FIGURE 2.19

Insertion loss and 20 GT/s PWF with and without channel.

any impedance discontinuity. The second case with "ripples" has two significant discontinuities from 80-mil via stubs placed within the transmission line.

The noise in the insertion loss profile may be extracted into an integrated parameter useful to quantify the transferred reflection noise level. Mellitz et al. (2011:9) repurposes the PWF defined by Healy in calculations for integrated crosstalk noise (σ_{ICN}) and applies it to a new parameter, integrated insertion loss noise (σ_{IILN}). These parameters represent the average total power from the noise source.

In the first step, the noise in the ripples is isolated from the insertion loss as the difference to an insertion loss fit over a specified frequency range. An insertion loss fit is defined by the IEEE 802.3ap Backplane Ethernet specification in Annex 69B (2007:172). Other methods are permissible, including one published by Mellitz et al. (2011:8) that includes feed-forward equalization. The least mean squares fit in the 802.3ap specification is defined in equations (2.14), (2.15), (2.16), (2.17), and (2.18) (IEEE 2007:172):

$$f_{avg} = \frac{1}{N}\sum_{n} f_n \tag{2.14}$$

$$IL_{avg} = \frac{1}{N}\sum_{n} IL(f_n) \tag{2.15}$$

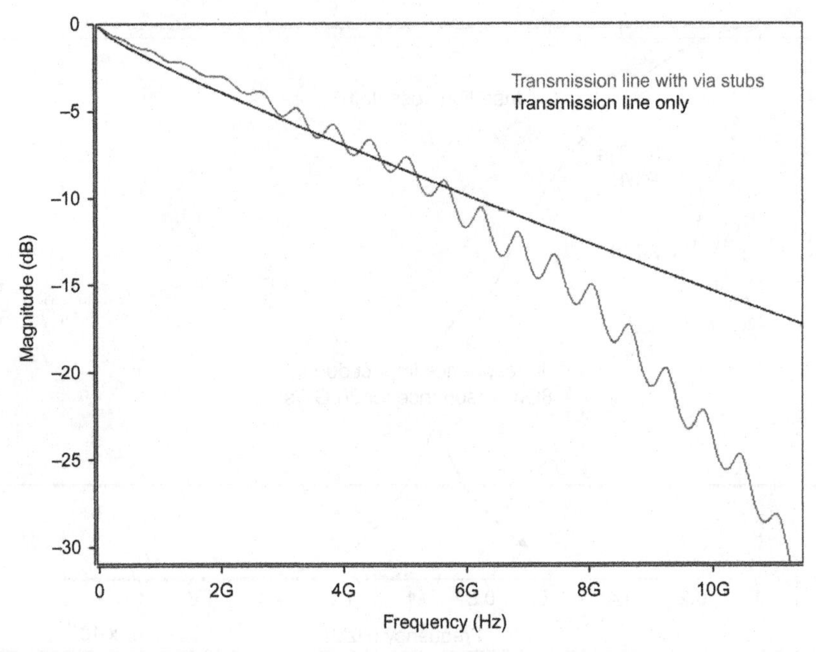

FIGURE 2.20

Insertion loss with and without reflective ripples.

$$m_A = \frac{\sum_n (f_n - f_{avg})(IL(f_n) - IL_{avg})}{\sum_n (f_n - f_{avg})^2} \tag{2.16}$$

$$b_A = IL_{avg} - m_A \times f_{avg} \tag{2.17}$$

$$A(f) = m_A f + b_a \tag{2.18}$$

The insertion loss noise (ILN) is calculated as the difference between the insertion loss and the insertion loss fit. The σ_{IILN} will be integrated over a specified frequency range to find the power and then changed to voltage according to Mellitz (2011:9) in equations (2.19) and (2.20):

$$ILN(f_n) = |sdd21(f_n)| - 10^{\frac{ILfit(f_n)}{20}} \tag{2.19}$$

$$\sigma_{iiln} = \sqrt{\frac{2 \times \Delta f \times \sum_n PWF(f_n) \times ILN(f_n)^2}{F2 - F1}} \tag{2.20}$$

For the responses shown in Figure 2.9, the σ_{IILN} is calculated to be 2.27 mV for the smooth transmission line and 21.6 mV for the transmission line with vias for an input data rate of 8.0 GT/s.

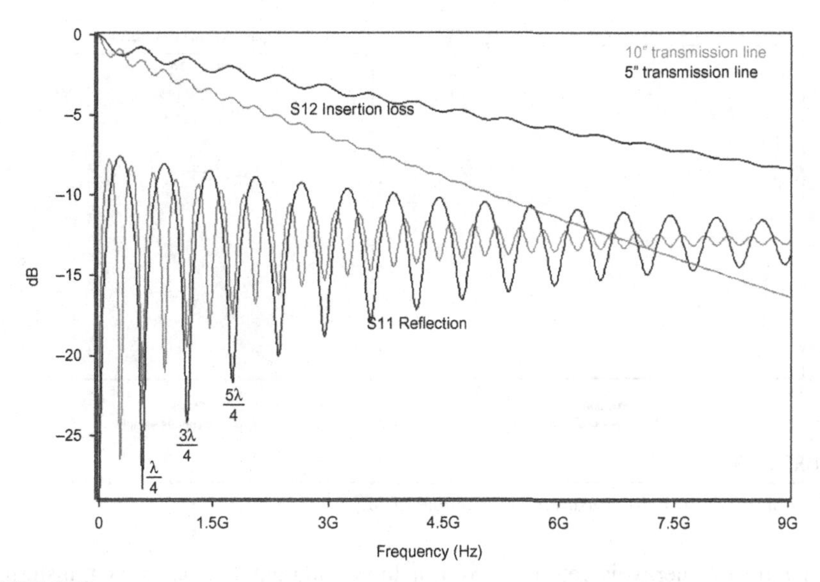

FIGURE 2.21

S_{11} and S_{12} for 5- and 10-inch transmission lines.

RETURN LOSS

Return loss is the ratio of the input signal to the reflected signal. The reflection is the result of an impedance discontinuity when transitioning between transmission lines or structures of different impedance. Return loss is a metric to describe how the impedance is matched between transmission lines and devices or systems. Return loss is commonly expressed in dB, being defined as shown in equation (2.21):

$$RL(dB) = 10\log_{10}\frac{P_r}{P_i} \tag{2.21}$$

where P_r is reflected power and P_i is the incident power. Frequency domain analysis of S_{11} from an S-parameter is commonly labeled return loss, though S_{11} is actually the magnitude of the reflection coefficient ($S_{11} = |\Gamma|$). Return loss and reflection coefficient are tightly related as shown in equation (2.22):

$$RL(dB) = -20\log_{10}|\Gamma| \tag{2.22}$$

and reflection coefficient Γ is a complex number describing the magnitude and phase of the reflection, as shown in equation (2.23):

$$\Gamma = \frac{V_r}{V_i} \tag{2.23}$$

The relationship between incident power and reflected power is illustrated clearly in Figure 2.21 for a 10-inch transmission line. At frequencies for which a

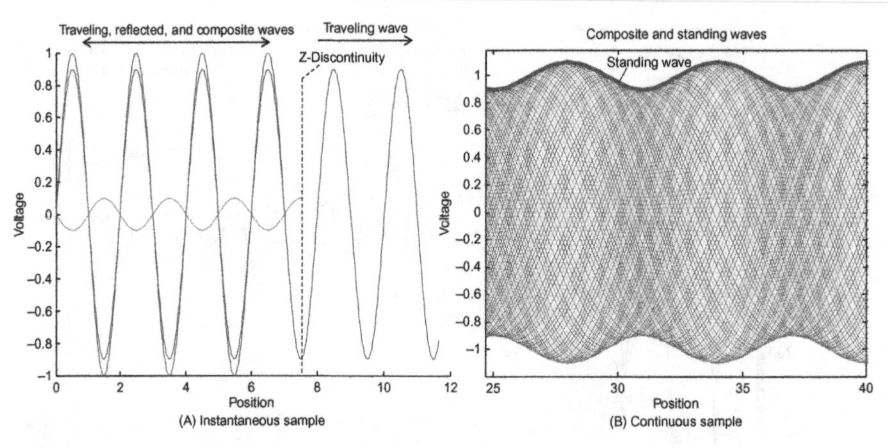

FIGURE 2.22

VSWR of ~1.2 and S_{11} maximum at multiple of λ/2.

high ratio of energy is returned (S_{11}), a lower amount of energy is transferred (S_{21}). The return loss also features "nulls" or frequencies where very little energy (−40 dB) is returned. Nulls in the return loss result from the transmission line length. As length increases, the gap in frequency between the nulls becomes smaller and maintains a half-wavelength separation.

S_{11} nulls

Nulls in S_{11} are created when transmission line length creates a certain phase relationship between the incident and reflected waves. This complex phase information within the reflection coefficient Γ is observable in the time domain when the standing wave ratio (SWR) approaches 1:1. SWR is measured at the input of a transmission line and is a ratio of the maximum voltage to minimum voltage on a standing wave, defined by equation (2.24). When the reflection coefficient Γ approaches 0 (nothing reflected), the ratio of input to output voltage becomes 1:1, and return loss in equation (2.22) becomes a small number, creating the null. So what creates an SWR of 1:1?

$$SWR = \frac{V_{max}}{V_{min}} = \frac{1 + |\Gamma|}{1 - |\Gamma|} \tag{2.24}$$

To illustrate this phenomenon, Figure 2.22 first demonstrates a typical case where 10 percent of the signal is reflected (a return loss of 20 dB) with an SWR of 1.1:0.9. In (A) we observe the instantaneous time sample of a wave traveling over a transmission line and meeting an impedance discontinuity. Ten percent of the traveling wave is being returned at the discontinuity with a reverse magnitude. A composite waveform is shown as the dashed line, created by the sum of the incident traveling wave and the reflected wave. The transmission line length (or length until the discontinuity in this illustration) is such that the reflection phase is misaligned

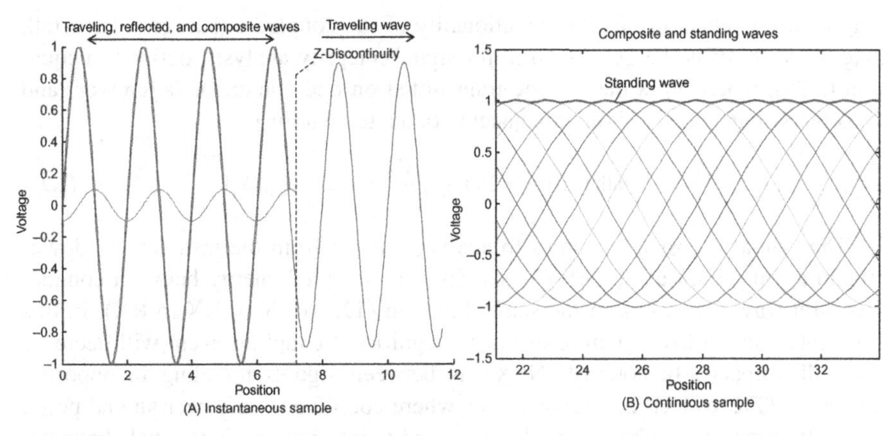

FIGURE 2.23

VSWR of ~1.0 and S_{11} Null at multiple of λ/4.

by half a wavelength and creates the lowest possible composite waveform maximum. These alignments occur at transmission line lengths in multiples of half-wavelengths. In (B) we observe the composite waveform at additional sample times. Over time the composite waveform changes. At each sample in time, the maximum voltage of the composite waveform is recorded. The maximum voltages create the standing wave shown in (B), which swings from 0.9 V to 1.1 V.

The condition creating the return loss null is illustrated in Figure 2.23. In (A) an instantaneous sample is shown that has met a discontinuity, though the reflection is now misaligned by a quarter-wavelength. As the reflected wave proceeds and passes through 0 V the incident wave is reaching a maximum of 1 V. The resulting composite wave (sum of incident and reflected) maintains a 1-V level, equal to the incident wave. The reflection coefficient Γ magnitude appears to approach 0 in equations 2.23 and 2.24 for this instantaneous sample. In (B) continuous samples are taken and the maximum of the composite waveform does not virtually see any variation over time. The ratio of the maximum to minimum values of the composite wave yields an SWR of 1:1, and the reflection coefficient magnitude is 0 at all samples in time.

CROSSTALK

Coupling strength between conductors is expressed in the frequency domain to characterize change in coupling across frequency. The crosstalk response of an entire system or component is a primary tool for the design engineer and is often used for component specifications. The concepts of power spectral density apply to crosstalk coupling as well. The magnitude of coupling between conductors is only relevant at frequencies in which spectral content is present. Two primary crosstalk expressions, far-end crosstalk (FEXT) and near-end crosstalk (NEXT), may be used depending on the directionality of the coupled conductors. Crosstalk magnitude in dB is the general form for signal integrity analysis, defined in equation (2.25), where V_t is the voltage transmitted on one conductor (aggressor) and V_r is the voltage received on an adjacent conductor (victim):

$$\text{Crosstalk (dB)} = -20\log_{10}\left|\frac{V_t}{V_r}\right| = -20\log_{10}|S_{ij}| \tag{2.25}$$

The primary crosstalk sources to analyze will be from the upstream or downstream agent within a link. FEXT describes the coupled energy between conductors with waves traveling in the same direction (TX to TX or RX to RX). In this case, increases in losses before or after a significant coupling event will decrease crosstalk impacts. In contrast, NEXT is between signals traveling in opposing directions (TX to RX). In circumstances where coupling occurs near an end point, crosstalk impacts may be severe. Near an end point, transmitting signals have not seen much attenuation and will create strong reverse waves onto receiving conductors.

Data rate and spectral content are relevant when drawing conclusions about the severity of crosstalk in the frequency domain. System crosstalk levels at the Nyquist frequency near 40 dB (1 percent or about 10 mV out of 1.0 V signal) and below are often tolerable in high-speed systems. Comparatively, system crosstalk levels of 30 dB (0.3 percent, or 30 mV) are significant—and may only be tolerable when interconnect insertion loss is very low. However, if a characterized component demonstrates 30 dB of crosstalk, it is not necessarily a severe case. The impact on the system performance depends on the placement of the component and the loss in the channel. FEXT through the component will be further attenuated in the channel. For example, a component with 30 dB of FEXT in a channel with 20 dB of insertion loss will observe a moderate 0.3 percent (about 3 mV) of coupled energy at the receiver from the component. If such a component supplies NEXT, the location is extremely important. Components in the middle of the channel will attenuate crosstalk levels before reaching a receiver, while placement near a device will lead to high crosstalk noise without the opportunity for attenuation.

Crosstalk sum

The summation of multiple crosstalk responses can be used to describe the coupling that results at the end of the interconnect on a receiving node. Coupling seen at any receiver is the culmination of multiple crosstalk sources. One common source is transmission line coupling from the nearest adjacent conductors. Further, vertical structures such as a ball grid array (BGA) or connector are an array of crosstalk sources, which may not be the same source as that seen by the PCB transmission line. Crosstalk sums are commonly used to observe the total amount of crosstalk noise in a system from these sources. The sum is also used for industry specifications on components (primarily connectors) to restrict the allowable noise.

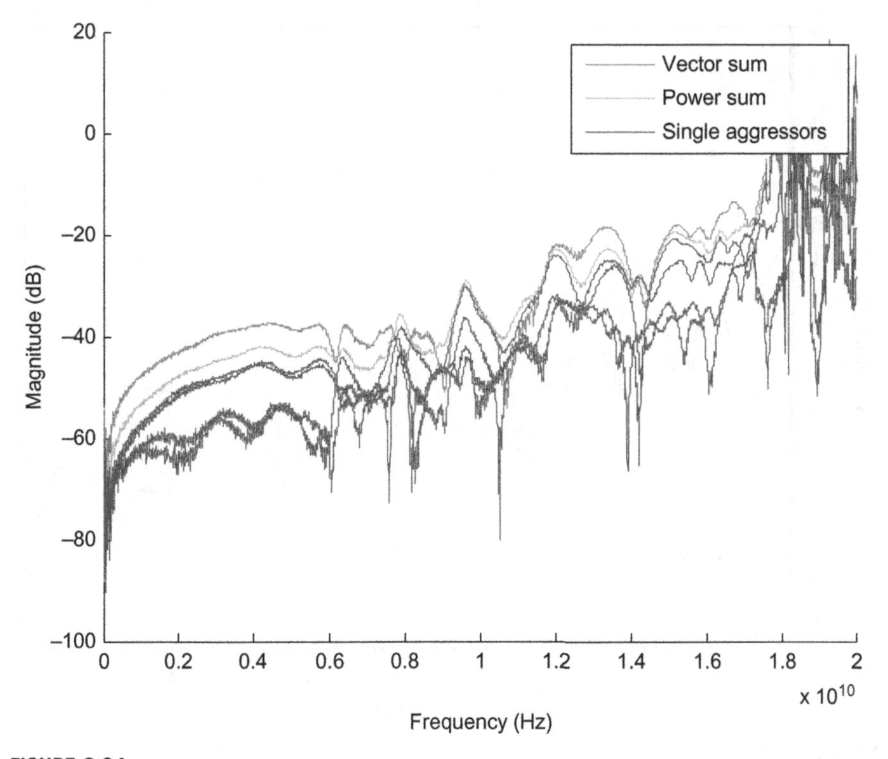

FIGURE 2.24

Vector crosstalk sum and individual responses.

A vector sum represents the impact from all the aggressors on the signal. Vector sums maintain the phase information, allowing for constructive cancellation between sources. Differential crosstalk responses, $DDXTALK_n$, are the complex S-parameter responses and may be NEXT or FEXT sources. An example of the use of a vector crosstalk sum is the connector DDNEXT requirement in the PCIe CEM Specification 3.0. The vector sum equation for n crosstalk sources is defined in equation (2.26):

$$\text{Vector Sum (dB)} = 20 \log_{10}\left(\left|\sum\nolimits_n DDXTALK_n\right|\right) \tag{2.26}$$

A power sum describes all the energy transferred, removing phase information to represent the worst case addition of the sources. The equation for the power sum of n crosstalk sources is defined in equation (2.27):

$$\text{Power Sum (dB)} = 10 \log_{10}\left(\sum\nolimits_n |DDXTALK|_n\right) \tag{2.27}$$

Power and vector sums may be composed of all FEXT, all NEXT, or a mixture of both. To represent the system-level impact, the directionality of each conductor should be properly accounted for. In Figure 2.24, a comparison is made

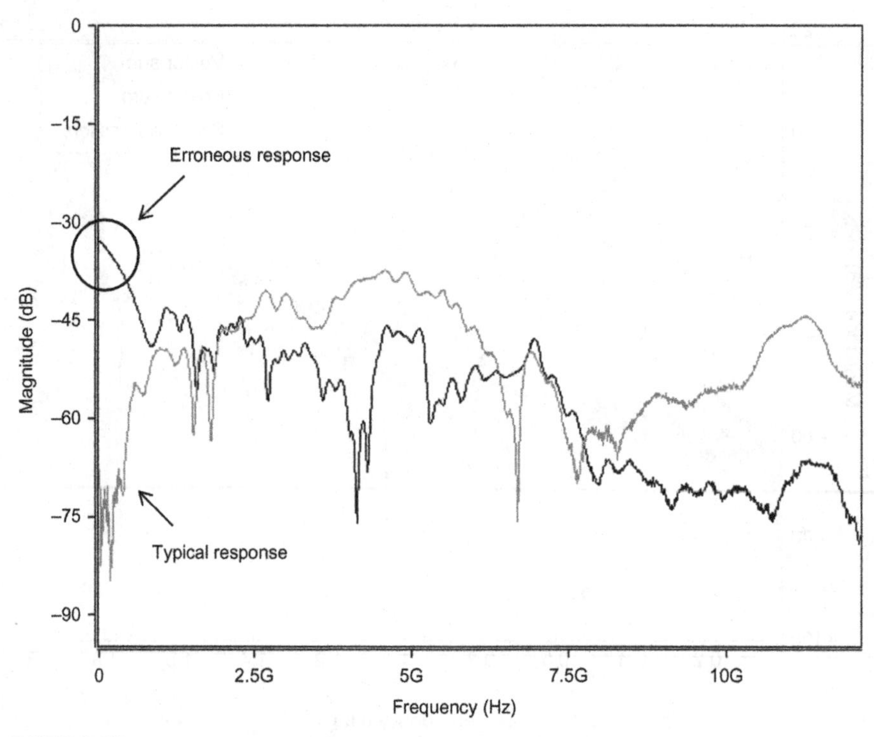

FIGURE 2.25

Typical and erroneous channel crosstalk measurement.

between different crosstalk sums on the four nearest aggressor measurements from a standard PCIe 3.0 connector. In this example, all lanes are treated as NEXT per the specification, though not all conductors contribute NEXT in the system environment. The figure demonstrates the significant difference between the two methods, with the vector sum demonstrating the worst case result.

Review of crosstalk in the frequency domain will help check for modeling or measurement mistakes. In Figure 2.25 a significant amount of coupling (−30 dB) is observed near DC in a VNA measurement. It is expected that as frequency decreases and approaches DC coupling becomes nonexistent, reaching the equipment noise floor (60 dB or below). The measurement in this figure is erroneous and should be completely discarded and re-measured.

INTEGRATED CROSSTALK

A power spectral–dependent crosstalk parameter, integrated crosstalk noise (σ_{ICN}), is introduced by Adam Healy (2007:2) and shown in equation (2.28). The parameter captures the average power for the specified operating frequency, giving significance to coupling at frequencies that matter the most.

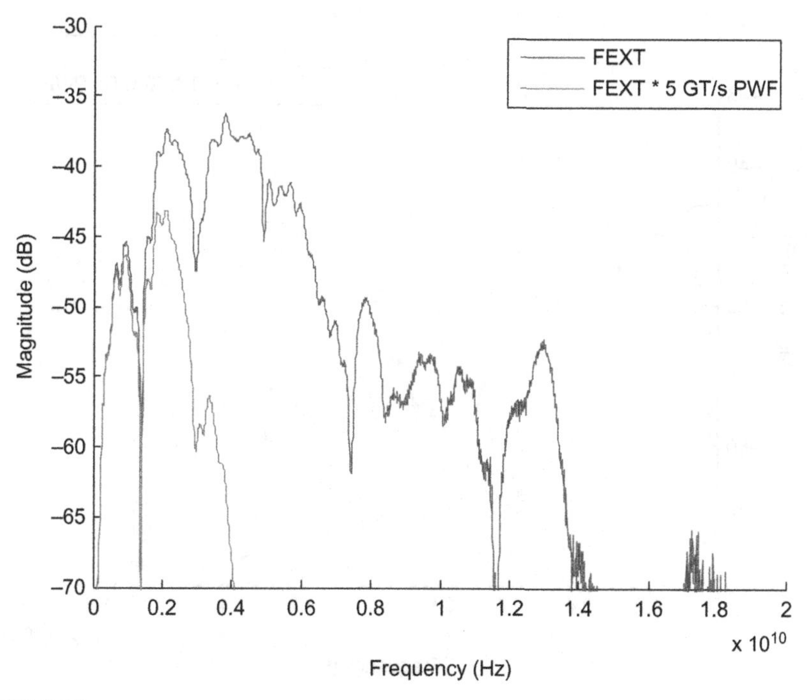

FIGURE 2.26

Typical crosstalk response multiplied with 5-GT/s PWF.

$$\sigma_{icn} = \sqrt{\frac{2 \times \Delta f \times \sum_n PWF(f_n) \times \sum_x A_x^2 \times |SDD21XTK_x(f_n)|^2}{F2 - F1}} \tag{2.28}$$

In Figure 2.26, a typical crosstalk response for a 20-inch channel containing two connectors is shown. With an attributable response of −40 dB of coupling at 5 GHz, the power weighting function used by σ_{ICN} can re-color the response with data rate dependency. An additional two responses are shown as the product of the crosstalk with the power weighting function at 5 GT/s in Figure 2.26 and 20 GT/s in Figure 2.27. The σ_{ICN} is 6.7 mV at 5 GT/s and 7.7 mV at 20 GT/s.

SIGNAL-TO-NOISE RATIO

The transmitted signal and noise sources in the channel may be combined to create a signal-to-noise ratio (SNR). An SNR created from integrated noise parameters σ_{ICN} and σ_{IILN} is proposed by Mellitz et al. (2007:10). Insertion loss fit at Nyquist and a peak pulse response height are at least two options for the transferred signal, S_{rx}. This SNR is purely an interconnect parameter, quantifying the usable signal at the end of the channel. While SNR is data rate dependent, it is not aware of silicon capability and it may be misleading to compare the

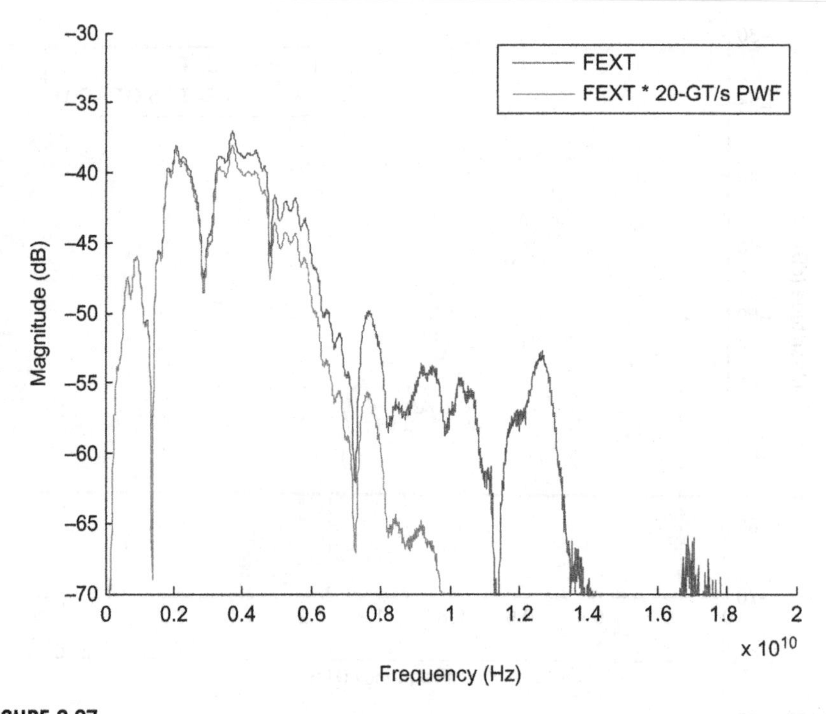

FIGURE 2.27

Typical crosstalk response multiplied with 20-GHz PWF.

interconnect SNR across interfaces that apportion the UI budget between silicon and interconnect differently (such as PCIe to USB). If desired, additional noise terms may be correlated and added to relate SNR metrics with circuit performance such as jitter, equalization capability, receiver noise, and alien crosstalk. SNR is defined in equation (2.29):

$$SNR = 20\log\left(\frac{S_{rx}}{\sqrt{\sigma_{icn}^2 + \sigma_{iiln}^2}}\right) \tag{2.29}$$

The SNR is an accurate way to compare trade-offs in interconnect choices without full time domain simulations that will derive eye height and widths. Design choices, such as PCB loss, signal separation, via discontinuity, stack-up size, connector choice, etc., are some of the many considerations that can be calculated quickly with SNR. It is found that under significant changes in the channel it is required to have additional parameters to maintain accurate SNR comparisons that properly track time domain results. Healy (2007:9) demonstrated this idea first in the use of σ_{ICN} by insertion loss.

SNR correlation to time domain eye height is demonstrated in the simulated results of Figure 2.28. The color contour is the simulated eye height for 8.0-GT/s

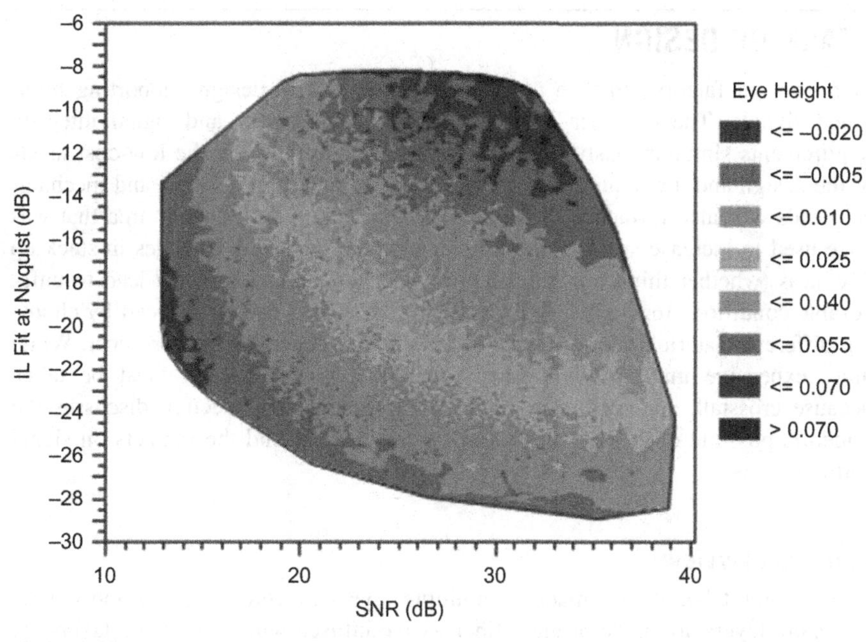

FIGURE 2.28

Eye height correlation to SNR and insertion loss.

Table 2.2 Range for Random Channel Variables

Parameter	Min	Max
Termination	80 ohms	120 ohms
Package Impedances	Low	High
Package Length	0.4″	1.4″
Trace Length	3	23
PCB Loss	0.6 dB/inch at 4 GHz	0.9 dB/inch at 4 GHz
Number of Vias	0	6
Via Stub Length	0 mils	97 mils
Crosstalk Spacing	9 times the height	13 times the height

PCIe 3.0. The vertical axis is insertion loss fit at Nyquist, which is added to assist in correlation over a wide variety of interconnect changes. An extremely wide variety of topologies to stress reflection and crosstalk noise terms were simulated to verify correlation. Larger eye height correlates to larger SNR values (to the right) and higher insertion loss (more positive, to the top). The parameters defined in Table 2.2 were randomized to create thousands of data points for correlation.

STACK-UP DESIGN

Many design factors must be considered in a stack-up design supporting high-speed signals. The designer needs to satisfy many power and signal integrity requirements simultaneously. Balance must be found between the low-cost needs of the design and the requirements for high-speed signaling. Chassis and mechanical needs will also influence the board size and may reduce layout area that was once used to increase spacing and minimize signal coupling. Changes in stack-up thickness (whether thinner or thicker) driven from form factor may lead to unfavorable conditions for high-speed signaling. Signal loss can be reduced by choosing different material and copper foil types, which generally increase cost. While more expensive materials will lower dielectric losses, caution must be taken because crosstalk and reflections may be increasing. This section discusses the tradeoffs possible during the stack-up decision process and the impacts on signal and noise.

Build-up overview

Printed circuit boards are made of multiple layers of core, prepreg, and copper foil. Core layers are made of glass fiber weave infused with resin. Core layers are already cured and hardened at high temperature and have a copper foil on each side. Prepreg, short for pre-impregnated, is a glass weave impregnated with resin that is not yet hardened. The unhardened prepreg will allow the resin to flow, filling in the gaps between the signals, when all the layers are heated and pressed together. Cores and prepegs come in various thicknesses, while copper foil is available in ½-and 1-ounce sizes, approximately 0.65 and 1.3 mils, respectively.

Buildup of the PCB begins from the inner layers. For example, a 4-layer stack-up like that shown in Figure 2.29 begins with the core, already cured and having copper foil on both sides. The copper layers on each side of the core will become metal layers 2 and 3. The outer copper foil on each side of the core is etched, and an oxide treatment is applied to improve adhesion to the prepreg

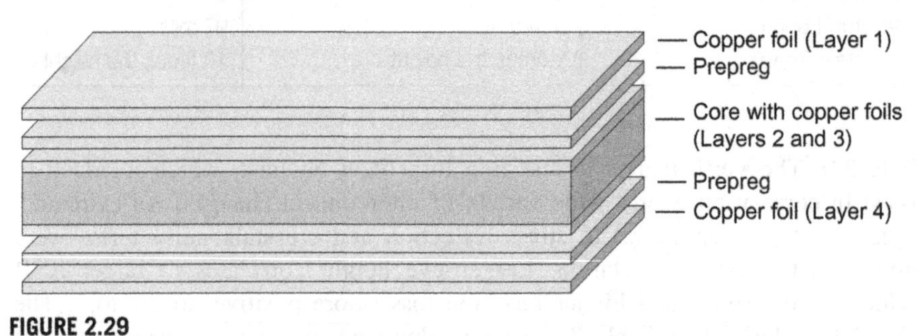

FIGURE 2.29

Build up for a 4-layer stackup.

Table 2.3 Impedance Requirements for Common High-Speed Signals

Interface	Impedance	Mode
PCIe 2.0, 3.0	85	Differential
SATA 2.0, 3.0	85	Differential
Clocks	85, 100	Differential
Ethernet 10 G	100	Differential
DDR DQ, DQS	40, 50	Zo, Single Ended
Miscellaneous Signals	50, 55	Zo, Single Ended
USB 2.0, 3.0	85	Differential

layers that are added to each side. This is followed by the laying down of copper foil and etching on the outside, creating metal layers 1 and 4.

IMPEDANCE TARGET (ROUTING IMPEDANCE)

The stackup needs to support the required impedance of each interface present on the design. The impedance requirement may come from a standardized interoperability specification or from a specific product design guide. An analysis with a field solver is needed to ensure that trace geometries will meet the required impedance on the layer(s) on which each interface will be present. Layers without particular interfaces are not required to support their impedance targets. In early design stages, an approximate impedance calculation may be performed with equations described earlier in this chapter. Table 2.3 provides examples of standardized impedances. The most common impedance targets are 40 and 50 ohms single ended and 85 and 100 ohms differential.

The primary parameters that affect trace impedance are the dielectric heights, dielectric constants, and trace width, height, and separation (if differential). While these are the primary controls for impedance, these parameters also may have an influence on signal loss and crosstalk that the design engineer must consider.

Optimal routing impedance

What is the optimal routing impedance? In years past, at lower frequencies, impedances of 50 ohms single ended and 100 ohms differential were the common targets for PCBs, having a history in coaxial cabling manufacturing for decades. More recently, interfaces have reduced the differential impedance requirement to 85 ohms, showing improved signaling results for package and board routing. When an industry specification is involved, it is expected that the investigation of optimal impedance through full link simulation has already been done to find the balance between power and reflections. While full link simulations provide the final measure of improvement, a first-order assessment of the possible optimal impedance can be found through viewing the impedance

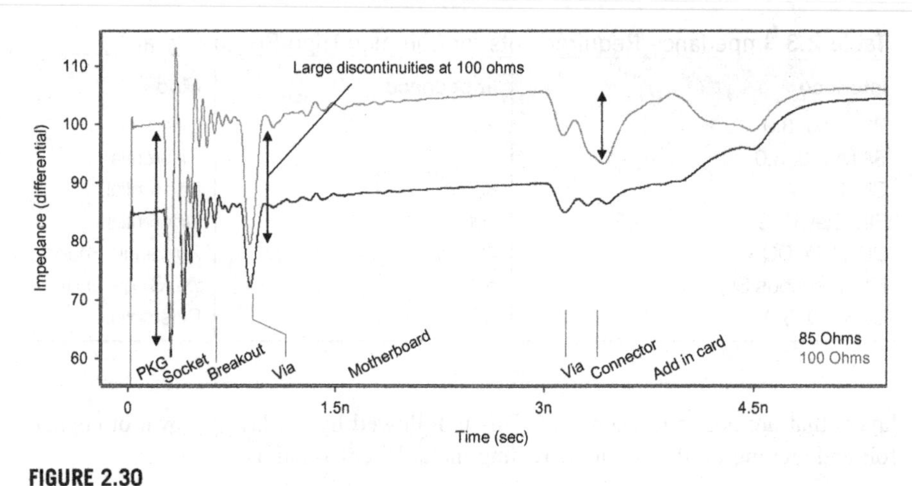

FIGURE 2.30

TDR comparison of PCIe channel at 85 and 100 ohms.

difference between the transmission line and the discontinuities in a time domain reflectrometery (TDR) response. Many interfaces have migrated to 85 ohms in order to more closely match the impedance of plated through hole vias in packages and PCBs, which may be as low as 70 ohms. As the mismatch and reflected voltage is reduced, the margins are improved.

As an example of 85-ohm improvement, Figure 2.30 shows a TDR response of a short PCIe channel with 85- and 100-ohm transmission line routing. The largest reflections for the 100-ohm transmission line are seen at the package via, motherboard via, and connector gold finger. Some items in the channel are dominated with inductance and may have higher impedances (approximately 100 ohms), such as sockets and connectors, creating a consideration that 100-ohm impedance may be optimal. However, the TDR response with these components in the channel, as well as support from full link simulation, can confirm whether 85 ohms is optimal—and in this example it is clear from the TDR that 85 ohms offers the least difference between transmission line and discontinuities. Often 100-ohm buffer terminations are discussed as a clause for 100-ohm routing. In such a case, the whole of the termination including the capacitance should be considered. For high-speed signals, the buffer parasitics (primarily capacitive) are present and lower the effective impedance seen at higher frequencies. Therefore, a 100-ohm resistive termination may appear electrically lower and have a smaller discontinuity. Still, the recommendation is to return to simulation with proper buffer modeling to understand any contribution to optimal routing (or buffer) impedance. If buffer termination has a negligible role on reflection and signaling margin, decisions should defer to voltage and power requirements. In the end, full link simulations demonstrate the total effect of all interconnect components.

PCB LOSSES

Signal loss through PCB materials is of high importance for high-speed interconnects. Designs with longer routing are often loss constrained, limited to a maximum length because of loss rather than other noise sources. Many material choices are available to the designer to achieve a desired loss target. To ensure that production loss is in agreement with pre-silicon and design expectations, several validation paths may be taken.

Traditionally, differential insertion loss measurements are taken with a 4-port VNA. In the laboratory, loss may be measured on a per-board basis with test structures or coupons placed on the edge of the board. In some cases, measurements are possible in the BGA areas for point-to-point nets. Often these measurements are difficult without easily accessible ground pins.

A newer approach was developed in 2009 by Jeff Loyer and Richard Kunze. Single-ended transmission to differential insertion loss (SET2DIL) is a means to make PCB loss measurements without the need for expensive VNA equipment. The process, now largely adopted by PCB vendors, uses TDR equipment on test coupons to determine whether PCB loss meets expectations. Detailed procedures and methodologies are available online in the 2010 DESIGNCON paper, "SET2DIL: Method to Derive Differential Insertion Loss from Single-Ended TDR/TDT Measurements."

In either approach, non–transmission line components in the measurement affect the accuracy of the insertion loss measurement for the PCB. The primary concern is the discontinuity from vias and via stubs located near the test structures. An assumption then can be made that shorter via stubs and lower frequency loss measurements will not see as much error introduced from the via structure. As a rule of thumb, we can assume that insertion loss measurements may be taken up to 5 GHz with a via stub of no more than 50 mils with a negligible effect on insertion loss calculations. High speeds and longer via stubs require mitigation—such as enlarged anti-pads to mitigate via effects on the test structure. More via mitigation techniques are discussed later in this chapter.

We may make a generalization and categorize PCB material loss performance into three groups: standard FR4, mid-loss, and low-loss materials. Standard FR4 dielectrics will see attenuation between 0.70 and 0.85 dB/inch at 4 GHz. In comparison, mid-loss and low-loss materials may be 0.6 to 0.7 dB/inch and 0.3 to 0.5 dB/inch, respectively. Cost increases vary greatly from vendor to vendor, though some general expectations may be set for improved materials at the time of publication: 10 percent or more for mid-loss and 30 to 70 percent for low loss. It should be noted that these loss expectations are discussed at nominal laboratory conditions. PCB materials vary in response to temperature and humidity. For more details, the role of environment on PCB loss is further discussed in Chapter 3.

When a supplier returns loss results that exceed the targets, what can be done? Some options are available to consider for a reduction in loss, in addition to the supplier's review for any manufacturing mistakes:

- Change dielectric material—lower loss tangents from a resin or glass change is a primary method for loss reduction.
- Change to a copper foil with lower roughness properties—reduced copper roughness improves high frequency performance.
- Increase trace width—conductivity is increased with wider traces. Stack-up changes may be necessary to maintain the same impedance.
- Increase dielectric thickness—thicker dielectrics decrease the dielectric loss. The most significant gains are available to microstrip routing.
- Choose core between signal and the primary reference plane—after etching, the widest surface of the trace faces the core. Skin effect resistance is minimized when more trace surface is provided on the face on which current will be flowing.

DIELECTRIC LOSS

Signal loss from dielectric materials is described by the dissipation factor. The factor is unit less and often called the loss factor, dielectric loss, loss angle, or loss tangent. It describes the amount of charge that is polarized and displaced within a dielectric material when an electric field is applied. When this happens, part of the applied field is reduced and converted to heat, and signal loss has occurred. The dissipation factor is provided on the data sheet from PCB vendors. Off-the-shelf and inexpensive FR4 materials, called "regular FR4" herein, range in dissipation factors from 0.017 to 0.022, while specialty low-loss materials can be as low as 0.003. When longer routing lengths are achieved through low-loss dielectric materials, some caution needs to be taken due to the higher contribution from conductor loss, return loss increase, and crosstalk increase from low-loss dielectrics.

Lower loss dielectrics

Signal attenuation can be reduced through a selection of lower loss materials at an increased cost rather than reducing the PCB routing length. Materials with lower attenuation are achieved through a change in the epoxy throughout the glass weave to achieve a lower dissipation factor. Lower dissipation factors, which correlate to higher cost, may be 0.003 or lower. It is possible to achieve total signal attenuation as low as 0.35 dB/inch at 4 GHz.

The total attenuation at a single frequency scales with length, so to double the length is to double the loss at a single frequency. However, care must be taken to consider not only the loss at one point, but rather the entire insertion loss profile

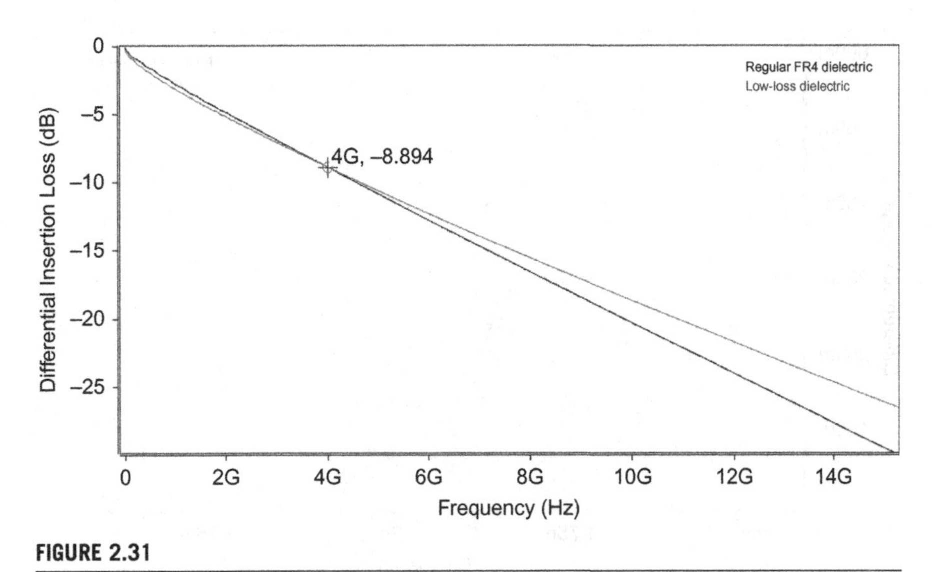

FIGURE 2.31

Insertion loss comparison for equal insertion loss at 4 GHz.

over frequency. In low-loss materials, the losses at lower frequencies are dominated by copper losses rather than dielectric losses and take a square root of frequency shape from the skin effect. As shown in Figure 2.31, two materials with the same insertion loss at one frequency, chosen at 4 GHz, can have a very different loss profile over frequency. For the simulation, the length for FR4 material is 10 inches and the low-loss material is 18.5 inches.

Signal content at lower frequencies will be more greatly attenuated due to the higher contribution of conductor loss with the lower loss material. For a signal operating at 4 GHz, most spectral content will be below 4 GHz, which will lead to a smaller pulse height than the high-loss material. The pulse response in Figure 2.32 demonstrates how this is realized in the time domain. For the purpose of illustration, the propagation delay is normalized, aligning the responses for wave shape comparison. In this example the lower loss material reduces additional RX gain to compensate for the signal loss if the length is not reduced.

The pulse response tail, or dispersion, is different between the low- and high-loss materials. Dispersion is the result of a change in propagation delay across frequencies. Lower loss materials generally observe a slower change in delay over frequency, reducing the differences in velocity and reducing the tail or residual ISI that signals must equalize.

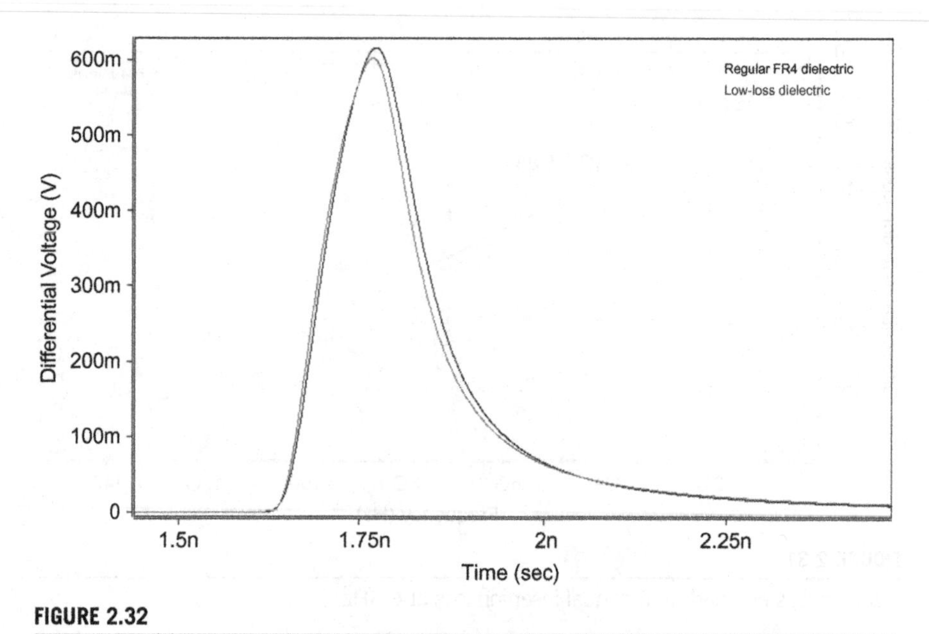

FIGURE 2.32

8-GT/s pulse response comparison for equal insertion loss at 4 GHz.

Some care must be taken for potential noise increases with low-loss dielectrics. Decreasing the loss in the PCB will yield stronger reflections that were otherwise attenuated, similar to the effect that may be seen on short channels. The effect seen is greater margin reduction from discontinuities in the channel. Crosstalk has a potential increase in low-loss dielectrics. Dielectric constants in standard FR4 dielectrics tend to increase with thickness, a characteristic that assists in reducing crosstalk. A survey of available low-loss materials does not exhibit this trend, leading to an increase in crosstalk levels in asymmetric stackups.

Care must be taken that link margins on low-loss stackups are sufficient through simulation. Back-of-the-envelope calculations, such as "30 percent loss reduction means 30 percent more length," should not be taken confidently without investigation into crosstalk changes and full link simulations.

Hybrid stackups

An alternative loss reduction option at a reduced cost is a blended stackup of both regular FR4 and mid- or low-loss dielectrics. A blended, or hybrid, stackup may replace some regular FR4 dielectrics with higher grade materials. Substitutions are possible for the outside layers and some or all of the internal prepregs and cores. Lower costs may be achieved through the substitution of all the prepregs or all the cores exclusively. However, costs and production capability will vary across vendors and stack-up proposals must be vendor-approved.

CONDUCTOR LOSS

On typical FR4 materials, conductor loss is a lower percentage contribution than dielectric loss to the total attenuation at higher frequencies. Conductor losses increase at a square root of frequency, while dielectric losses increase linearly, making dielectric loss the primary concern. However, when dielectric losses are intentionally made very low, conductor loss becomes increasingly important. It is worthwhile to review the parameters available that influence conductor loss and their effects on total attenuation.

High-frequency skin effect losses are documented in textbooks, describing loss as a function of frequency when the current is no longer uniform throughout the conductor. The nonuniform current meets a greater resistance, leading to greater signal loss. These skin effect losses begin when the skin depth is less than the conductor thickness. The skin depth is defined in equation (2.30):

$$\delta = \sqrt{\frac{\rho}{\pi f \mu}} \tag{2.30}$$

where f is frequency, μ is the permeability of the conductor, and ρ is the resistivity of the conductor.

In designs where loss is a primary concern, trace width is often considered as an opportunity to improve losses through wider traces. A comparison is offered to clearly demonstrate the improvement in loss from trace width alone on regular FR4 and again on low loss. For the comparison, the stack-up heights and dielectric constants are the same between regular and low loss. The loss tangent decreased from 0.022 to 0.003 to create a low-loss model. An initial trace and spacing of 4 mils that yields 85 ohms is taken. The second model increases trace width to 6 mils and increases spacing to 8 mils to keep impedance constant at 85 ohms.

In Figure 2.33 and Table 2.4 it can be seen that the percentage of total attenuation improved from wider traces is two times more significant on low-loss material. On low-loss material, conductor loss is contributing a much greater percentage of the losses. Insertion loss is improved 7 to 8 percent on regular FR4 materials and 16 percent on low-loss materials.

Surface roughness

The skin effect descriptions are useful until the current depth is the same height as the microscopic features of the conductor surface. At this point, generally around 1 GHz, it is no longer accurate to assume the conductor has a smooth surface. At higher frequencies the current depth becomes the same order of magnitude as the surface imperfections. The larger these imperfections (the rougher) become, the more the loss increases. A unique mathematical model is necessary for these features and is described in Chapter 3. It is necessary to discuss the benefit of various copper types as a stack-up design enabler for increased routing length in high-speed signaling.

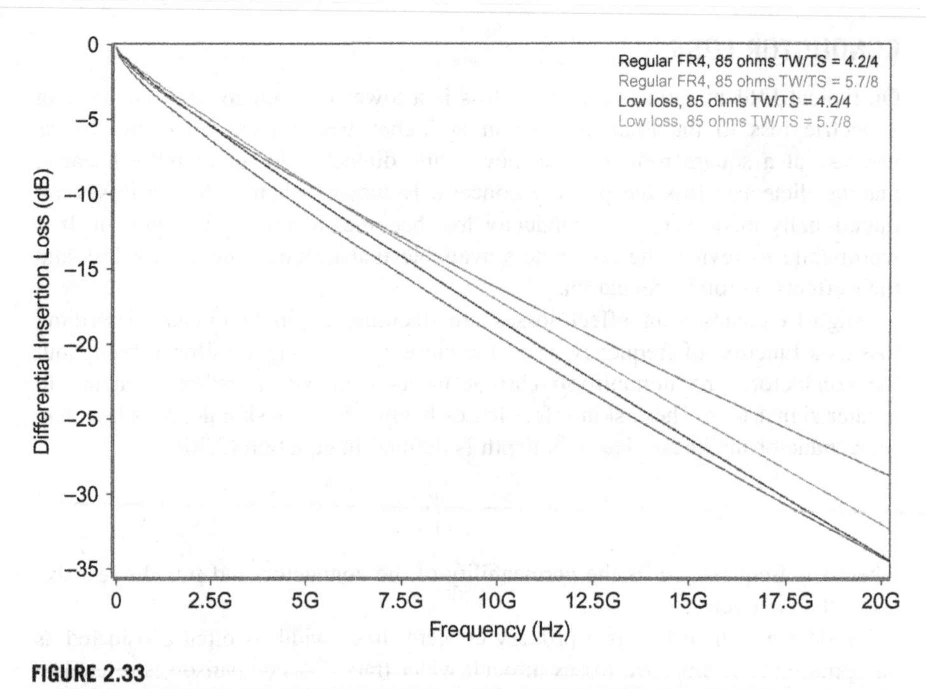

FIGURE 2.33

Loss sensitivity to trace width on regular and low-loss dielectrics.

Table 2.4 Insertion Loss Improvement from Increased Trace Width

	4-Mil Trace Width	6-Mil Trace Width	Improvement
Regular-loss PCB	0.812 dB/in at 4 GHz	0.746 dB/in at 4 GHz	8.1% at 4 GHz
	1.497 dB/in at 8 GHz	1.388 dB/in at 8 GHz	7.2% at 8 GHz
Low-loss PCB	0.505 dB/in at 4 GHz	0.420 dB/in at 4 GHz	16.8% at 4 GHz
	0.882 dB/in at 8 GHz	0.733 dB/in at 8 GHz	16.8% at 8 GHz

Note: *Trace spacing adjusted to maintain 85-ohm differential impedance in both models. Geometries for 4 mils are TW/TS = 4.2/4 and 6 mils are TW/TS = 5.7/8. Loss tangent is 0.022 for regular PCB and 0.003 for low-loss PCB. One-oz copper is modeled with a core thickness of 4 mils, prepreg of 14 mils, and dielectric constants of 4.1.*

A primary method for copper adhesion to the dielectric has been through the intentional roughening of the copper. PCB suppliers desire to meet minimum peel strength; however, it is at the cost of electrical performance. If smoother copper is required, then suppliers are required to find alternative methods to promote adhesion of the copper to the dielectric—perhaps an aggressive oxide treatment. It is common to find higher costs for smooth copper on microstrip layers, which require the highest peel strength.

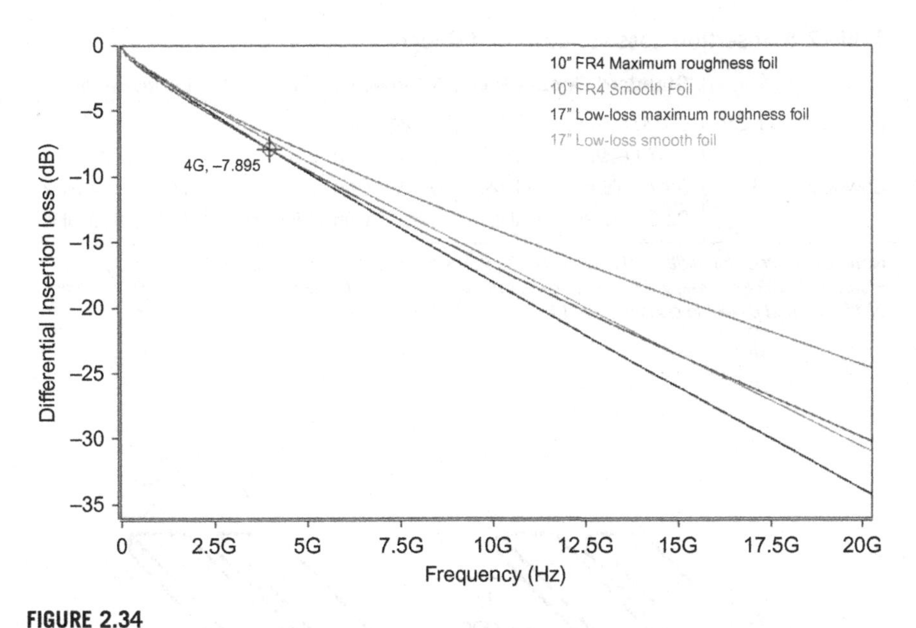

FIGURE 2.34

FR4 and low-loss SDD12 for rough and smooth surfaces.

Details about the copper roughness may not always be available from suppliers. Referencing the data sheet from a copper foil supplier may not be sufficient because these specifications describe the material prior to adhesion treatment. The Institute of Printed Circuits (IPC) has begun the work toward standardization of copper roughness profiles, but there is more work to do. IPC document 4562A has begun to identify some of these profiles.

The insertion loss benefit of copper smoothness depends on the operating frequency and chosen dielectric loss. As frequencies increase and skin depth decreases to the size of the copper imperfections, the benefit of smooth copper becomes more significant. Figure 2.34 and Table 2.5 show the difference of moderate and smooth copper on regular FR4 and low-loss materials. Regular FR4 and low-loss materials are modeled at 10 and 17 inches, respectively, to achieve the same insertion loss as 4 GHz under maximum roughness conditions. The comparison is created when both materials change to a smooth copper foil selection. The copper roughness is modeled using the Huray roughness model and recommended settings described in Chapter 3 for high and low copper foil roughness, 79 and 50 spheres, respectively. As can be seen in the figure, the lower loss material containing a stronger contribution from conductor loss has a greater benefit. The regular FR4 material observes an 8.3 to 9.3 percent improvement, while the low-loss material achieves a 14.2 to 16.8 percent loss improvement.

Table 2.5 Insertion Loss vs. Improved Copper Roughness

	Standard Copper Foil	Improved Copper Foil	Improvement
Regular-loss PCB	0.790 dB/in at 4 GHz	0.724 dB/in at 4 GHz	8.3% at 4 GHz
	1.474 dB/in at 8 GHz	1.336 dB/in at 8 GHz	9.3% at 8 GHz
Low-loss PCB	0.470 dB/in at 4 GHz	0.403 dB/in at 4 GHz	14.2% at 4 GHz
	0.831 dB/in at 8 GHz	0.691 dB/in at 8 GHz	16.8% at 8 GHz

Note: *Geometries for width and spacing are TW/TS = 5/6.5 mils. Loss tangent is 0.022 for regular PCB and 0.003 for low-loss PCB. One-oz copper is modeled with a core thickness of 4 mils, prepreg of 14 mils, and dielectric constants of 4.1.*

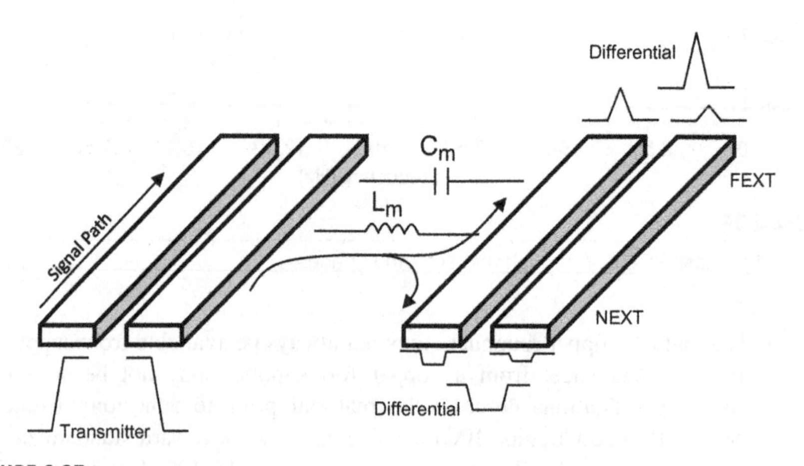

FIGURE 2.35

FEXT and NEXT diagram.

CROSSTALK MITIGATION THROUGH STACKUP

This section discusses what opportunities the design engineer has to reduce crosstalk before the stackup is finalized. Typically, dielectric thickness is the only variable considered to play a role in crosstalk when stack-up decisions are being made. However, this section considers additional options that are available during the stack-up decision-making process.

Crosstalk on differential signals arises from mutual capacitive and inductive coupling between signals as illustrated in Figure 2.35. It is important to know the origin of the transmission line crosstalk, which affects its polarity, so that efforts may be taken to mitigate it. When the aggressor has a positive rising edge, the NEXT will be positive voltage, while FEXT may have a positive or negative voltage. When mutual capacitance dominates, the far-end response will be the same sign as the derivate of the aggressing signal. For example, a rising edge from the aggressor will lead to a positive crosstalk response at the far end of the victim. It will be the opposite for the inductively coupled case. Most PCB transmission lines are dominated by inductive coupling. Innovative and discrete structures may be designed that will increase the capacitive or inductive coupling in order to compensate for the crosstalk present on the transmission line. The entire far- or near-end coupling can be described in equations (2.31) and (2.32). Coupling is described as ratio of inductance, mutual (L_m) to self (L_s), and modified by the capacitive ratio, mutual (C_m) to self (C_s):

$$NEXT \cong \frac{L_m}{L_s} + \frac{C_m}{C_s} \tag{2.31}$$

$$FEXT \cong \frac{L_m}{L_s} - \frac{C_m}{C_s} \tag{2.32}$$

Stripline dielectric

The crosstalk on a stripline transmission line is dependent on the dielectric constants through which the forward and reverse waves travel. The tribal knowledge in the SI community is to design with symmetric (physically) and homogeneous (electrically) dielectric stackups for crosstalk reduction. Symmetric striplines have more tightly coupled fields and less fringing than stackups with distant secondary planes (asymmetric). Uniform dielectric material above and below conductors will minimize crosstalk. Selection of a prepreg with a dielectric constant different from the core is expected to increase crosstalk levels. Modeling and simulations confirm these design rules of thumb. However, some recent measurements have suggested that traditional modeling approaches have shown some possible gaps in modern guidance. The results discussed suggest that symmetric heterogeneous stripline may have more crosstalk than asymmetric and that some combinations of non-homogeneous dielectric stackups can offer a slight crosstalk reduction.

The finding comes out of questions related to the material composition in the signal layer. The substance between the signals may almost uniformly consist of resin, making a uniform dielectric an incorrect assumption. This is caused by the stiffness of the glass, and the phenomenon can be confirmed in cross-sectioning. Application of a lower dielectric constant between the signals has shown some correlation improvement. Questions remain concerning the thickness of the resin layer and whether modeling is always necessary to achieve correlation (more details on modeling in Chapter 3). The performance of crosstalk correlation is highly recommended to ensure that expected coupling is achieved.

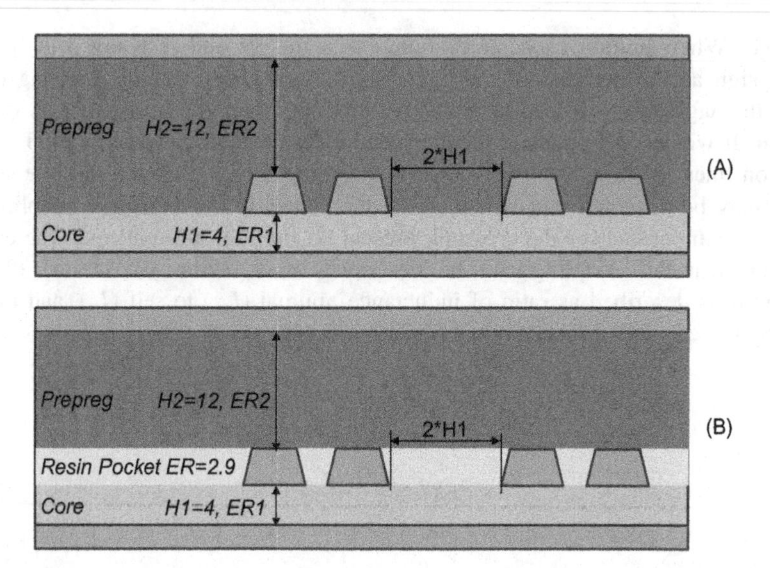

FIGURE 2.36

Asymmetric stripline stack-up parameters.

Table 2.6 Asymmetric Stripline Coupling for Various Dielectrics

Er1	Er2	(a) Integrated Crosstalk	(b) Integrated Crosstalk
4.0	4.0	4.65	5.95
3.7	4.0	8.91	3.29
4.0	3.7	6.16	7.31

Note: *Integrated crosstalk is the forward wave taken over a 5-inch length. The calculation for integrated crosstalk is derived earlier in the chapter, in the Time Domain section.*

To demonstrate the sensitivity to homogeneous and non-homogeneous dielectrics, simulation results for three stack-up scenarios are provided for symmetric and asymmetric stripline stackups. Non-homogeneous experiments apply a 0.3 (8 percent) difference in dielectric constant. This difference represents a worst-case difference in a typical stack-up design—many designs are less than this amount. The symmetric simulated stackup is shown in Figure 2.36. The signal separation is set to two times the core thickness, slightly aggressive from a standard of three times separation.

The simulated results over a 5-inch transmission line are reported as integrated crosstalk in Table 2.6. The traditional design in (a) demonstrates the lowest crosstalk when the design is homogeneous. Dielectric deviations with 8 percent higher or lower dielectric constants lead to crosstalk increases. The resin model (b) indicates higher crosstalk than otherwise modeled in homogeneous stackups. An increase in dielectric constant in the thicker dielectric demonstrates a crosstalk

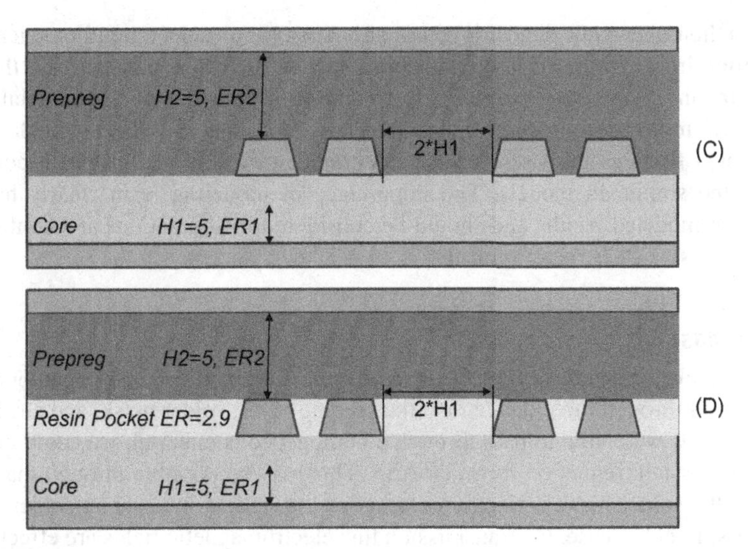

FIGURE 2.37

Symmetric stripline stack-up parameters.

Table 2.7 Symmetric Stripline Coupling for Various Dielectrics

Er1	Er2	(c) Integrated Crosstalk	(d) Integrated Crosstalk
4.0	4.0	1.70	6.39
3.7	4.0	2.36	5.74
4.0	3.7	2.26	5.64

Note: *Integrated crosstalk is the forward wave taken over a 5-inch length. The calculation for integrated crosstalk is derived earlier in the chapter, in the Time Domain section.*

reduction, even lower than the homogeneous dielectric. The high crosstalk case of lower dielectric constant in the thicker dielectric is unlikely in stack-up designs and is offered for informational purposes only.

Simulated results for a symmetric stripline in Figure 2.37 are shown in Table 2.7. Similar to the asymmetric stripline, the homogeneous dielectric begins with low crosstalk and increases when the signal layer is modeled as resin. The increase in crosstalk in the symmetric stripline is much greater (3.75×) than that seen in the asymmetric case (1.28×). In all three dielectric combinations, the new modeling approach demonstrates increased crosstalk in the symmetric stripline stackup.

A comparison of the symmetric and asymmetric rules reveals that the symmetric homogeneous model had 2.7× lower crosstalk than the asymmetric homogeneous model and may actually be much closer to the same coupling at 1.07× more. The guidance from the simulation in order to reduce crosstalk is the following: symmetric and asymmetric models may be regarded nearly equivalent under a homogeneous dielectric constant. Attempt to select non-homogeneous dielectrics in a symmetric stripline and a higher dielectric constant for thicker layers in an asymmetric stripline. It has been observed that as material loss decreases, fewer design choices for dielectric constant will be available. The unexpected findings put stress on the importance of PCB-level crosstalk correlation to the simulated models. The importance of modeling resin clearly has an impact on modeled results and should be considered in simulations and confirmed through passive PCB measurements.

Solder mask

Microstrip electromagnetic fields experience much variability in propagation delay as they pass through three dielectrics: the prepreg layer, solder mask, and air. If the electric fields were able to pass through a homogeneous material, we could expect to see a crosstalk reduction for microstrip. This may be possible through management of the solder mask dielectric constant or through an increase in solder mask thickness. In either case, the transmission line electromagnetic fields are effectively passing through materials and experience a smaller disparity in dielectric constant.

The contour in Figure 2.38 shows how the far-end crosstalk, measured as the integrated crosstalk noise, may change with solder mask dielectric constant and thickness. The highlighted region demonstrates where crosstalk is at a minimum, the potential optional design location for microstrip. While design may be easy in simulation, it may be more difficult practically to control solder mask properties in manufacturing.

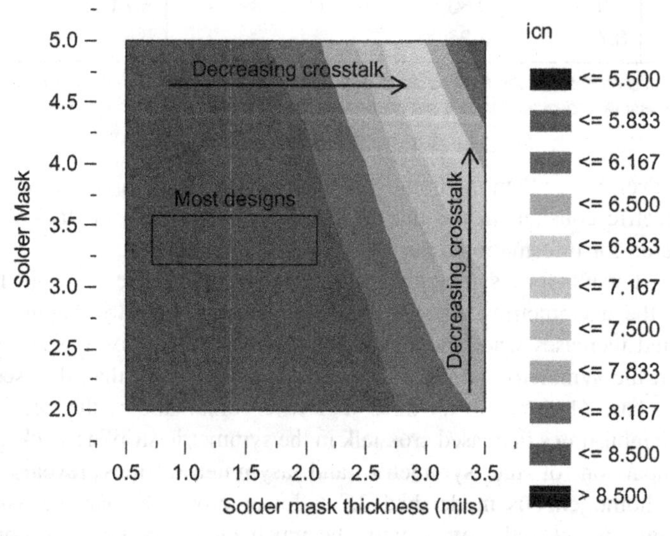

FIGURE 2.38

Five-inch microstrip route with 9-hour separation.

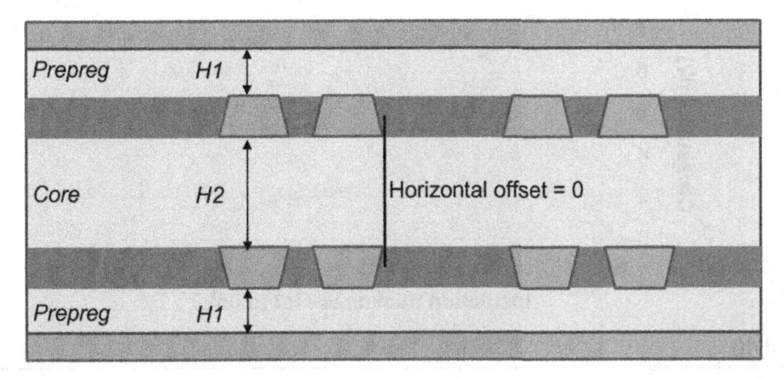

FIGURE 2.39

Dual-stripline Cross-section.

DUAL STRIPLINE

A PCB stackup with dual striplines, as shown in Figure 2.39, reduces the cost of the board by removing the number of isolating metal reference layers, without sacrificing signal layers. The higher signaling density provided by striplines can also help enable designs of smaller form factor and more features. The benefits of using dual stripline make it very popular in computer system designs across different business segments.

A key issue in a dual stripline is the inter-layer crosstalk. Without the reference layer in between, signal nets on the two adjacent stripline layers are directly coupled. The potential impact due to the increased crosstalk is significant unless properly managed. The most common techniques to reduce the inter-layer crosstalk are to increase the thickness of the insulation dielectric layer and/or to make an effort to route trace orthogonally on adjacent layers. A thicker dielectric layer between the stripline layers, however, will increase the total thickness of the board and, subsequently, exaggerate the via stub effects for the signals that have a long stub of the transition via. In addition, it may have an impact on the form factor, cost, weight, etc. Routing the signals on two adjacent layers orthogonally can minimize both the coupled loop area of the magnetic flux and the overlapping area for the coupled electrical field. The routes of the signal traces, however, are primarily determined by the component placement. Orthogonal routings may not always be practically feasible in an actual design.

It is crucial for a successful dual-stripline design to minimize the inter-layer crosstalk and balance of all the design constraints. This requires the careful planning of stackup and component placement and application of routing techniques to mitigate the inter-layer crosstalk.

PCB stackup

The thickness of the insulation layer is directly correlated to the crosstalk strength. Figure 2.40 shows the simulation result of the differential far end inter-layer crosstalk

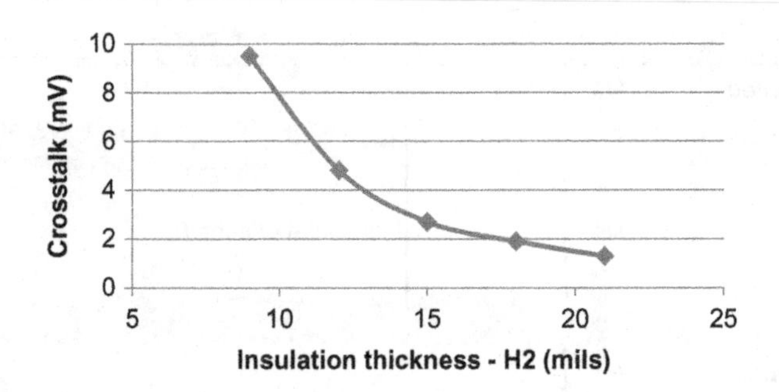

FIGURE 2.40

Inter-layer crosstalk vs. thickness of the insulation.

as a function of the insulation thickness. The two differential pairs are directly overlapping each other on different layers. The distance of the traces to the primary reference plane is 4 mils, and the length of the traces is 10 inches. As shown in Figure 2.40, the inter-layer increases as the insulation thickness decreases, and the increase in speed is faster than linear. A sufficiently large separation distance between routing layers is very effective in reducing the inter-layer crosstalk. It is a design trade-off at the system level how thick the insulation should be.

Besides the insulation thickness, the distance to the primary reference plane, H1, as given in Figure 2.39, is another key parameter in the stack up selection. A smaller H1 will increase the coupling strength between a signal trace to its primary reference plane, and, hence, potentially reduces the inter-layer crosstalk. The smaller H1 will also lead to a thinner trace width to maintain the same impedance target. Thinner traces will reduce the electrical coupling strength but increase the conduction loss at the same time. Overall, a smaller H1 will enable an even more compact stripline design, but a design trade-off between crosstalk and loss will be needed, which typically requires full link simulation to verify.

In general, the ratio of the height to the primary reference plane to the insulation thickness correlates to the crosstalk strength. This ratio can be used as a key design parameter in the selection of dual-stripline stackup. In a typical design, a minimum ratio of three times the dielectric height as recommended. When the ratio is too small, besides the increased inter-layer crosstalk, the impedance of the signal traces will be strongly influenced by the traces on the adjacent layers, especially when the relative physical position between them varies significantly over the course of the routing.

Another important aspect of a dual-stripline stackup is the arrangement of prepreg and core layers. In Figure 2.39, the two signal layers are on the two sides of a core dielectric layer. However, this may not always be the case. The two adjacent signal layers can either center a core or a prepreg dielectric layer, as shown in Figure 2.41. The arrangement of core and prepreg is often not freely selectable and largely depends on the total layer count in the stackup and the layers to be used as stripline layers. The two different constructions have an impact on the performance as well as the modeling of the dual striplines.

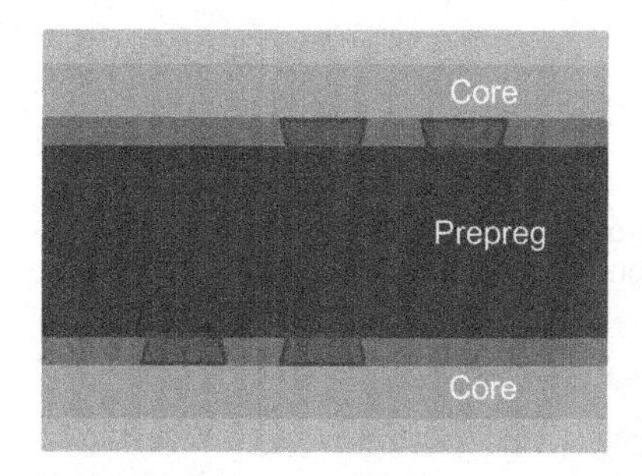

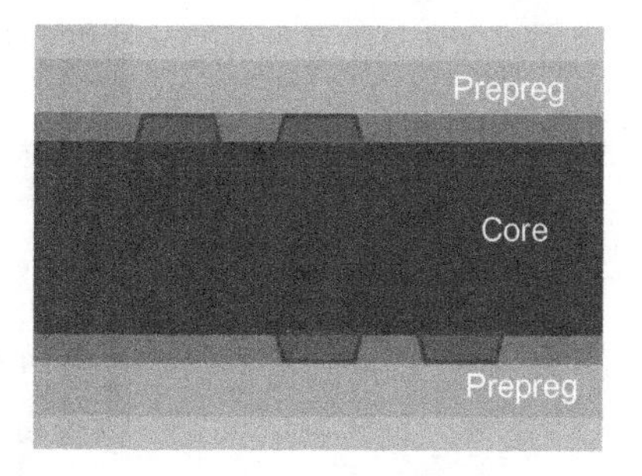

FIGURE 2.41

Dual-stripline stack-up construction.

Due to the over-etching effect in the PCB manufacturing process, the shape of the trace cross section is trapezoidal. The wider side is the surface that is next to the core dialectical layers, and the narrow side is the one next to the prepreg layer. In a dual-stripline design, whether the two signal layers are centering a core or prepreg makes a difference. When they are centering on a prepreg layer, the narrow sides of the traces are facing each other, and the wider sides are facing the primary reference, which may help reduce both the electrical coupling and conduction loss, though the significance of the impact depends on the geometries.

On the other hand, the dual stripline centering on a core layer has much tighter inter-layer misregistration control, and the tolerance is typically two to three times smaller than that of centering on a prepreg layer. When the inter-layer misalignment needs to be tightly controlled, the core-layer center dual-stripline construction is preferred.

In either case, the transmission line models should comprehend the exact construction, material properties, and geometries of the dual-stripline design.

Angled routing

Crosstalk between two coupled transmission lines is accumulative. When the insulation layer is not thick enough and the inter-layer crosstalk is considerably high, a long parallel coupled segment may result in a significant accumulation of the inter-layer crosstalk in phase. The impact can be much bigger if the signals are interleaved, e.g., the signaling directions are opposite. It is highly recommended to avoid interleaved routing of dual stripline, especially when the coupling location is close to the receiver of either of the signals. The most common mitigation technique is to avoid parallel routing by routing traces on different layers crossing over each other with angles, and, in most cases, an orthogonal cross-over is preferred, because it reduces both the electrical and magnetic coupling strength.

Figure 2.42 illustrates the angled cross-over between dual striplines. The red and blue differential pairs are routed on the two adjacent layers, and the reference layers are hidden. To eliminate the ambiguity, the angle is defined between the signaling directions of the two traces. When the angle is 0 degrees, the two traces are

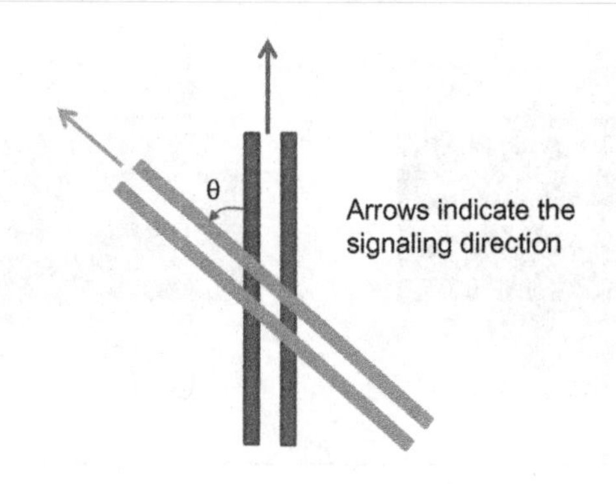

FIGURE 2.42

Cross-over angles.

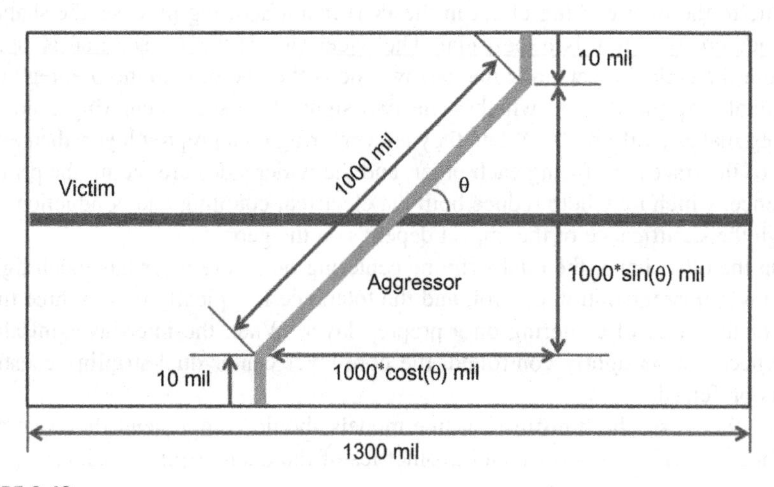

FIGURE 2.43

An illustration of a 3D model of angled dual-stripline cross-over.

in parallel and non-interleaved, and when the angle is 180 degrees, the two traces are in parallel and interleaved. However, for the case in between 0 and 180 degrees, the definition of FEXT and NEXT between the coupled traces becomes blurred. In general, the crosstalk in an acute angle cross-over is considered far end and, otherwise, near end. Although the illustration is drawn in differential signaling, the same idea applies to the single-ended interfaces.

Three-dimensional full-wave simulations are needed to study the crosstalk of an angle-end cross-over. To investigate the crosstalk, a 3D model setup was built as shown in Figure 2.43.

The first step is to construct a simplified dual-stripline stackup in a given 3D solver. The stackup, material properties, and geometries are provided in Figure 2.44. After that, traces are built parametrically to simulate the ideal angled

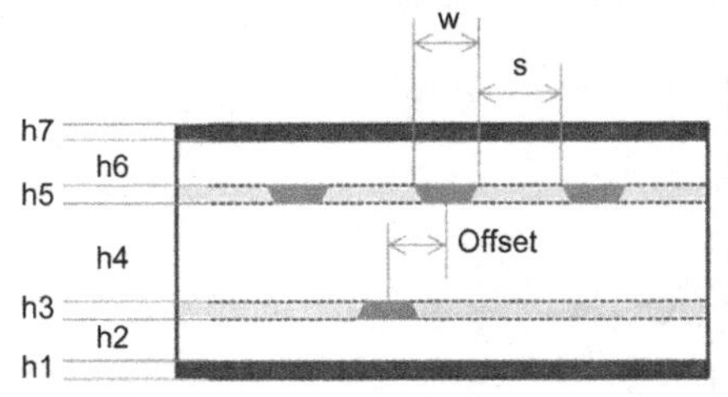

h1 = 1.3 mil	h6 = 4 mils
h2 = 4 mils	h7 = 1.3 mil
h3 = 1.3 mil	w = 4 mils
h4 = 12 mils	s = 12 mils
h5 = 1.3 mil	Offset = 0 mil
Dielectric: FR4 with Er=4.0 Boundary: Radiation	

FIGURE 2.44

Model assumptions: stackup and geometries.

Cross-over angle (degrees)	Differential coupling strength (dB) @ 4.8 GHz		
	h4 = 5 mils	h4 = 14.7 mils	h4 = 25 mils
0	–25	–42	–55
15	–58	–76	–90
30	–72	–106	–100
45	–80	–99	–105
60	–86	–96	–104
90	–93	–99	–99
120	–75	–83	–91
135	–71	–79	–80
150	–60	–67	–71
165	–42	–48	–53
180	–14	–29	–39

FIGURE 2.45

Inter-layer crosstalk vs. cross-over angle.

cross-over, and the crossing angle (θ) is swept in this simulation. During the sweep, the trace length of the aggressor and victim is kept unchanged to minimize the impact of the trace loss to the crosstalk result. This requires the adjustment of the size of the reference planes. For example, in Figure 2.43, the length of the victim is 1300 mils. The length of the aggressor is 1020 mils, including two 100-mil lead-in traces. As a result, the width of reference plane is 1300 mil and the length of the reference plane is a function of the cross-over angle, $1000*\sin(\theta)$. The size of the 3D model is adjusted for each angle value, and the results will be processed and the extra trace length will be re-embedded. The end result is the crosstalk between the two 1-inch dual-stripline traces crossing over each other with different angles.

The result of the differential 3D modeling is summarized in Figure 2.45. For the easy interpretation of the data, only the coupling strength is reported at the Nyquist frequency of a high-speed differential interface running at 9.6 Gbps. The data clearly show the trend that a larger insulation thickness and bigger cross-over angle will lower the inter-layer crosstalk. It is also shown that the interleaved

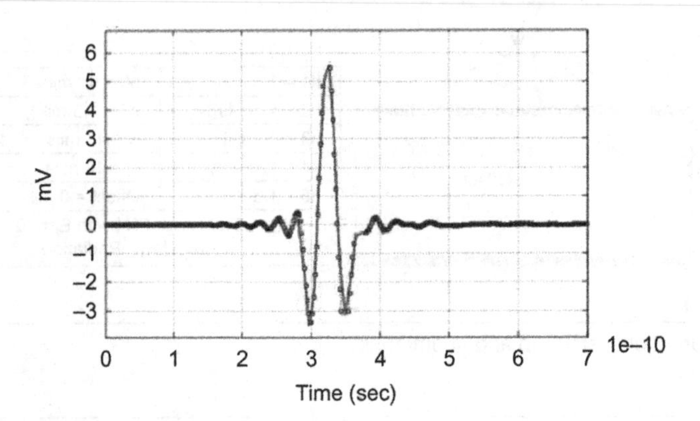

FIGURE 2.46

Inter-layer NEXT in an angled cross-over.

coupling is often stronger than the non-interleaved one. Please note that the 3D modeling result may have a monotonic trend as the angle changes. This could be due to the fact that the data are reported at a single frequency, as well as the numerical uncertainties, especially when the coupling strength is very small (<-90 dB).

In the layout practice, the selection of a minimum allowed cross-over angle depends on the stackup and trace geometries, particularly the thickness of the insulation layer. With a very large separation distance, for instance, when the ratio is greater than 10, a small angle or even parallel routing would have a very small impact from the inter-layer crosstalk. For the more common designs that have a ratio greater than three of the insulation thickness to the distance to the primary reference plane, a minimum cross-over angle of 30 degrees would be a rule of thumb. Many factors, however, may contribute to the overall inter-layer crosstalk, such as the disparity of the signal strength of the aggressors and victims, the location of the cross-overs, the number of the cross-overs, etc. It is highly recommended to run full-link simulation on the dual striplines to make design decisions at the system level.

The inter-layer NEXT waveform in the time domain from a differential cross-over is given in Figure 2.46. The peaking point in the waveform correlates to the physical cross-over point. It is interesting to see that the crosstalk waveform starts long before the cross-over, and the width of the crosstalk waveform is electrically much bigger than the cross-over geometries. This is because in the neighborhood of the cross-over, the two traces are close enough for the crosstalk to noticeable. The smaller the cross-over angle is, the larger the neighborhood will be for the traces to separate far enough. This also provides a different explanation as to why a large angle is preferred.

This leads to an important routing rule when the angled cross-over is applied. When the dual stripline needs to turn back and forth when crossing over the traces on the adjacent layer, the traces should be routed far enough after a cross-over

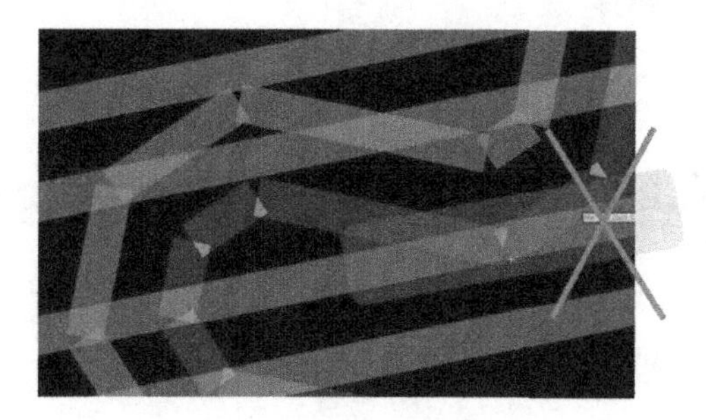

FIGURE 2.47

Avoid "quick turns" in angled cross-overs.

before turning back. The so-called quick turns as illustrated in Figure 2.47 should be avoided. If traces on one routing layer turn quickly back and forth over the traces on the other layer, the total overlapping area increases. As a result, crosstalk also increases accordingly.

In an actual design, it is very unlikely that dual-stripline cross-over occurs just once as that in the 3D model of Figure 2.43. Typically, a dual-stripline net will cross over many other signal nets. The overall crosstalk will be an aggregative effect of all the cross-overs. Fortunately, because of the randomness of the locations of the cross-overs and the randomness of the bit patterns driven on the signal nets, the crosstalk from different aggressors is often not additive in phase. The statistical impact of the crosstalk from multiple aggressors that have uncontrolled phase relationships reduces significantly the probability of hitting the worst-case inter-layer crosstalk.

Key design considerations when angled cross-over dual striplines are used are the component placement and dual-stripline routing layer assignment. If the two adjacent layers are assigned to interfaces routed between different components that will cross each other, the angled routing will be a natural fit. Another important design consideration is the cross-over location. For example, when a PCIe port crosses over a DDR channel, we should avoid locations where the signals are too asymmetrical—for instance, the DDR signals are too strong (close to the TX) and PCIe signals are too weak (close to the RX). Although the coupling strength is the same, a strong signal attacks a much weaker signal and will result in a much larger impact on the link performance. The worst-case scenario happens when the traces are interleaved on the adjacent layers in the pinfield. However, an interleaved cross-over may not be as bad if the cross-over happens in the middle of the channels of the two interfaces when both sides have similarly strong signals.

When traces on the two adjacent layers need to be routed in the same direction—for example, traces of the same PCIe port—parallel routing may be unavoidable. Parallel dual striplines can be closely coupled to each other and have large ICN. For such cases, zigzag routing can be a mitigation option, as illustrated in Figure 2.48 below. For each cross-over, the cross-over angle should be large enough and no quick turns allowed. It is also recommended not to have

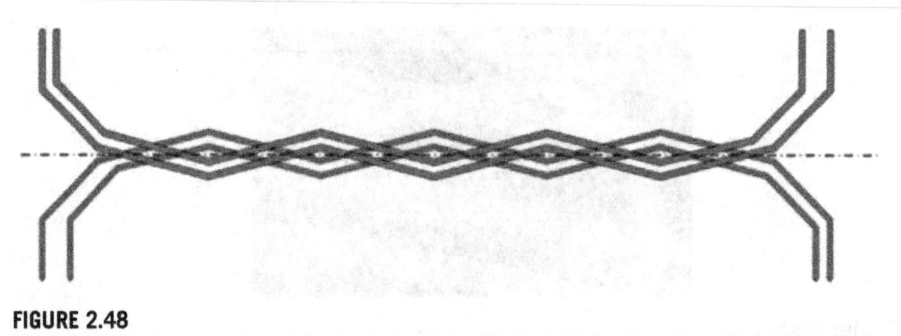

FIGURE 2.48

Zigzag routing.

the cross-over points uniformly spaced, since the periodic zigzag trace pairs may cause the in-phase accumulation of crosstalk.

A zigzag routing scheme can be combined with the routing technique for mitigating a fiberweave effect. If traces on each layer are routed with an angle of no less than 15 degrees in opposite directions, dual striplines will cross-over each other with an angle larger than 30 degrees. This mitigates the impact of both the ILC and fiberweave effects. Note that the zigzag routing increases the routing length and potentially the layout complexity.

Parallelism

Sometimes parallel dual striplines are evitable. For parallel dual-striplines, it is expected that the crosstalk will hit the maximum when the traces are completely overlapping each other, and the crosstalk will decrease to zero when the inter-pair spacing is sufficiently large. The crosstalk properties between these two extreme points need to be studied, especially for differential to differential crosstalk where four wires are involved. The inter-pair spacing for dual striplines is defined as the center-to-center pair spacing, e.g., the distance between the geometry center lines of the two different pairs, as shown in Figure 2.49. When the pairs are overlapping each other, the center-to-center spacing is 0, and center-to-center is (2W + S) when the edge-to-edge spacing is 0.

Figure 2.50 shows the simulated FEXT amplitude versus center-to-center spacing with various insulation layer thicknesses (H2). It is interesting to see that there exist minimum inter-layer crosstalk points. The x-axis is the center-to-center spacing between differential dual-stripline pairs in mils, and the y-axis is the peak crosstalk amplitude in mV for the coupled 10-inch pairs. The curves of different colors correlate to the different insulation layer thicknesses. When the center-to-center spacing is 0, e.g., the two differential dual-stripline pairs are overlapping each other, the inter-pair is greatest, as expected. As the spacing increases, the peak inter-layer crosstalk amplitude decreases and eventually crosses zero. After the zero-crossing point, the inter-layer crosstalk will increase with an opposite

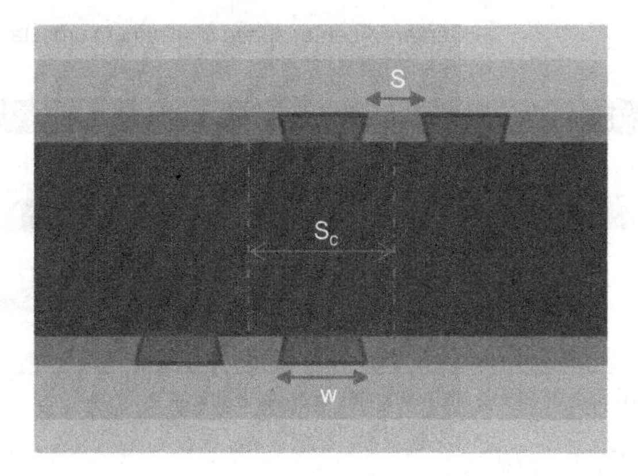

FIGURE 2.49

Center-to-center pair spacing.

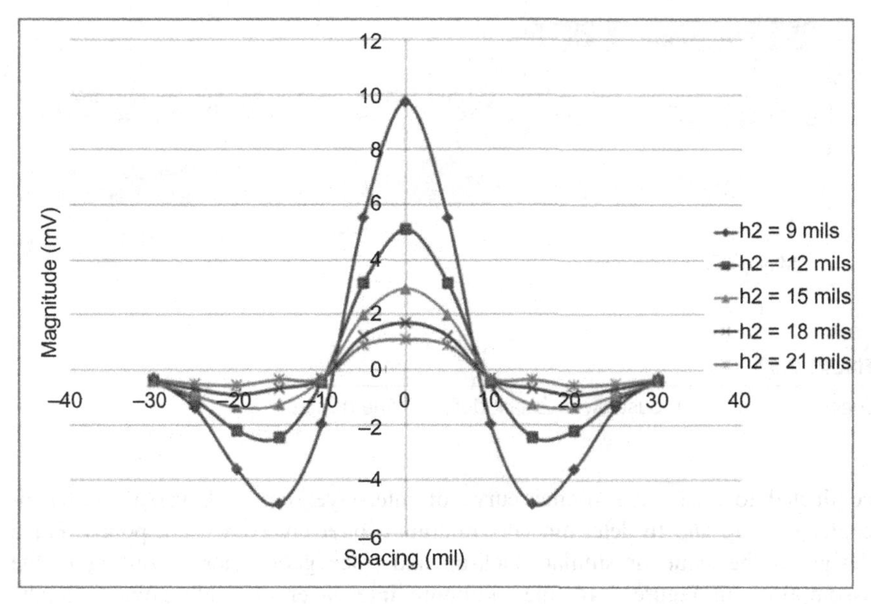

FIGURE 2.50

Differential far end inter-layer crosstalk vs. center-to-center pair spacing.

polarity. Note that the minimum crosstalk points are relatively stable with regard to the thickness of the separation dielectric layer.

The existence of minimum inter-layer crosstalk points makes it possible to route differential dual striplines in parallel with controlled crosstalk. The exact location of the minimum inter-layer crosstalk points depends on the stackup, material properties, and trace geometries. Therefore, simulations should be

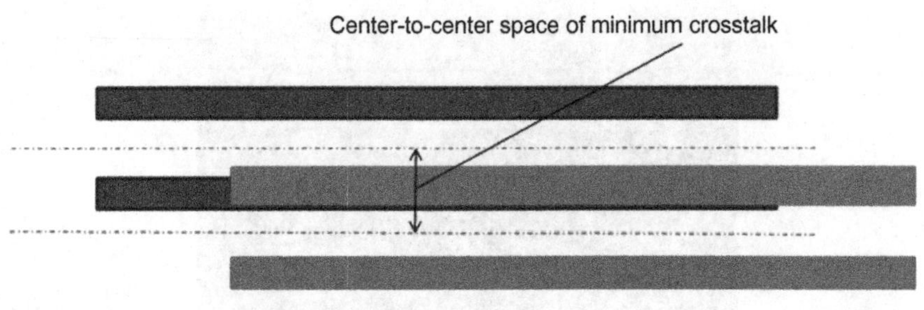

FIGURE 2.51

Routing differential space along the minimal crosstalk line.

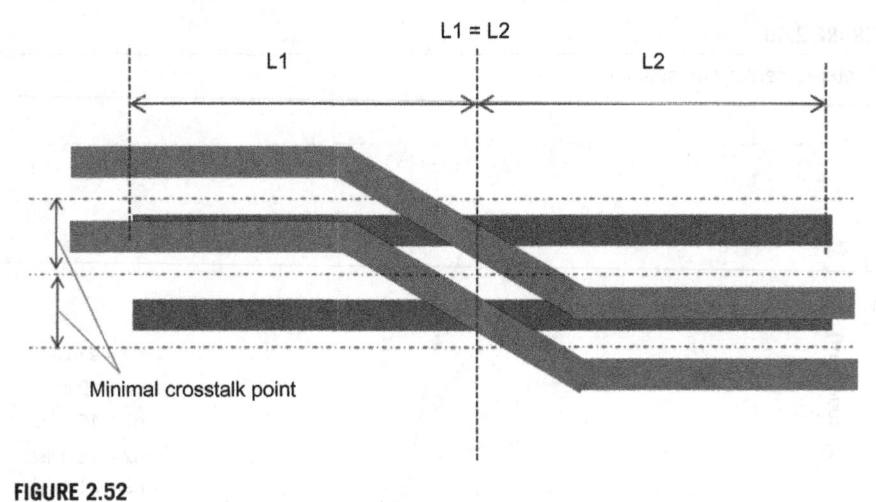

FIGURE 2.52

Jogging scheme for robust low-crosstalk dual-stripline design.

conducted to create the similar curve of inter-layer crosstalk versus center-to-center spacing and to determine the minimum inter-layer crosstalk points. For a design of the same or similar stack-up and trace geometries according to the assumptions in Figure 2.44, the minimum inter-layer crosstalk point is at the center-to-center spacing of ~10 mils. Theoretically, if a dual-stripline pair is routed on the minimum inter-layer crosstalk point of another dual-stripline pair, inter-layer crosstalk will be minimized, as shown in Figure 2.51. When the high-volume manufacturing tolerance is considered, the actual location of the dual striplines may shift away from the desired point. As a result, the inter-layer crosstalk may increase considerably. The tolerance can be quite large when the two adjacent stripline layers are centering on a prepreg layer.

The following wiring scheme, the "jogging scheme," is recommended for robust, parallel dual-stripline designs, as shown in Figure 2.52. For two coupled

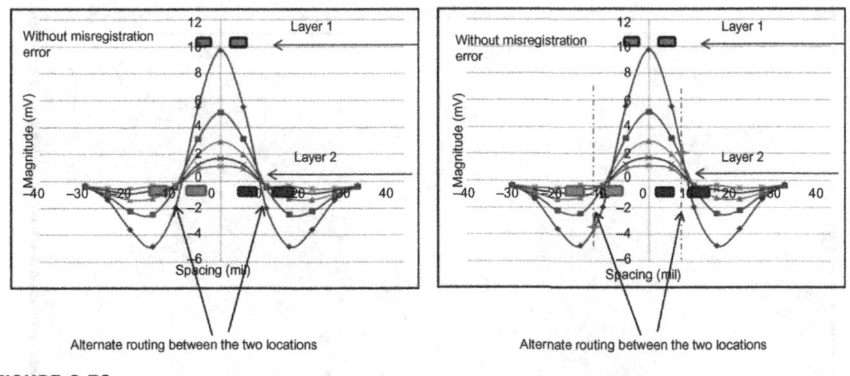

FIGURE 2.53

The jogging scheme in the presence of layer-to-layer misregistration.

parallel dual-stripline pairs on different layers, one of the dual-stripline pairs alternates between the two minimum inter-layer crosstalk points of the other. The total routing length along each of the two minimum inter-layer crosstalk points should be equal. It is okay to jog a pair more than once for more routing flexibility. However, it is better to minimize the number of jogs, because each jog will increase the trace length and add additional phase skew between the two pairs, which may decrease the effectiveness of crosstalk mitigation.

The reason for using the jogging scheme is illustrated in the plot (Figure 2.53). When there is no registration error, both jogging and no-jogging schemes can minimize crosstalk. When there is a misregistration error, inter-layer crosstalk increases as the traces shift away from the desired location. However, the inter-layer crosstalk picked up at different jogging segments has opposite polarity and will cancel each other out. As a result, the overall increase of the inter-layer crosstalk is small. The jogging scheme robustly reduces the inter-layer crosstalk between parallel differential dual striplines as shown in Figure 2.53.

Note that the jogging scheme will still have residual inter-layer crosstalk due to the errors in the prediction of the minimum inter-layer crosstalk points as well as the layout implementation. It is recommended to either simulate or budget for the residual error in the full-link simulation.

For differential near end inter-layer crosstalk, the crosstalk curves remain similar in shape. The amplitude, however, is larger than the FEXT for the same stackup, and the location of the minimal crosstalk points is different too. The same mitigation schemes may be used as illustrated in Figure 2.51. However, the jogging scheme will not work for the inter-leaved dual-stripline design. Because the NEXT does not accumulate in phase, the negative and positive crosstalk will not add up and cancel each other, and instead, a positive and negative peak will be created.

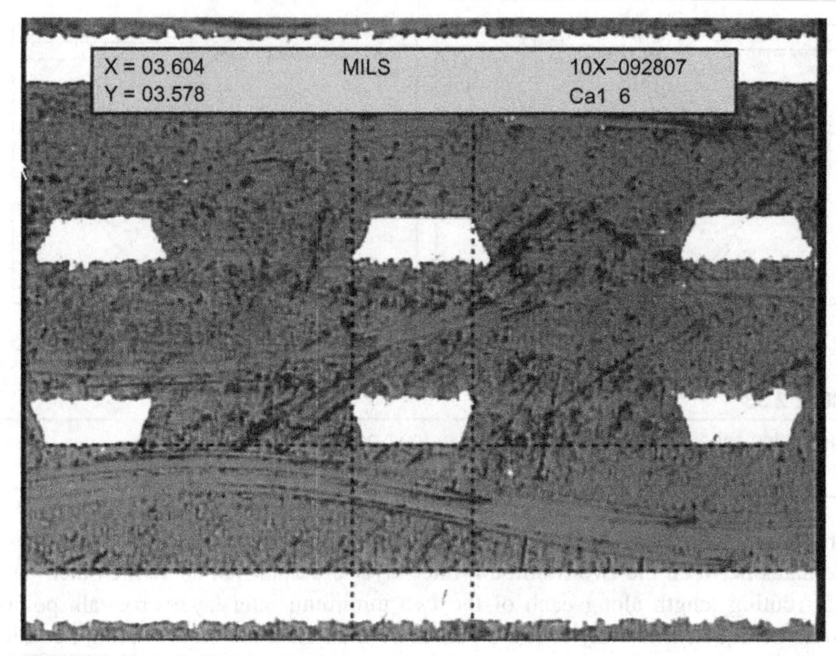

FIGURE 2.54

Cross section of DDSL implementation.

DENSELY BROADSIDE COUPLED DUAL STRIPLINE

An alternative layout technique is densely coupled dual stripline (DDSL), also called broadside coupled stripline. In DDSL, coupling is increased horizontally, between signals, and vertically, between layers. Here, the differential pair is established broadside. Utilizing the DDSL technique can minimize NEXT and FEXT through a deliberate increase of the coupling between traces. The design can be tuned for a minimal amount of crosstalk. Further, the broadside coupling can be applied to improve routing density. By shifting the alignment between N and P of a differential pair, one can tune to optimize FEXT or NEXT. Figure 2.54 shows a cross section of the compact dual stripline.

From Figure 2.54 we observe a key manufacturing variation. The trace width is different between the top and bottom traces in the picture. This is due to differences in etching of the copper. This possibility should be modeled in analysis to account for worst-case impedance, crosstalk, and common-mode possibilities. In this design, the wide side of the trace faces inward within the differential pair. This is obtained with a core construction in the middle. The core in the center and wider trace provides more differential coupling, has advantages of reduced crosstalk, and reduces registration shift.

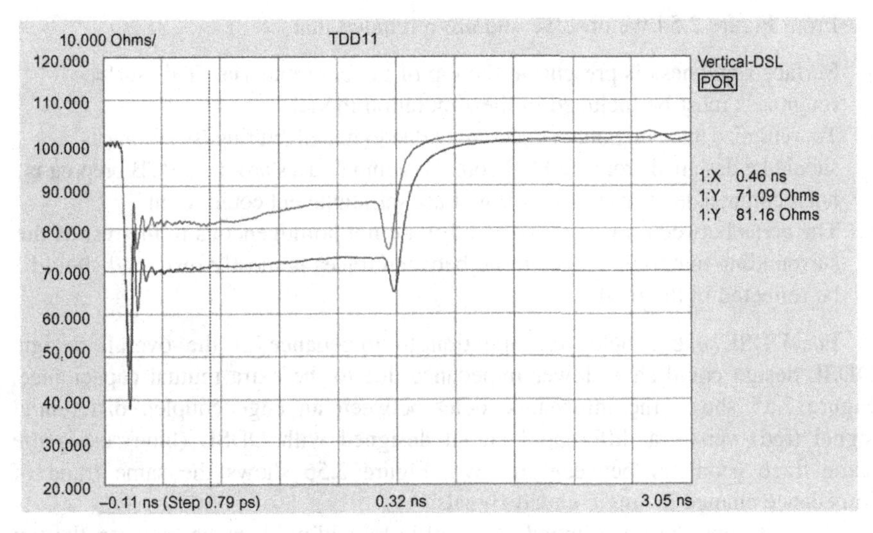

FIGURE 2.55

TDR of differential signal with DDSL and edge coupled.

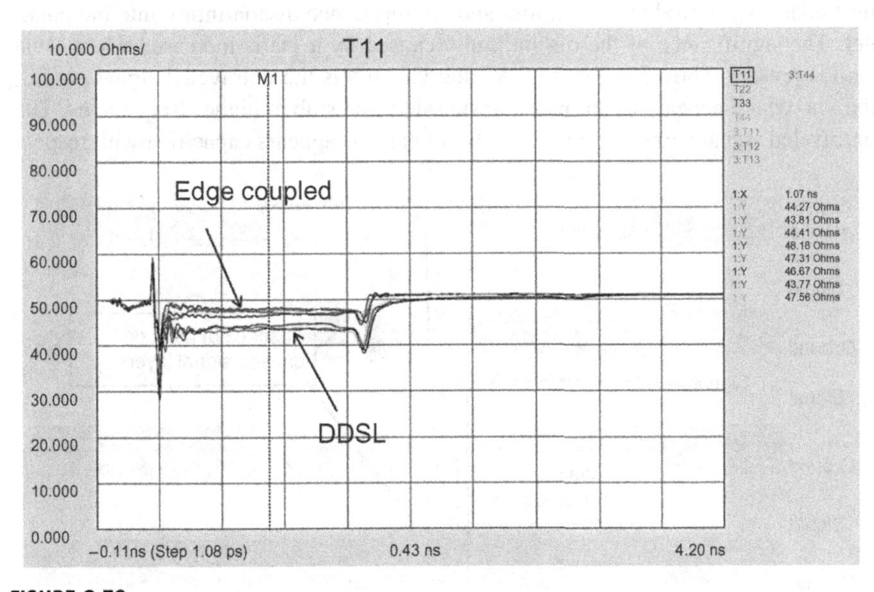

FIGURE 2.56

TDR of single-ended signal with DDSL and edge coupled.

From Figure 2.54 we observe and are reminded that

- Surface roughness is present on the top of the core material. This surface roughness must be included in the simulation model.
- To minimize trace alignment uncertainty due to manufacturing issues, traces should be designed around a PCB core. Designing traces around a PCB prepreg is not recommended because significant trace misalignment could result.
- The areas between traces (horizontally) are not homogeneous to the rest of the surrounding material. The material between traces is mostly resin and should be reflected in the model.

For DDSL one should pay attention to impedance of the overall design. DDSL design could show lower impedance due to the extra mutual capacitance. Figure 2.55 shows the impedance delta between an edge coupled differential signal (red) versus a differential signal designed with DDSL (blue) using the same trace geometry between the two. Figure 2.56 shows the same trend of impedance change in single-ended signals.

One can tune the trace impedance profile by adding distance between the top and bottom traces. Care must be taken to ensure crosstalk does not become a factor as you pull the traces apart.

VIA STUB MITIGATION

Plated-through-hole (PTH) vias used for layer transitions at a BGA, connector, or mid-board will introduce signal loss and an impedance discontinuity into the channel. The significance of the discontinuity caused by a via is increased when a via stub is present. Shown in Figure 2.57, the via stub is the untraveled signal path of the via, which appears as an open transmission line stub at higher frequencies. The untraveled signal barrel for typical PCB dimensions appears capacitive with respect

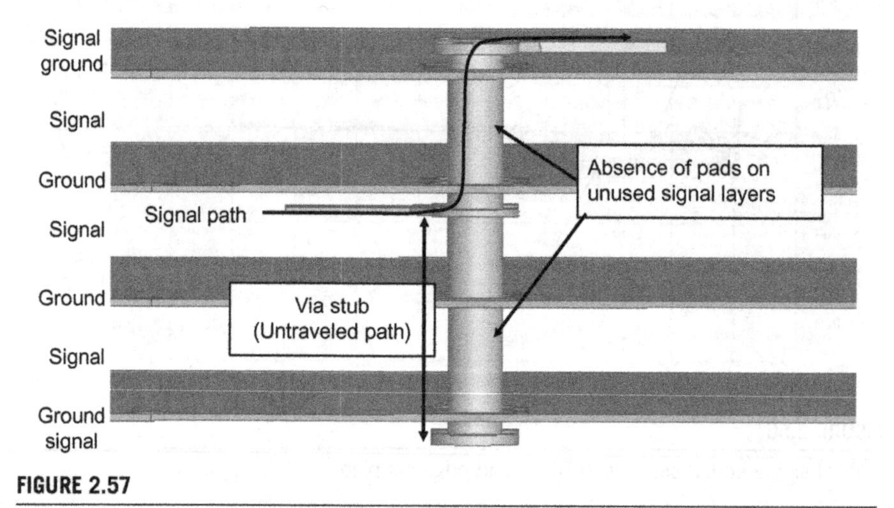

FIGURE 2.57

PTH untraveled signal path described as "via stub."

to typical transmission line routing at 100 or 85 ohms. The magnitude of the discontinuity is a function of the via stub length, drill size, pad size, and separation from planes (anti-pad).

The choice of routing layer affects the length of the via stub discontinuity. Routing on the lower signal layers is preferred but not always possible on high-speed designs. A design with many critical high-speed signals may need every layer for high-speed routing. Adding PCB layers to route all high-speed critical needs on lower stripline layers is often cost prohibitive compared to other techniques. In Figure 2.58, the insertion loss is demonstrated for vias that transition from layer 1 of a 14-layer 0.093-inch stackup to various internal layers, from layer 3 to layer 14. The layers compose a range of via stubs ranging from 0 (layer 14) to 80 mils (layer 3) in length. The vias are shown for common 10-mil and 28-mil drill sizes. Signal layers not connecting to the via do not have pads, a practice used to help decrease the discontinuity without any cost or impact. In the figure, the large Q nature of the via means that an impact is observed at frequencies much lower than the resonant frequency. The longest via stubs have the greatest impact on lower frequencies, with a resonant frequency as low as 9 GHz for the 10-mil drill size.

The 28-mil drill size is often used for press fit connectors like PCIe and has a much greater capacitance and therefore more signal loss than the standard 10-mil drill used for layer transitions on BGA breakouts. The responses shown for a 0.093-inch thickness are fairly indicative of the effect of the same length stub in a stackup with a different layer count. For example, the discontinuity of a 50-mil stub

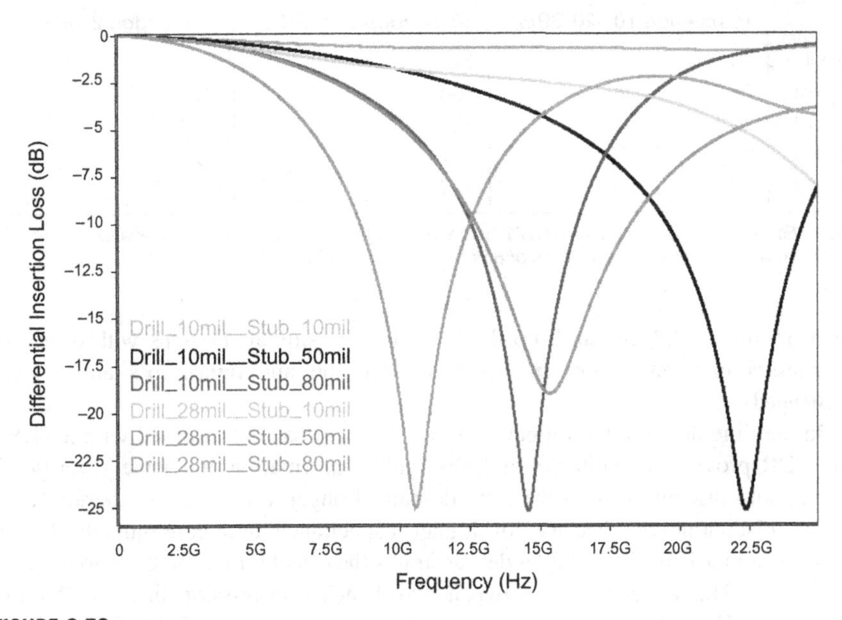

FIGURE 2.58

Insertion loss for various via stubs.

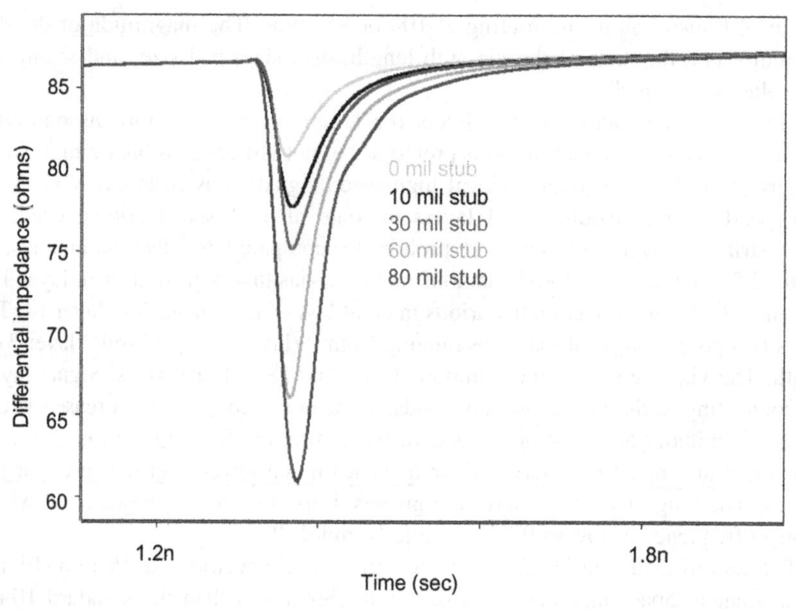

FIGURE 2.59

TDR of various via stubs.

Table 2.8 Impedance Discontinuity for 10-Mil Via Drill Size (ohms)

	15 ps edge 10–20 GB/s	30 ps edge 6–0 GB/s	50 ps edge 2–6 GB/s
0 mil	80.8	81.5	82.1
10 mil	77.7	78.5	79.4
30 mil	75.1	76.2	77.4
60 mil	66.1	67.6	70.1
80 mil	61.1	62.5	64.7

***Note:** For dielectric constant of 4, 20-mil pad size and 10-mil anti-pad. Minimum impedance discontinuity is observed after 4 inches of 85-ohm transmission line.*

for a 62-mil and 93-mil stack-up thickness is very similar. Designs with different thicknesses may vary most in crosstalk—insertion and return loss are largely unchanged.

In the time domain, the reflection from via stubs can be observed with a TDR. The TDR provides quantitative and observable information about the severity of the via stub discontinuity in the time domain. Longer via stubs will demonstrate greater discontinuities because of higher capacitance and consequently lower impedance in a TDR. Figure 2.59 demonstrates the effect on vias with various stubs by a TDR. The via is placed between two 4-inch transmission lines at 85-ohm impedance. The lowest observed impedances are summarized in Table 2.8.

Table 2.9 Eye Height for 10-Mil Via Stub (mV)

	8.0 GT/s	16.0 GT/s
T-Line only	0.83 UI	0.64 UI
T-Line with via	0.81 UI	0.45 UI

Note: *A different t-line length is used at 8.0 and 16.0 GT/s to achieve 15 dB at Nyquist for a fair comparison. No additional noise sources are added to the simulation.*

Does the via respond differently due to higher edge rates? To review the impact of the via, impedance is measured by TDR and cataloged in a table for three different data rates. A constant transmission line length of 4 inches is placed before the via being tested. The results in Table 2.8 reveal that faster edge rates lead to only minimal change in reflection magnitude. If the change in impedance is slight, then what is the cause for significant eye closure at faster data rates?

The via stub becomes more detrimental at higher data rates. A simulation demonstrates the impact at 8 GT/s and 16 GT/s in Table 2.9 using the same device under test (DUT) used in Figure 2.59. The simulation includes reasonable equalization—two taps of transmit de-emphasis and a receiver CTLE. The results show that at 8 GT/s the via degrades the eye opening 2 percent while at 16 GT/s the impact is a larger 19 percent.

The cause for reduced eye opening is found in the reduced unit internal and unchanged reflection duration. The pulse response observed at the end of a line is shown in Figure 2.60. Three waveforms are shown with 0-, 30-, and 80-mil via stubs. The reflection occurred at a delay of three times the distance to the discontinuity, plus the delay from the discontinuity to the receiver.

An overlay of the bit length for 5 GT/s and 20 GT/s signaling helps illustrate why discontinuities are an increasing problem for higher data rates (see Figure 2.61). At 5 GT/s, the duration of a reflection will last for approximately 2 bits. This is contrasted at 20 GT/s, where approximately 8 bits may be affected. When superimposed, a single bit may be influenced by the summation of eight individual reflections. Some options are available to reduce the impact of via and via stub reflections. The goal of optimization is to remove the reflection seen in TDR or the pulse response, providing the least amount of distortion possible.

Impedance optimization

The geometries related to the PTH via are the direct cause of the impedance discontinuity. It is possible to tune some of these parameters in order to match the transmission line impedance leading into the via, thereby eliminating the discontinuity. This process must be done uniquely for each PCB layer transition and for each drill size. Impedance optimization for PTH vias can be a very resource-intensive task, requiring much 3D simulation. Limitations on how parameters can be tuned are bounded by manufacturing capability and layout area. As the

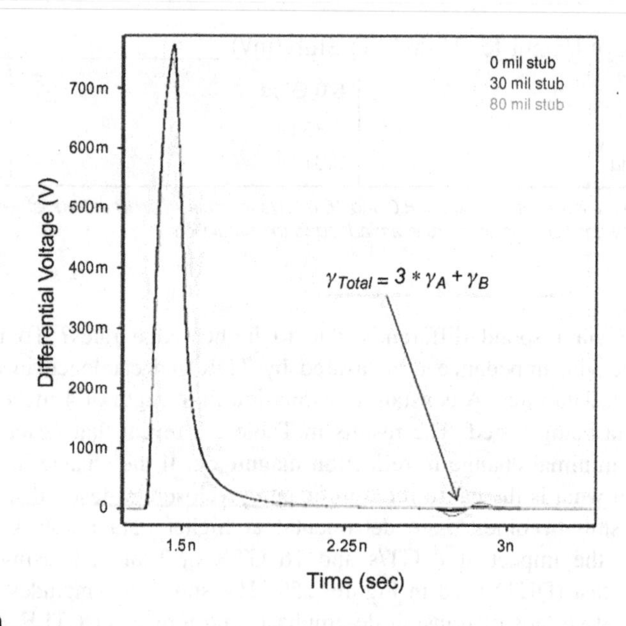

FIGURE 2.60

Pulse response with reflections from via stub.

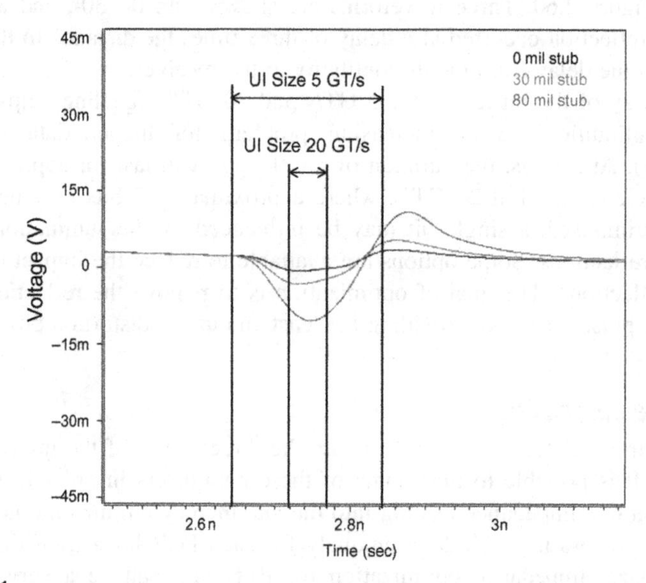

FIGURE 2.61

Zoom of pulse response reflection caused by various via stubs.

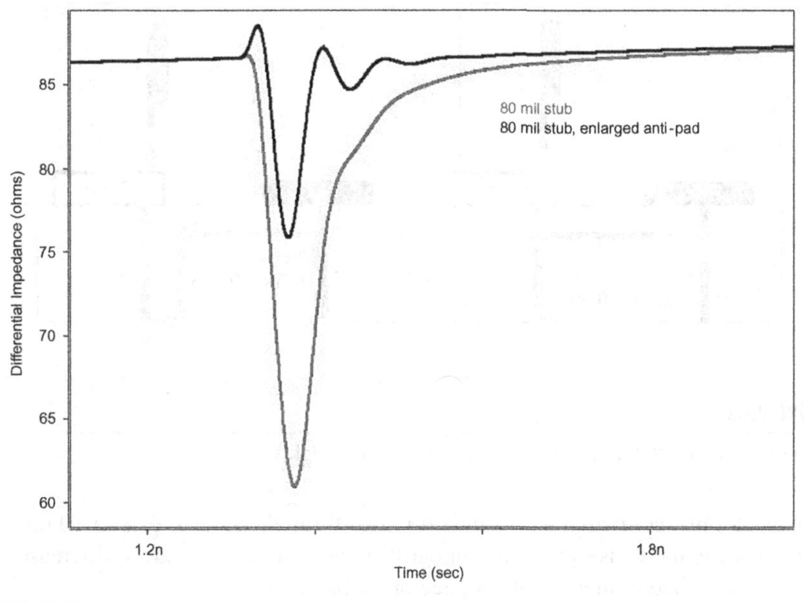

FIGURE 2.62

TDR response of 80-mil via stub with standard and enlarged anti-pad.

structure is dominated by capacitance, the mitigation approach should be to consider what structural changes will decrease capacitance or add inductance to the via. The parameter ranking in order of importance is as follows:

- Significant: Via stub length
- Significant: Anti-pad clearance
- Significant: Drill size
- Moderate: Pad size
- Low: Total board thickness

One of the primary parameters for optimizing impedance is the anti-pad size. Anti-pads may be increased or joined, on some or all metal layers, to decrease the capacitance. The disadvantage of increased anti-pads is the removal of the metal layer, which may be needed as the reference for signal routing. This is particularly problematic in congested areas like BGA and connector pin fields. Figure 2.62 shows the TDR response for an 80-mil via stub with a nominal anti-pad and an enlarged anti-pad. Observing the impedance, the discontinuity is significantly improved, giving the appearance of a 10-mil via stub. A compromise may be found by increasing or joining the anti-pad on some but not all layers. While pad size has a significant influence, it generally cannot be reduced further due to manufacturing requirements. Drill diameters are the result of mechanical requirements and cannot simply be reduced. For example, some press fit connectors require a drill size as

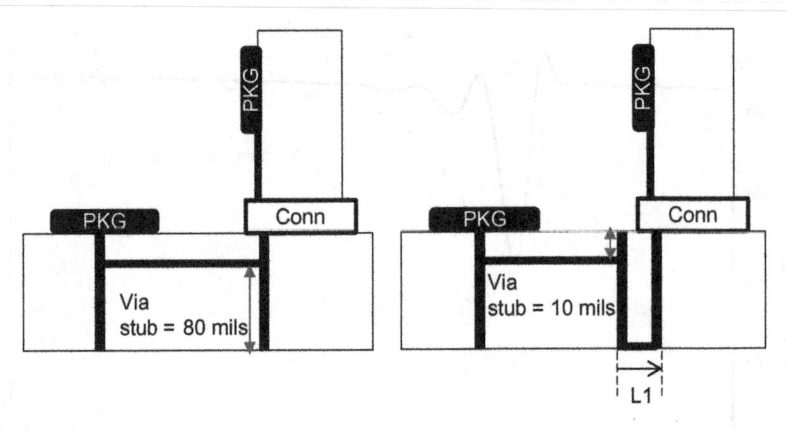

FIGURE 2.63

Original topology (left) and u-turn via implementation (right).

large as 28 mils, compared to a standard-plated through-via drill of 8 to 10 mils. If a large drill is the cause for signal integrity issues, the options are to increase the anti-pad size, choose an alternate connector, or back drill.

U-turn via

An approach to reducing the system return loss is the use of the "U-turn via." This method is patented under U.S. patent #WO2006050286A2. This method introduces an additional via to redirect the signal path and reduce the via stub. In Figure 2.63, left, is the original topology with an 80-mil via stub. On the right, an additional via is used to direct the signal path through the bottom of the otherwise 80-mil stub. In the end, the net effect of two vias with no stub is better than one via with a long stub. This technique is most useful before connector vias where the larger capacitance from larger drills (≤15 mils in diameter) creates more discontinuity than normal transition vias (~10 mils). At speeds of 10 GT/s and above, via mitigation techniques such as U-turn become critical in enabling the use of upper stripline layers.

Figure 2.64 demonstrates a practical implementation of a U-turn via before a 28-mil-diameter PCIe connector via/pin. The transmission line between the added via and original via is shown as L1. The length of L1 is important to minimize the discontinuity between the vias. If the length of L1 is electrically long compared to the data rate, the reflection from the trace will contribute to more ISI in the eye opening. A maximum length for L1 to ensure the most benefit from the U-turn via is 500 mils for 8GT/s, with preference at 300 mils.

Back-drilling

Back-drilling can be a relatively inexpensive means to reduce via stubs compared to other high density interconnect (HDI) techniques that require sequential

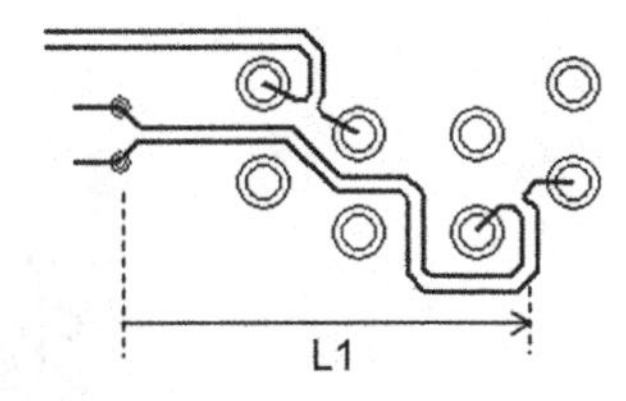

FIGURE 2.64

Layout example of U-turn implementation.

buildups. In this method, boards are sent through the drilling process for a second time, incurring a cost proportional to the number of drills required. In the drilling process, the via stub is virtually removed, leaving a minimum possible via stub length of 20 mils: the industrial capability today. While 20 mils may be electrically negligible for edge rates associated with 10-GB/s signaling, it will be more significant at higher data rates. It is expected that today's achievable via stub tolerance will be almost 50 percent lower by 2020.

The disadvantage of back-drilling is the destruction of the routing channel where the drill has passed through and consideration must be taken as to which layers must not be broken when drilling. For example, it may be advantageous to reduce a highly reflective 80-mil via stub to a less reflective 40-mil via stub, more than the minimum of 20 mils, if it preserves upper routing channels.

When the back drill is to be performed, copper traces require an increased clearance from the via drill. In dense areas, such as BGA pin fields, the back drill can completely eliminate the possibility to route any traces between the vias when pitch is less than 48 mils (10-mil drill). The destruction of the channel is caused by a larger drill and the associated uncertainties shown in Figure 2.65 and Table 2.10. The minimum back-drill size is 8 mils larger than the original drill to compensate for the uncertainty of the precise drill location when the boards are aligned to be processed for drilling, thus ensuring that the via has been completely removed. An additional guard band of 9 mils (E) will be needed between the larger back-drill size (B) and trace due to layer-to-layer registration error (±5.0 mils), minimum dielectric (3.5 mils), and trace etch variations (0.5 mils).

Differential signal escape in small pitch situations (<48 mils) is possible when placement of signal pins in the package is carefully planned. An illustration of this design is shown in Figure 2.66. Placement of critical signals that will break out on upper PCB layers and require back-drilling should be placed inward of or behind, while signals that break out on lower PCB layers that do not require back-drilling are placed on the outside area of the BGA. Breakout channels are still destroyed when back-drilling is performed; however, pin-out is designed such that no signals requiring an escape in upper layers are located behind the back-drill region.

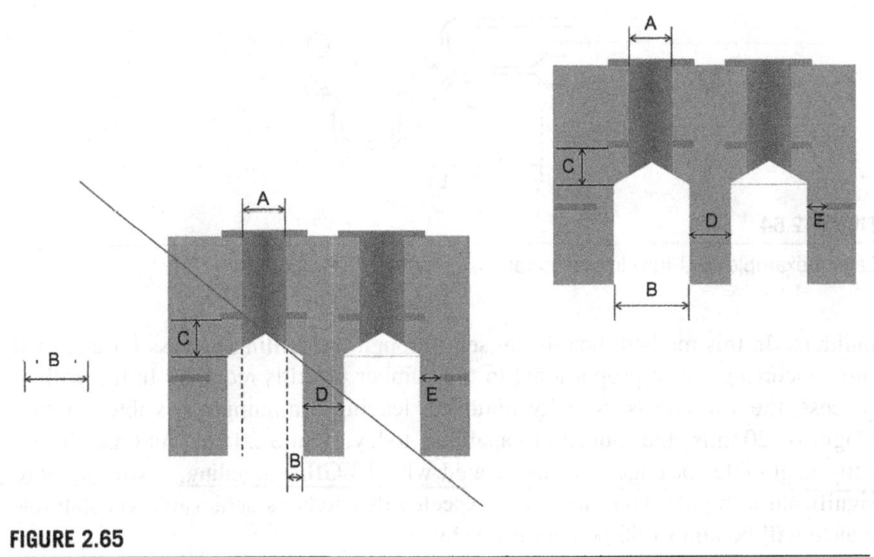

FIGURE 2.65

Back-drill parameters.

Table 2.10 Industry Back-Drill Capability

Parameter	Designator	Capability (mils)
Minimum Drill Diameter	A	10
Minimum Back-Drill Diameter	B	A + 8
Depth Uncertainty + / −	C	10
Minimum Drill Web	D	10
Minimum Drill to Copper Trace	E	9

Note: *Industry capabilities taken at the time of printing in 2015.*

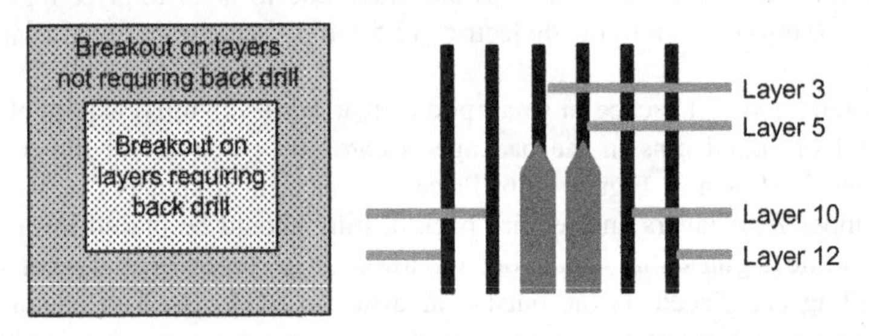

FIGURE 2.66

Top (left) and side (right) views of back-drilling method underneath BGA.

Figure 2.66 illustrates the same drill depth for all vias, but additional processes may be completed to perform multiple drill depths at additional cost; however, this is still expected to be cheaper than other HDI methods.

Connectors with through-hole mounting technology cannot be back-drilled due to the solderwick located on the backside of the board, necessary for mechanical reliability. Press fit mounted connectors may be back-drilled up to 20 mils below the end of the pin. While via stubs may not be greatly reduced for press fit connectors, the pads on the bottom layer are removed for a significant gain in via barrel capacitance.

Blind and buried via

Blind, buried, and microvias are a means to reduce the impact of vias by eliminating the unused signal path. Often referred to as high-density interconnect (HDI), these methods are a significant cost after the PCB construction. A buried via, not connecting to external layers, is an increased cost because the layers connected must be completely built up and plated before the outer layers are introduced. A blind via is drilled, penetrating only one or two layers. Blind vias connect to an external layer. A microvia is a type of blind via that is laser-drilled. The laser-drilled layers are assembled with a laser-drillable prepreg. Often available in the same 106, 1080, and 2116 styles, these prepregs flatten out the concentrated glass weaves, which are more difficult for the laser to penetrate, ensuring more consistent burn-through depths for each microvia. The resin-to-glass ratio is not changed for these materials, yielding the same effective electrical parameters with the added benefit of uniformity and therefore a decreased fiber weave impact. Blind and microvias are smaller, 100–200 μm depending on the vendor. When considering use of blind and microvias, an aspect ratio of 1:1 or less is required, 0.7:1 being preferred.

A variety of microvias exist, as shown in Figure 2.67. From left to right, the illustration shows (3) through (7) as an external microvia, internal microvia, staggered microvia, stacked microvia, and stepped microvia. The internal microvia (4) demonstrates an optimization that may improve microvia performance by voiding the ground plane above and below the microvia. This optimization, depending on the stackup, may help bring microvia impedance closer to the routing impedance. Optimization can be determined with 3D simulation. The same optimization may be used for blind and buried vias.

PCB LAYOUT OPTIMIZATION

This section provides guidance for optimization of the layout for signal integrity performance. Discussions include crosstalk, return path, length matching, and power integrity. To begin, the "top 10 rules" for high-speed design are provided and are applicable to nearly any differential interface.

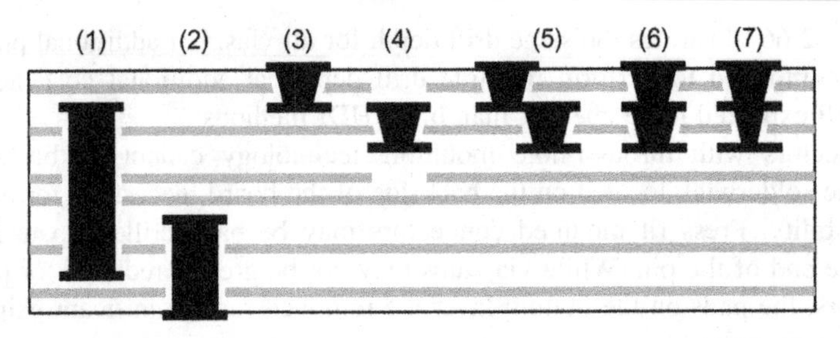

FIGURE 2.67

HDI structures: (1) buried via, (2) blind via, (3) microvia, (4) internal microvia, (5) staggered microvia, (6) stacked microvia, and (7) stepped microvia.

1. Choose a routing layer that minimizes PTH via stub length. Give precedence to reduced PTH via stub lengths for press fit and through-hole mount connectors with larger PTH via drill and pad diameters.
2. Adding ground PTH vias near the signal transition when changing layers also changes the reference plane.
3. Minimize stripline crosstalk with a separation of at least 3 times the dielectric thickness of the primary reference plane.
4. Minimize microstrip crosstalk with a separation of at least 9 times the dielectric height for signals above 5 GT/s, and 7 times for signals below 5 GT/s.
5. Increase pair-to-pair spacing for interleaved routing, especially when near end points.
6. When signals are serpentine, maintain a self-spacing equal to or greater than that which would be used for a crosstalk aggressor.
7. Remove the ground plane directly beneath the edge pads on plug-in cards and the mounting pad of surface-mount connectors.
8. Avoid routing traces between PTH vias carrying the positive and negative signal of a differential pair in BGA and connector pin fields.
9. Do not route signals of plane splits in the primary reference plane, which shall be ground. Avoid routing over plane splits to loosely coupled secondary reference planes.
10. Place DC blocking capacitors in a line symmetrically.

LENGTH MATCHING

Routing of differential pairs inevitably results in length mismatches between the signals within a differential pair that must be compensated. As signals are routed, bends add positively or negatively to the accumulating skew. In all cases, the outside signal in the bend has a longer length than the inside signal, creating a length mismatch as indicated in Figure 2.68(A). If another bend is taken in the opposite direction, the accumulated skew will be negated. Corners routed at 90 degrees are generally held as the least favored routing type for signal integrity as they add the

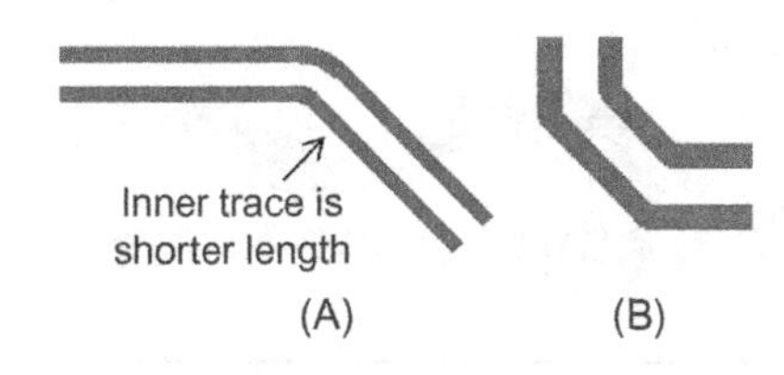

FIGURE 2.68

Bends in routing.

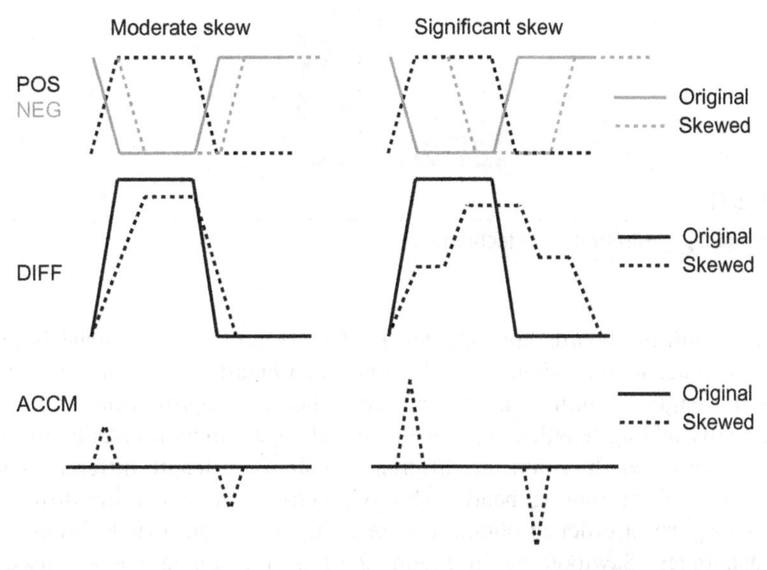

FIGURE 2.69

Graduated effect of intra-pair skew.

most routing length mismatch and theoretically the greatest capacitance. It is advised to add mitering to the bends to reduce the capacitance in the bend, as shown Figure 2.68(B). However, 90-degree bends are permissible and according to Montrose [4, p. 4] are indistinguishable in time domain and frequency domain measurements when edge rates are above 15 ps.

The primary reason to maintain skew in differential pairs is to maintain low EMI, one of the favorable advantages of differential signaling. When electric fields no longer cancel from equal and opposite currents, common-mode EMI will become a serious problem. A second issue is AC common-mode increase (ACCM) due to delay differences within the differential pair. Any delay difference within the differential pair when received at the differentiator will lead to energy being converted to common mode that would have otherwise been part of the differential signal. The loss in the differential signal is observed as a rise time or signal height reduction, as illustrated in Figure 2.69.

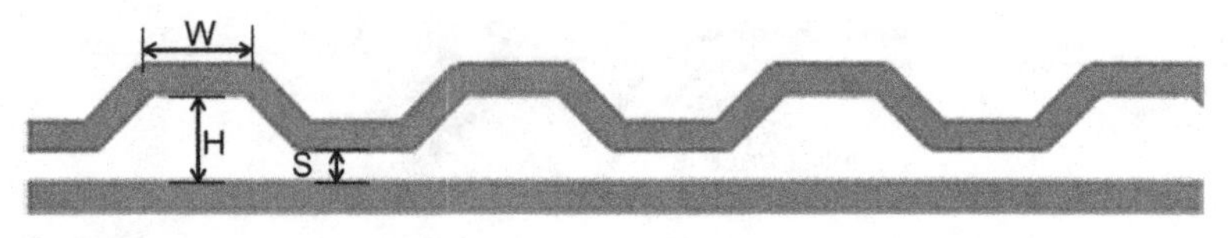

FIGURE 2.70

Length matching with sawtooth technique.

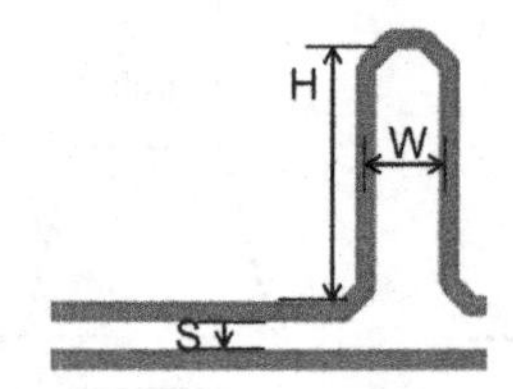

FIGURE 2.71

Length matching with trombone technique.

When multiple boards are included in the topology, caution must be exercised not to accumulate significant skew between boards. A common target is a matched length within 5 mils on each board (approximately 0.85 ps). Intentionally adding length to one of the signals is a common practice to maintain a matched length within the differential pair when length differences have been created from routing bends. The symmetry requirement for differential pairs is forgone in order to obtain a more important length match through several techniques. Sawtoothing in Figure 2.70 is a common method used for length matching that may increase running length by about 30 percent for the duration of the sawtooth. In order to manage the impedance discontinuity, rules must be followed: sawtooth length *W* may be up to three times the trace width; and sawtooth separation may be 2–3 times the original separation. Trombones in Figure 2.71 may also be used for length matching. Trombones uncouple the differential pair and have the disadvantage of symmetry, exposing the differential pair to common-mode increases. The advantage of trombones is the ability to match length in a relatively discrete location, without increasing the trace pitch and width of a bus.

Connections at devices, capacitors, and connectors are other locations where skew is introduced. As signals escape from components, differential symmetry is preferred. Some pin and pad orientation relationships with the escape routing may lead to skew, as shown in Figures 2.72 and 2.73. These cases intentionally preserve symmetry at the expense of matched length, which could have been done with asymmetrically angled routing. Length compensation should be performed as soon as possible as demonstrated on the right side of Figure 2.72.

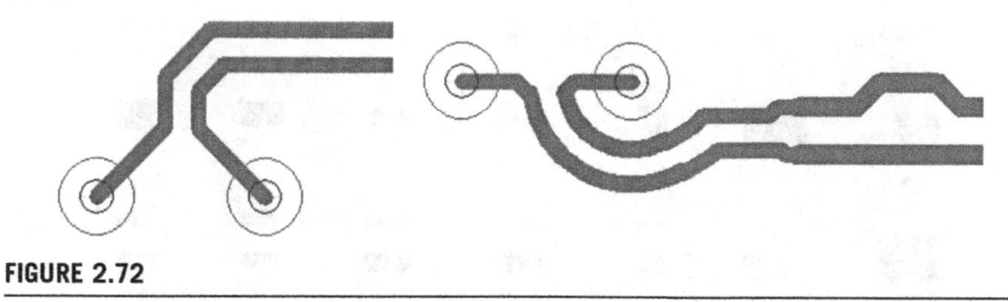

FIGURE 2.72

Symmetric routing examples at pins and vias.

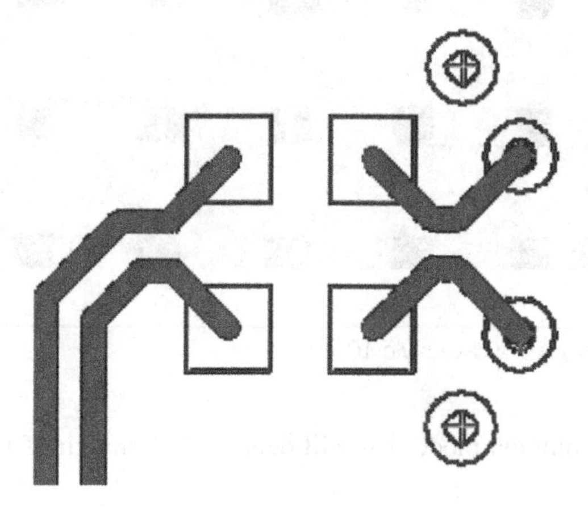

FIGURE 2.73

Symmetric routing examples at pads and vias.

FIBER WEAVE EFFECT

Signal traces placed in the PCB may lie in parallel to the glass weave in the dielectric and can result in significant signal skew. An example of what this may look like on glass weave style 106 is shown in Figure 2.74, with a differential pair superimposed on top of the glass weave. When one signal within a differential pair lies on top of a glass bundle and the other lies between bundles, the signals within a differential pair will have different propagation rates. Signal velocity is related to the dielectric constant, which is uniquely different between glass (3.7 and higher) and epoxy (approximately 3.0), by equation (2.33) for velocity, where c is the speed of light in a vacuum:

$$v = \frac{c}{\sqrt{\varepsilon_r}} \tag{2.33}$$

Loyer [5, p. 5] suggests the worst-case ε_r disparity between signals is 0.8, leading to a skew rate of 60 ps every 4 inches. Signal skew from glass weave, similar to intra-pair length matching, can lead to significant EMI issues and

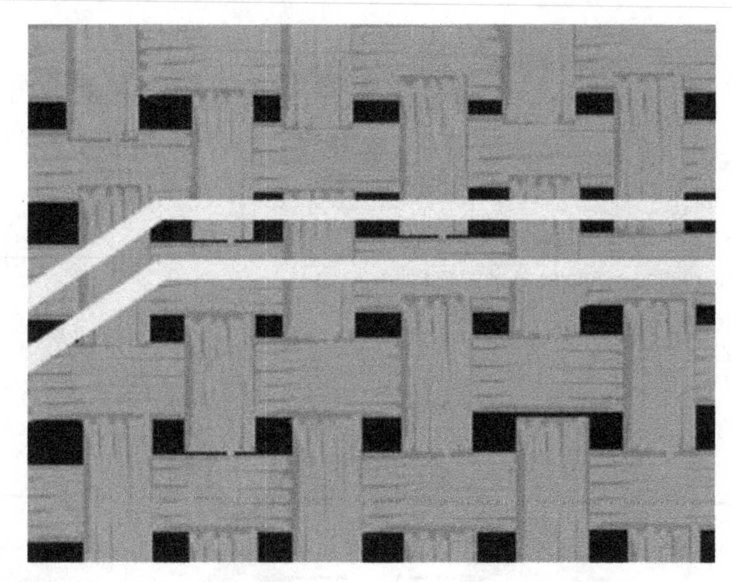

FIGURE 2.74

Differential pair over glass weave style 106.

increases in AC common mode that will degrade link margin if it is not properly managed.

In order to mitigate the effect, design the board such that traces will be routed 2 degrees relative to the weave and that no routing is parallel to the weave for a significant amount of time. However, fiber weave may vary by as much as 5 degrees from the board edge. A minimum of 7 degrees is required to mitigate the effect. At 7 degrees, signals will see an averaged dielectric constant (approximately the value quoted by board manufacturers). Effective solutions to compensate for fiber weave are the following:

- Lay out the board so that routing is at an angle—however, this may be a challenge for CAD designs.
- Rotate the board design by at least 7 degrees.
- Use a denser weave than 106 or 1080 glass (2216 or 7268); this can help mitigate the effect.
- If using a stripline, select different weaves for above and below the trace. Symmetric stripline stackups (same thickness above and below the signal) will see an averaging of the dielectric constants.
- Choose spread glass dielectrics (1067, 1086) that have a "flattened" glass weave to reduce dielectric constant difference between P and N traces.
- Route traces in zigzags. In order to route traces in a coordinal direction, frequent zigzags may be performed at +7 and −7 degrees from the board edge. Zigzags should traverse at least three times the glass pitch.

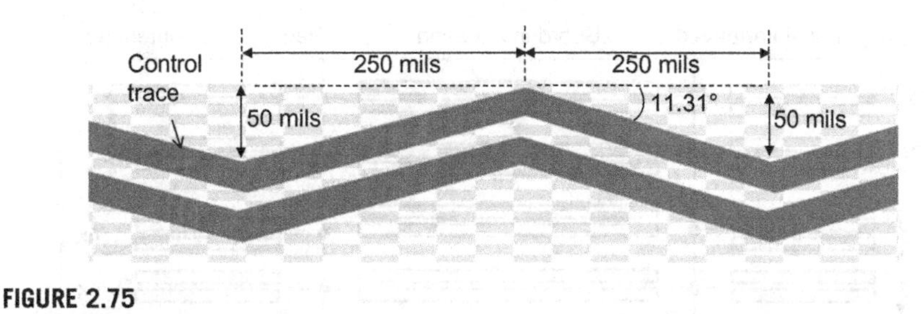

FIGURE 2.75

Zigzag routing stays on grid at 11.31 degrees (units in mils).

An implementation of traces routed in zigzag is shown in Figure 2.75. Manual implementation of the zigzag in CAD tools can be easier when traces remain on grid. In the example, the control trace remains on a 5-mil grid size with 11.31-degree routing at the specified lengths.

These options come with an implementation cost. A 10-degree design rotation generates unused panel space, often increasing the number of panels necessary for a desired product volume. Zigzag of traces that would otherwise be parallel or orthogonal to the board edge consumes layout area that could be reduced or used to implement features.

CROSSTALK REDUCTION

Once the stackup is finalized the only variable seemingly available to the design engineer to reduce crosstalk is signal separation. However, this section explores many additional opportunities to reduce crosstalk in the layout. Refer to the stack-up section of this chapter for stackup-based crosstalk reduction techniques.

Interleaving

Interleaving signal routing on the PCB may be an opportunity to reduce the total crosstalk seen in a system. Care must be taken to understand the crosstalk tradeoff between noninterleaved and interleaved options. It is recommended to complete system simulation to understand the impact. Interleaving signals present a risk of significantly high NEXT. This method can be successful when care is taken to significantly increase separation between signals near the I/O or to interleave signals for only part of the total interconnect. One possibility may be to route packages non-interleaved, while using an interleaved motherboard route as shown in Figure 2.76. The ability to choose an interleaving route is highly dependent on the package pinout, which if already defined may restrict the layout decision. If interleaved routing is chosen, the end result may be microstrip signals that see less crosstalk and improved margins or a design that has saved board real estate by allowing microstrip traces to be placed closer together.

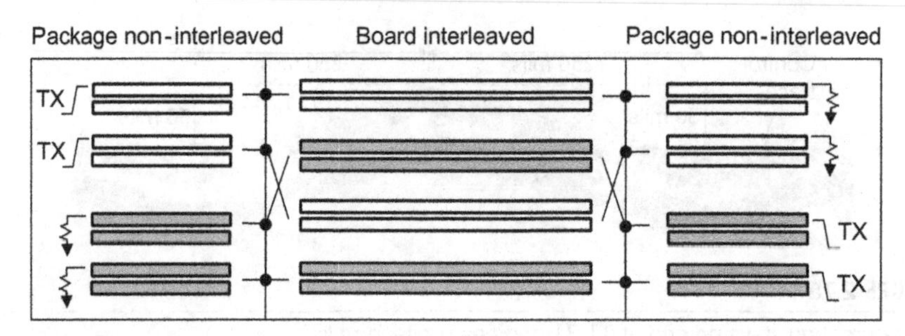

FIGURE 2.76

PCB interleaved routing.

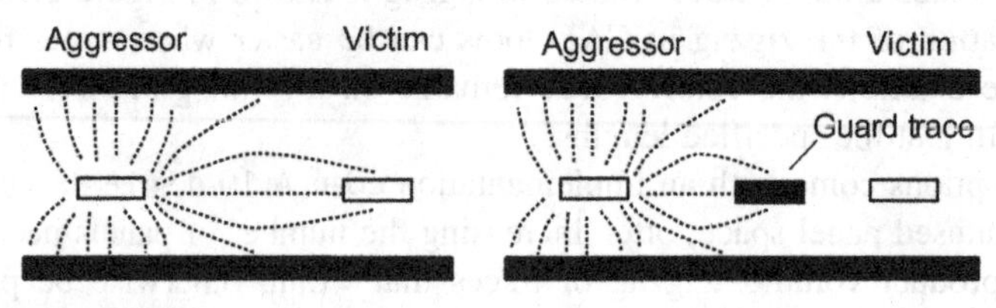

FIGURE 2.77

Electrical field with and without guard trace.

Guard trace

Placement of a guard trace between signals may be used to reduce coupling between signals (see Figure 2.77). The guard trace provides a return current path, taking electric fields that may have induced current as crosstalk onto a victim. The guard trace will have a slight impact on impedance, which may require a slight adjustment of trace width and spacing. Care must be taken to frequently connect the guard trace to the ground plane with ground vias to avoid return path issues that may actually increase crosstalk from a long return path loop. In order to validate the crosstalk reduction and return path, lab testing with all signals active is a must-do.

Signal-to-ground ratio

The number of signals to the number of ground pins is the signal-to-ground ratio. The S:G ratio may be used to describe connectors, BGA pin-outs, and sometimes PCB via transitions. The S:G ratio provides a quantitative description of the isolating ground pins between signals pins. Increasing the ground pins and reducing the S:G ratio will reduce the design's crosstalk. The isolation created by ground pins help reduce crosstalk and provide a return path for the current on a designated ground pin rather than induced current on the neighboring signal via. Placement of grounds between signals will provide shielding for crosstalk signals. In connector and BGA designs the number of available pins is limited and ground

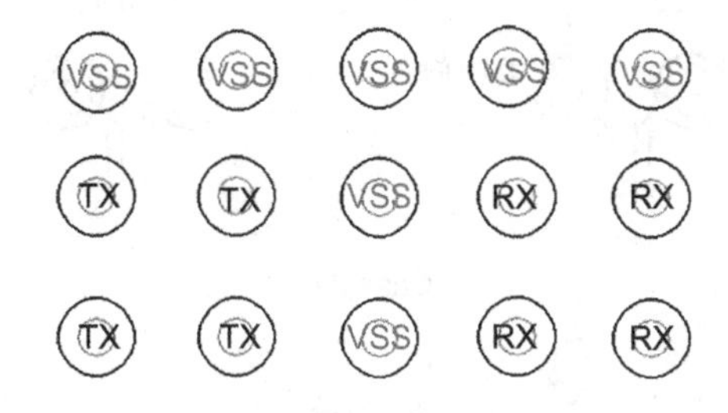

FIGURE 2.78

S:G ratio of 2:1 with isolation for TX to RX signals.

isolation cannot be supplied for every trace. One possible tradeoff is a lower S:G ratio for TX-to-TX signals and a higher S:G ratio for TX-to-RX signals.

Placement of ground between TX and RX signals as in Figure 2.78 is used to minimize the amount of NEXT when significant physical space cannot be created between the signals. Physical separation is often the solution for connectors. In BGA pin-outs, however, tightly spaced pin arrangements do not lend themselves to an increased distance to adjacent pins. The importance of isolation between TX and RX signals depends on the crosstalk magnitude and the component location within the interconnect. The highest priorities for TX to RX isolation are in package design and package escapes on the PCB. Connectors and vias in the middle of a lossy interconnect are more tolerant of NEXT due to higher attenuation before reaching a receiver.

Ground placement

Layout constraints may restrict the number of grounds available or the placement of grounds. Three examples are modeled in Figure 2.79. In case 1 and case 2 the ground vias are located within 50 mils of the signal vias. The distance is measured as the air gap between the nearest point of via pads. Case 3 locates ground vias closely between the signal traces.

FEXT simulation results are shown in Figure 2.80. It is seen that cases 1 and 2 offer similar crosstalk performance for ground vias encircling the signal vias. A crosstalk reduction is offered in case 3, demonstrating a change from −42.9 dB to −48.5 dB at 4 GHz. While there is crosstalk improvement from case 3, there are other factors to consider. A rigid placement requirement for ground vias between signals will eliminate the routing channel between the signals that was otherwise usable on other layers. Close placement of the ground via can have an influence on differential impedance and differential to common-mode conversion, which can only be assessed through simulation.

Orthogonal placement

Placement of an aggressor that is equal distance from both the positive and negative signals of a differential pair will greatly reduce the crosstalk between an

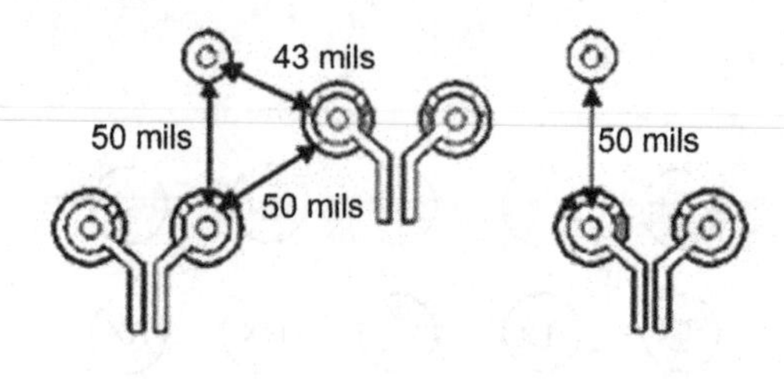

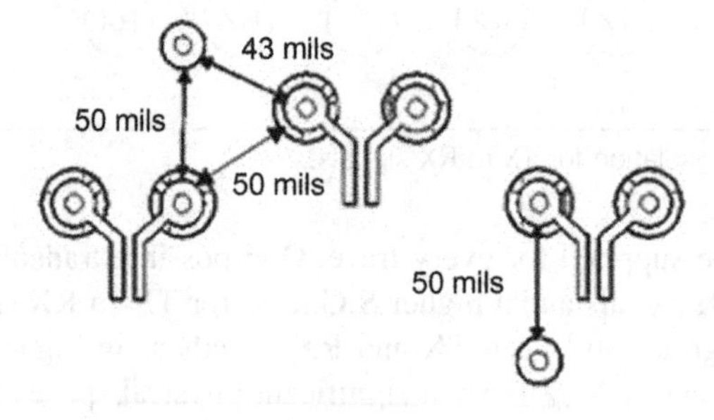

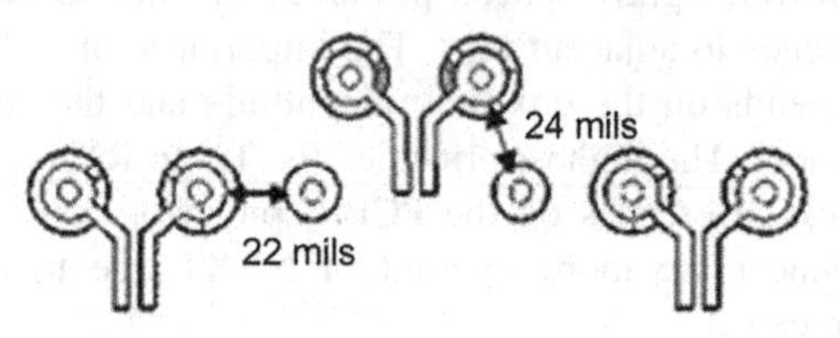

FIGURE 2.79

Layouts for ground placement review.

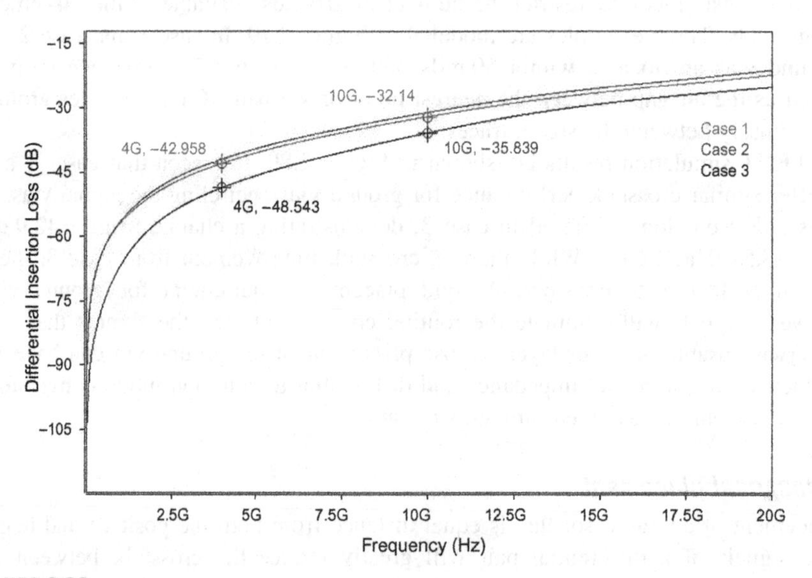

FIGURE 2.80

Simulated FEXT between vias.

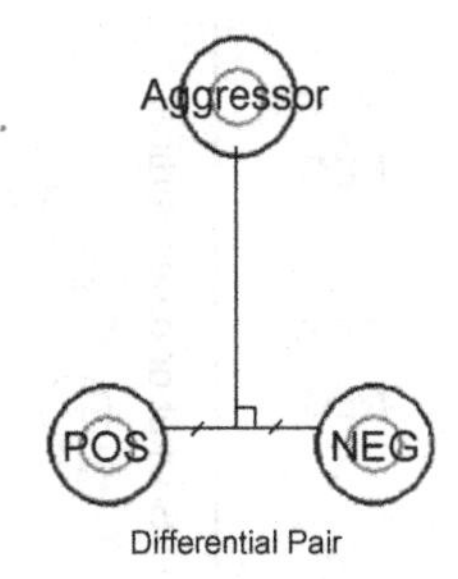

FIGURE 2.81

Differential pair via with an orthogonally placed aggressor.

aggressor and an odd mode differential victim (see Figure 2.81). In an orthogonal arrangement, the magnetic fields seen at the aggressor are equal and opposite. Ideally the arrangement would yield a net induced current of zero. In reality, the crosstalk is not zero. The effectiveness of orthogonal placement is dependent on differential pair skew and AC common mode. Minimization of the running differential pair P/N skew will maximize the effectiveness of the orthogonal placement. Accumulated running skew must be compensated before an orthogonally placed aggressor. Standard 5-mil length matching is sufficient. Transmitter edge rate and swing differences between positive and negative signals lead to unequally induced current and result in crosstalk. Differences in the positive and negative signal are measured as AC common mode (ACCM). There is no recommendation of the maximum ACCM required to minimize crosstalk. Simulation analysis with accurate transmitter models is recommended to assess the accumulated crosstalk.

Component (vertical to horizontal) cancellation

FEXT introduced from vertical interconnect components may be cancelled by accumulating transmission line crosstalk in the opposite direction in the interconnect. To take advantage of a crosstalk cancellation opportunity, the differential crosstalk polarity (positive or negative) of a vertical component must be determined to be opposing the positive crosstalk of the transmission line. Polarity is defined as the sign on the differential voltage induced on a victim by the differential rising positive edge of an aggressor. Many factors determine the crosstalk polarity of a vertical component: via or pin placement, S:G ratio, and via thickness. Open field transmission line polarity is almost always positive for an alternating routing pattern P1-N1-P2-N2-P3-N3; however, minimal spacing in escape regions may be subject to different behavior. Open field stripline routing has one polarity exception. Polarity may change from positive to negative in the less common stackup of asymmetric stripline, where dielectric constant of the thicker dielectric is much greater (more than 20 percent) than the thinner dielectric layer.

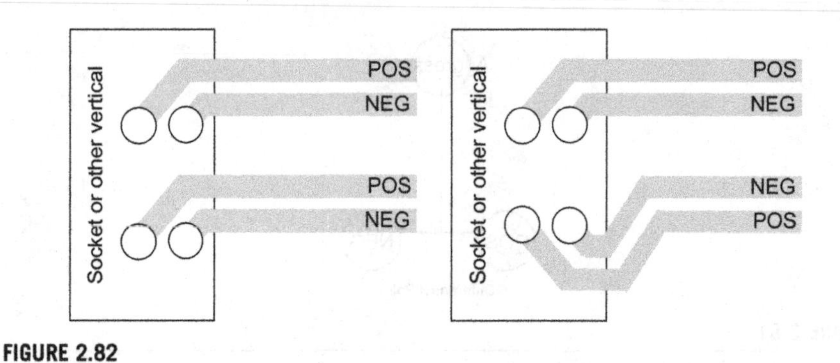

FIGURE 2.82

Left: positive t-line crosstalk polarity. Right: negative t-line crosstalk polarity.

The differential routing pattern P1-N1-P2-N2-P3-N3 is required to maintain positive differential crosstalk in the transmission line. This layout is shown on the left of Figure 2.82. An alternative routing of P1-N1-N2-P2-P3-N3-N4-P4 can strategically create negative differential crosstalk in transmission lines to cancel a vertical component that may have positive differential crosstalk. To employ this best practice of crosstalk cancellation, layout review of the vertical patterns and transmission line patterns is required.

If the routing pattern for signals is unknown, crosstalk cancellation cannot be relied upon. This may be the case with third-party devices or modules. Routing that utilizes polarity reversal does not negate the crosstalk cancellation effects, which are purely a spatial relationship between vertical and horizontal interconnect and are not dependent on the polarity connection of each lane.

An example of crosstalk cancellation is shown in Figure 2.83. In the figure, the crosstalk responses are the result of an aggressor pulse response. The socket (black color) generates a negative polarity response indicated by the negative voltage of the first peak resulting from the first edge of the aggressor. The simulation isolates the socket as a device under test, and no other interconnect component is included in the response. The microstrip routing (blue) of 10 inches at 9× height spacing is also isolated and demonstrates a positive polarity. The microstrip routing follows a P1-N1-P2-N2-P3-N3 pattern as shown on the left of Figure 2.82. The combined response of the socket and 10 inches of microstrip routing is shown in orange. The result is a reduction in crosstalk voltage. If the routing model were connected as shown on the right of Figure 2.82, a worst-case crosstalk scenario could occur where socket and microstrip crosstalk have a negative polarity. The result is constructive interference and the worst-case crosstalk scenario.

The waveform in orange is the composite response of both the socket coupling and the PCB microstrip routing. The composite crosstalk response is smaller than either of the responses alone. It can be expected, before basing a design on cancellation, that there are variations in the constructive cancellation amount in real

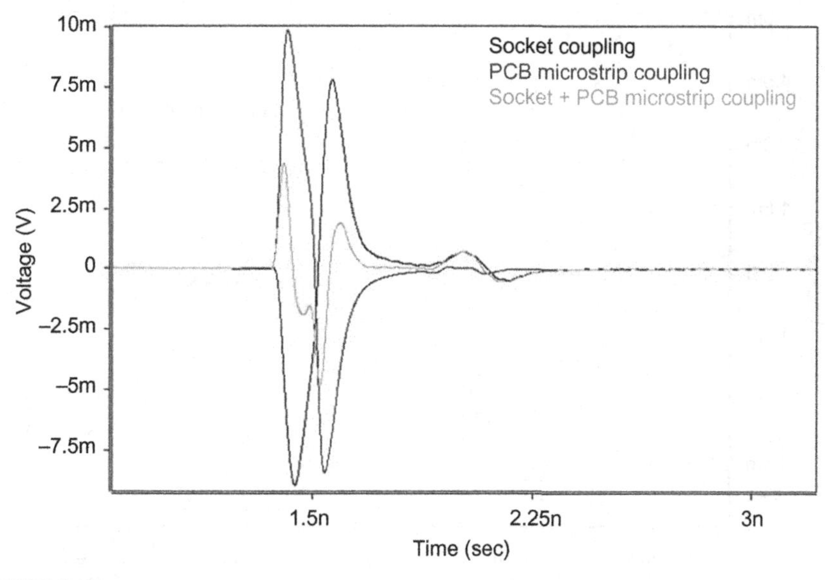

FIGURE 2.83

Far-end component crosstalk cancellation.

systems. Variations to consider are the initial phase relationship between PCB aggressors, the running skew between aggressors, and other dielectric property variations that may be required to generate sufficient (and opposite) PCB crosstalk. Figure 2.84 demonstrates a simulation waveform including the socket-to-PCB cancellation effect and a corresponding measurement. It can be observed that the non-idealities of the real system have led to changes in the crosstalk response.

NON-IDEAL RETURN PATH

High-speed signals require a continuous return path adjacent to the signal path. Return current travels the path of least inductance and flows on the ground or power plane directly below high-speed conductors. Lack of a sufficient return path, or interrupts in the return path, will lead to greater EMI emissions, increase transmission line impedance (appearing as an inductive spike), and reduce edge rates, causing lower margins. When return current deviates from a path directly beside the signal and the path is shared with other conductors, fields cross, creating a crosstalk increase. The signal impact scales with the distance or length in which the return current deviates from the signal path.

Routing over splits in reference planes should be avoided for high-speed signals. Return current must travel around splits as shown in Figure 2.85. If splits are on a secondary reference plane that is three times the height of a primary

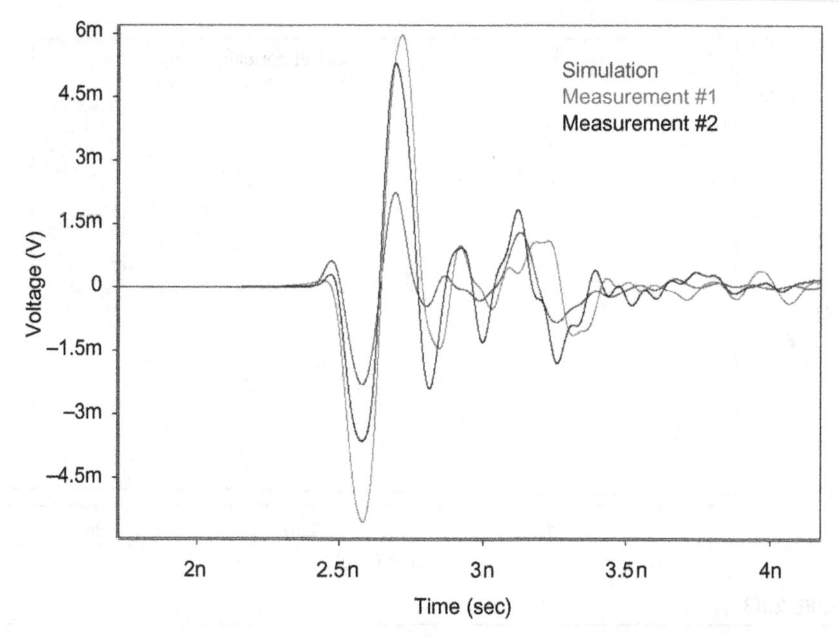

FIGURE 2.84

Simulated and measured constructively cancelled crosstalk response.

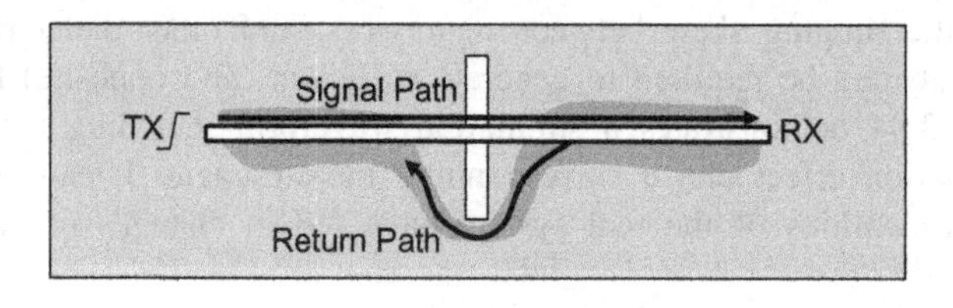

FIGURE 2.85

Split reference plane.

reference plane, then effects of plane splits are known to be negligible and can be tolerated without a noticeable signal integrity impact. Lower speed signals (less than 1 GHz) may route over significant splits or change power planes if the planes are connected by capacitors, but this is not acceptable for higher speed signaling where a continuous reference plane is required. In these cases the capacitors serve as a return path for AC current between the planes.

When a layer change occurs, the return current that was flowing on the nearest plane requires a short path to the new reference plane. Additional vias are added to "stitch" to ground planes together for the return path. Illustrated in the leftmost signal transition of Figure 2.86, a ground via provides the path for the return current. When a layer change occurs but the primary reference plane does not

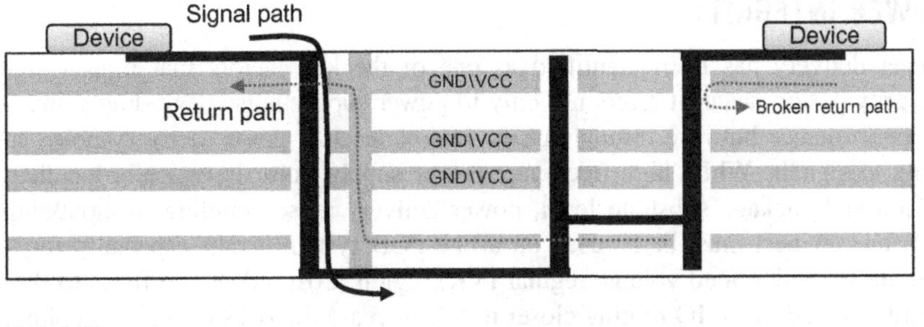

FIGURE 2.86

Reference plane change.

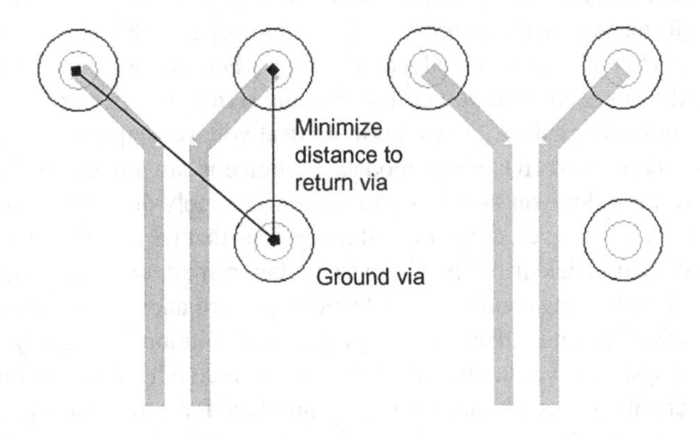

FIGURE 2.87

Return path via placement.

change, stitching vias are not required. The identification of the primary reference plane is required to make a decision that stitching vias are not required. A review of the height of the dielectric layers is needed to determine where the return current is expected. For example, a 1:3 dielectric height ratio will comfortably ensure that nearly all current will be on the nearest reference plane. Symmetry stripline stack-ups will share current on both planes, and would require stitching vias.

The best signal integrity practice is to minimize the distance to stitching vias and supply one stitching via for every differential pair. A distance of 50 mils is preferred, as shown in Figure 2.87. For lower speed signals (less than 8 GT/s), longer distances are acceptable with negligible impact. Symmetrical placement of ground vias to reduce common mode is also desired. Use of one via for two differential pairs may be permissible, but simulation is recommended to assess the longer return path. Post-layout simulation tools can provide analysis for return path issues, identifying signal loss and crosstalk concerns.

POWER INTEGRITY

Power delivery has been identified as one of the key factors that impact signal integrity performance. It refers not only to power supply noise degrading signal jitter performance but also throughout the interconnect as power delivery noise coupling to signals. While high-frequency power supply noise is typically handled at silicon and package substrate level, power delivery noise coupling to input\output (IO) interconnect must be focused on at the PCB system levels, especially for IOs that are located around voltage regular (VR) region. Cost reduction efforts to shrink board size and move IO routing closer to VR increase the risks of noise coupling.

Silicon and packaging design engineers have put a significant amount of research effort into how to link power delivery performance to signal integrity. Its coverage ranges from analysis methodologies and designing test structures for more accurate characterization of PDIO performance, to the best design rules to balance between cost and performance. A lot of progress has been made in this area, especially on analysis methodology and cost reduction designs. For example, methodology has been changed from decoupled power integrity and signal integrity analysis to power and IO co-simulation methodologies. Further, ideal voltage supply buffer models are changed to power-aware IO driver models. A tremendous amount of engineering work has also been done on how to optimize power supply design for good-enough IO design at Intel. However, there are still many gaps that need to be filled regarding more accurate and efficient modeling and simulation methodologies, including the worst yet realistic loading conditions, and precise performance specifications.

Besides effective and sufficient power delivery solutions to supply power for signals, good system-level interconnect co-designs including IO interconnects and power delivery networks are also equally important for a reliable system. There are many cases that demonstrate that without considering overall system-level interconnect designs, the system will not be stable and is vulnerable for fails. Those cases may be divided into two typical categories. One is IO signaling routed too close to VR, resulting in a significant amount of coupled noise. The other category is an improperly designed power delivery network that causes higher system-level noises that couple to IOs.

Although a classic motherboard-level voltage regulator with buck converter type topology only switches at 100's of KHz, its sharp rising and falling edge still contains high-frequency energy. Figure 2.88 illustrates a typical spectrum of the voltage waveform at a VR phase node. One may observe that there is noticeably high-frequency energy that can be coupled to signals.

Therefore, certain design rules between power and signaling need to be defined in that vicinity besides signaling itself. For example, Intel Corporation's published platform design guidance (PDG) typically lists general design guidance to lower the risks of noise coupling from VR to signaling. Industry-wide, more efforts are still ongoing to establish a reliable and efficient methodology that is capable of capturing noises from VR with different transient behaviors. With recently developed methodologies that allow full switch VR models to predict

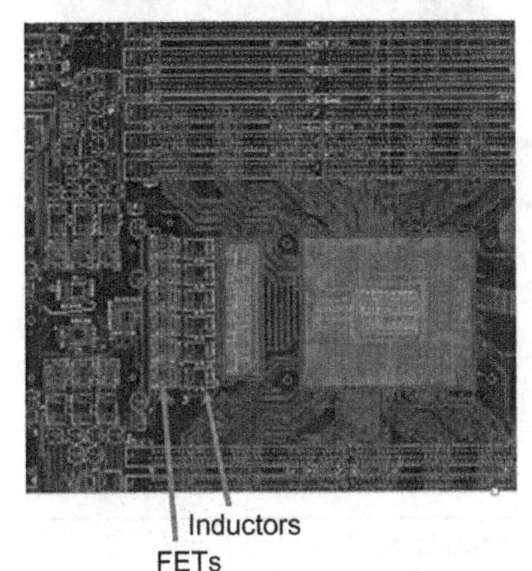

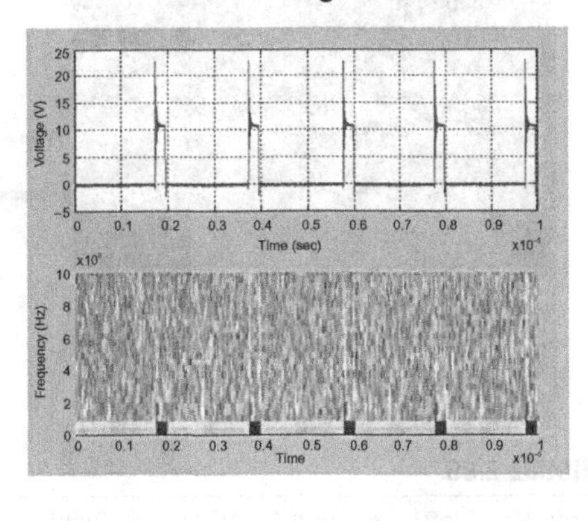

FIGURE 2.88

Left: placement of VR, IO routing, and CPU. Right: voltage waveform at VR phase node and its short-time FFT spectrum (STFFT).

system-level power delivery performance, a methodology to capture VR noises coupling to IO should be feasible.

Related to power delivery noise coupling, decoupling capacitors work with a very different noise generation and coupling mechanism than direct VR noise coupling to IO. It is well known that in order to effectively design a power delivery network by using decoupling capacitors, capacitor placement is critically important to ensure the smallest loop inductance. It is not only for a good power delivery solution but also for a minimized noise level.

Figure 2.89 illustrates typical mistakes in layout. On the left side of the layout, the power network is surrounded by ground vias. On the right, the power network is far from ground vias. If there are no good ground return vias for the power network, a big loop antenna will be set up and unexpected noises will be coupled to IOs. It is very noticeable that with the smaller return loop, field strength could be minimized and therefore less noise coupled to the signaling that is routed at the region, as shown in Figure 2.89.

In summary, power delivery directly impacts signal integrity from power supply noise and power network coupling noises. Fully independent design practice with respect to power delivery or signal integrity is not sufficient to guarantee a healthy system. Accurate and efficient methodologies are needed to incorporate both power network and signal network in order to capture their couplings.

REPEATERS

When a topology exceeds specifications, such as failure to meet eye-opening requirements, a solution may be to add a repeater to the link. Essentially,

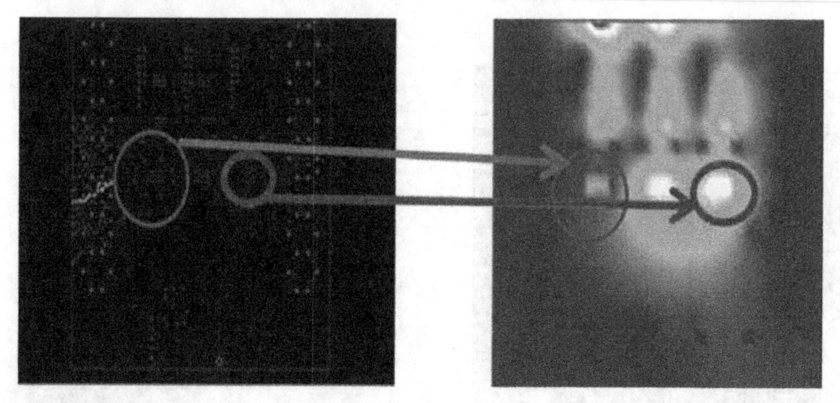

FIGURE 2.89

Left: Poor (left) and good (right) layout. Right: Corresponding field distributions.

Table 2.11 Repeater Comparison

Feature	Re-Timer	Re-Driver
Latency	Cycle increased	Sub-nanosecond analog latency
Protocol Aware	Sometimes	No
Random Jitter	Incoming jitter is terminated	Incoming jitter is usually attenuated, but some is added by the repeater itself
Reference Clock Input	Yes	No
Validation Time	Moderate	Significant
Interoperable	Yes	Not guaranteed unless active devices in a channel are defined by a specification
Power Consumption	More	Less
Cost	More	Less
Layout Area	More	Less

repeaters are added to compensate for excessive channel loss. Ideal repeaters are meant to be protocol-transparent, such that upstream and downstream agents are unaware of the addition of a repeater and require no protocol change for successful communication. Repeaters have two categories: re-drivers, which recondition signals, and re-timers, which terminate and retransmit. Each configuration offers some advantages and disadvantages, as shown in Table 2.11. The following questions will help determine the best choice for each design:

- What kind of repeater is necessary—re-timer or re-driver?
- What kind of signal integrity issue is the link experiencing? Is a repeater the solution?

- Is protocol awareness a requirement for this interface?
- Is interoperability with any end point a requirement?

Introduction to re-timers

Acting as an end point, the re-timer terminates and samples the incoming data, converting from analog to digital bits. Once the data have been sampled, all incoming jitter is no longer carried forward. The new transmission from the re-timer will carry only the jitter inherent of its own transmitter. This separation simplifies SI analysis into two separate channels—before and after the re-timer. It is assumed that an electrically compliant re-timer can support the same solution space as any other device, theoretically doubling the potential channel length when strategically located near the middle of the interconnect. The device may be protocol-aware of simple states, such as idle, or complex states, such as training routines. The importance of protocol awareness depends on the interface, and in some cases it is required in order to guarantee interoperability and specification compliance. For example, 10G-KR and PCIe 3.0 have a protocol for complex equalization training routines, while USB and SATA do not. Use of a re-timer that is not protocol-aware or specification-compliant on a protocol-aware interface will lead to significant validation time and invalidates the assumption of open system interoperability. A properly chosen re-timer that meets the requirements of a given interface will significantly reduce the validation time.

Introduction to re-drivers

Re-drivers recondition incoming signals to compensate for channel loss. Peaking filters like CTLE, or even FFE, are common means to amplify and shape the incoming signal. Re-drivers (or the amplifiers they comprise) are two types: limiting (high-nonlinear), and linear.

Limiting re-drivers have a very high-gain amplifier that produces a full swing output amplitude once the input exceeds a very low threshold. Limiting re-drivers reshape the signal significantly, essentially producing an output close to a square wave. One of their disadvantages is that all uncompensated incoming intersymbol interference (that is, residual ISI after their input CTLE) is turned into deterministic jitter at the output due to the slicing effect of the limiting amplifier. One of the main disadvantages of limiting re-drivers is that the block changes in the incoming signal's equalization, preventing it from reaching the output. Hence, it is inappropriate for use in busses whose protocols require adaptive TXLE training, such as PCIe3 and 10G-KR. This renders them only useful for simpler busses, such as USB3 and SATA. When used for these two busses, limiting re-drivers could affect a channel extension (in total dB of loss) of about 50 percent at most and 40 percent more conservatively.

The second type of re-driver, the linear re-driver, is the most prevalent in the 10G-KR and PCIe3 domains. These devices use a linear amplifier, whose compression point does not set in until the output reaches a level close to the full output swing. They are meant to maintain the linear relationship between the incoming and outgoing amplitudes, although they do reshape the signal, due to the effect of their equalization (CTLE, and sometimes also TXEQ). They are an attempt by semiconductor vendors to allow re-drivers to pass TXLE changes from input to output, albeit with slight nonlinearity, and more importantly the imperfect equalization of the channel extension renders them noncompliant to the specification.

Training sequences that request a certain equalization level will receive a reshaped waveform after the re-driver, interfering with receiver calibration. This result is a delivered signal other than intended or required. For this reason, it may not be possible to achieve interoperability compliance on some interfaces.

Modeling and simulation

Repeater models may be available in spice or IBIS-AMI formats. An analysis for interconnect using a re-timer will require two separate simulations as the signal is completely terminated and retransmitted by the re-timer. In an analysis with a re-driver, the signals pass through the vendor model in the interconnect simulation. Signals passing through a re-driver can observe an increase in data-dependent, uncorrelated deterministic and random jitter. It is possible, but unlikely, that re-driver simulation models include the added device jitter. It is important to obtain accurate models at manufacturing corners, in order to evaluate the change in linearity and output capability over process, voltage, and temperature. The availability of modeled and validated samples in manufacturing corners should be a consideration in the health assessment when reviewing the simulated or measured margin available at the end device.

PCIe considerations

Design for PCIe 1.0 and 2.0 have seen successful implementations with re-timers and re-drivers. The minimal protocol and training requirements on these two interfaces lead to success with moderate validation efforts using re-timers or re-drivers, as few equalization options are available. PCIe 3.0 (Gen3) introduced two new features that limit the design options for repeaters. Gen3 adds pre-emphasis equalization and 10 required equalization levels (presets) in the PCIe CEM 3.0 specification that must be achieved for interoperability for compliance.

Meeting these presets is a concern for re-drivers because of the coloring of the equalization of the signal. Using a linear re-driver allows the TXLE adaptation to pass from TX to RX. It is uncertain that a re-driver exists that compensates for a channel extension so perfectly that the duo of channel extension plus re-driver is transparent enough to guarantee that the entire equalization space required by the specification can be passed on from TX to RX as expected by the specification. Using a limiting re-driver does not allow TXLE to pass from TX to RX as expected and breaks down the adaptation. The link may be broken or never established, thus requiring that adaptation be disabled.

If adaptation is disabled, it must be disabled in the firmware for both devices in the link. The re-driver filter settings must be optimized in the laboratory. This restricts the use of the re-driver on links operating only to down devices that have been properly configured. Open slots available to any plug-in card are problems for re-drivers as the adaptation protocol cannot be disabled a priori and the re-driver filter will not be uniquely optimized to every end device available.

REFERENCES

[1] Rao F, Hindi S. Frequency domain analysis of jitter ampification in clock channels. In: Electrical performance of electronic packaging and systems. 2012. p. 51–54.

[2] Chaudhuri S, Anderson W, Bryan J, McCall J, Dabral S. Jitter amplification characterization of passive clock channels at 6.4 and 9.6 Gb/s. In: Electrical performance of electronic packaging, 2006 IEEE. IEEE; 2006. p. 21–24.

[3] Lee BT, Mazumder M, Mellitz R. High speed differential I/O overview and design challenges on Intel enterprise server platforms. In: Electromagnetic compatibility (EMC), 2011 IEEE international symposium on. IEEE; 2011. p. 779–84.

[4] Montrose M. Time and frequency domain analysis for right angle corners on printed circuit board traces. In: Proceedings of the IEEE international symposium on EMC. 1998. p. 551–56.

[5] Loyer J. PCIe 2.0 signal integrity considerations (Fiberweave Effect), [Online], 2007. Available: <http://www.pcisig.com/developers/main/training_materials/get_document?doc_id=f6a703c9dccb60d489bf1233e1eddc02a1b9be59>

[6] Huray PG, Hall S, Pytel S, Oluwafemi F, Mellitz R, Hua, D, et al. Fundamentals of a 3-D 'Snowball' model for surface roughness power losses. In: Signal propagation on interconnects. 2007. p. 121–24.

restricts the use of the [illegible] operating [illegible] devices that have [illegible] properly configured. Open [illegible] available [illegible] any plug-in card are problematic for [illegible] drivers [illegible] protocol cannot be disabled [illegible] and the [illegible] will not [illegible] optimize [illegible] to everyone [illegible] available.

REFERENCES

[1] [illegible] S. [illegible] domain analysis of [illegible] amplification in [illegible] channels [illegible] electrical [illegible] packaging and systems 2012. p. [illegible]
[2] Chandrasekhar S, Anderson [illegible], [illegible] McCall [illegible] [illegible] 64 and [illegible] electrical performance of electronic packaging. 2006 IEEE [illegible] 2006. p. [illegible]
[3] Lee [illegible], [illegible] M, Mellitz R. High-speed [illegible] serial I/O [illegible] and [illegible] challenges [illegible] platforms [illegible] EMC 2011 IEEE international symposium on IEEE. 2011. p. 779–84.
[4] [illegible] M. Time and frequency domain analysis [illegible] printed circuit board traces. In: Proceedings of the IEEE international symposium on EMC. 1998. p. 541–56.
[5] [illegible] J. PCIe 3.0 [illegible] integrity considerations [illegible] 2010 [illegible]. Available: [illegible]
[6] [illegible], Pytel S, Olowofela [illegible] Mellitz R, [illegible] D, et al. Fundamentals [illegible] model for surface roughness power losses. In: [illegible] 2007. p. 1–24.

CHAPTER

Channel modeling and simulation

3

Far better an approximate answer to the right question, than the exact answer to the wrong question, which can always be made precise.
— John Tukey

The quality of simulation models is becoming increasingly important with high-speed digital circuits. Models that exhibited certain amounts of undesired behavior that were once undetectable or even tolerated are now no longer acceptable. Customers demand quality and accuracy. Product managers require accurate pre-design risk assessments for more challenging interconnects. How can we ensure models are developed and simulation is executed to the best known methods? In this chapter, best known methods for the development of transmission lines, transmitters, and 3D models such as connectors, plated through hole (PTH) vias, and packages are discussed. Finally, simulation methods available in the industry to carry out analyses, such as comparative study, linearity review, bit error rate analysis, and volume defects per million analyses, are compared to determine the best approaches to satisfy the needs.

TRANSMISSION LINES

This section provides an introduction to the top issues with transmission line modeling for high-speed serial links. Important topics include model quality in causality and frequency dependence, surface roughness, temperature sensitivity, and high-volume model variations. It is through these guidelines and recommendations that transmission line models may be developed that correlate to measured performance.

CAUSALITY

High-speed circuits built on FR4 dielectrics have required a change from single-frequency models to frequency-dependent models in order to achieve accurate correlations. Single-frequency models may be sufficient for modeling insertion loss in the frequency domain due to the relatively flat response of dielectric loss tangent over frequency and the relatively small bearing of dielectric constant (which varies highly with frequency) on insertion loss magnitude. However, model accuracy is significantly lost when translating from frequency domain to time domain. The use of a single-frequency model lacks the description of varying propagation delay over frequency or dispersive properties that result from polarization of various molecule types that compose printed circuit board (PCB) dielectrics. Strict enforcement will be discussed that must be made to ensure that frequency-dependent models are causal. The principle of causality states that no effect may happen prior to a cause and may be described as shown in Equation 3.1:

$$h(t) = 0, \quad when\ t < 0 \tag{3.1}$$

Several textbooks and publications describe and build on the widely accepted frequency-dependent model known as the Djordjevic-Sarkar model. Djordjevic et al. [1, p. 662] offers causal closed-form equations based on an infinite pole Debye model describing the dielectric real and imaginary properties with correlations to measured data. Causality is achieved by the enforcement of analytical relationships between dielectric properties, known as Kramers and Kronig relations. These relationships include real and imaginary permittivity, inductance and resistance, and capacitance and conductance. Djordjevic calculates permittivity not by the general form in (3.2) but through a Hilbert transform in (3.3) [1, p. 664]:

$$\varepsilon(\omega) = \varepsilon' - j\varepsilon'' \tag{3.2}$$

$$\varepsilon(\omega) = \varepsilon' - jHilbert[\varepsilon'] \tag{3.3}$$

Djordjevic evaluates (3.3) in a closed-form equation to relate the real part as an analytic solution of the imaginary, thereby guaranteeing a causal relationship between the real and imaginary permittivity in (3.4):

$$\varepsilon_r(\omega) = \varepsilon'_{inf} + \frac{\Delta\varepsilon'}{m_2 - m_1} \frac{ln\dfrac{\omega_2 + j\omega}{\omega_1 + j\omega}}{ln10} - j\frac{\sigma}{\omega\varepsilon_0} \tag{3.4}$$

CHECKING FOR MODEL CAUSALITY

The use of noncausal models in time domain simulations can lead to erroneous results. The magnitude of the error depends on the ability of a simulator to tolerate or correct noncausality. S-parameters may be checked for causality using the Kramers and Kronig principles. A variety of commercial tools use one common method of verification to compare the difference in the real permittivity of the model and the real permittivity derived from the imaginary permittivity using the Hilbert transform. The difference between the real permittivity results are compared to an error threshold. Models that are determined to be noncausal may be corrected through the truncation and enforcement of the model data to maintain the requirement in (3.3). However, it must be cautioned that causality enforcement should only be performed in the case of small noncausality. Use of causality correction breaks the energy balance in the model. If correction is done by discarding the imaginary and recalculating it with (3.3), the imaginary/delay will be correct but the real/loss will be incorrect. If possible, it is better to generate a new physically correct model. When enforcement is performed, inspection of the time domain responses must be done to ensure a desirable response has been achieved.

Review of the time domain response can reveal causality or stability issues present in the model. Causality concerns are not limited to simulated models but are also in measured models that are susceptible to numerous noise sources. While visual inspection cannot determine whether a model is purely causal, it can reveal acausality and the magnitude of concern. Figure 3.1 and Figure 3.2 illustrate two examples: a noncausal measurement and a noncausal simulation model. The provided example is severely noncausal due to the energy appearing at the system output before the propagation delay of the channel. The next section discusses the requirements for creating causal and frequency-dependent models.

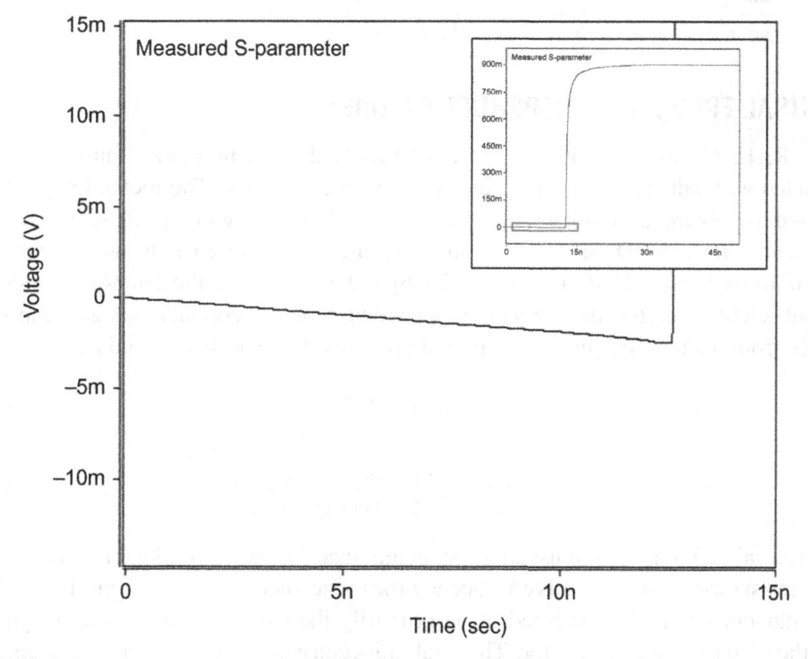

FIGURE 3.1

Noncausal step response from measured S-parameter.

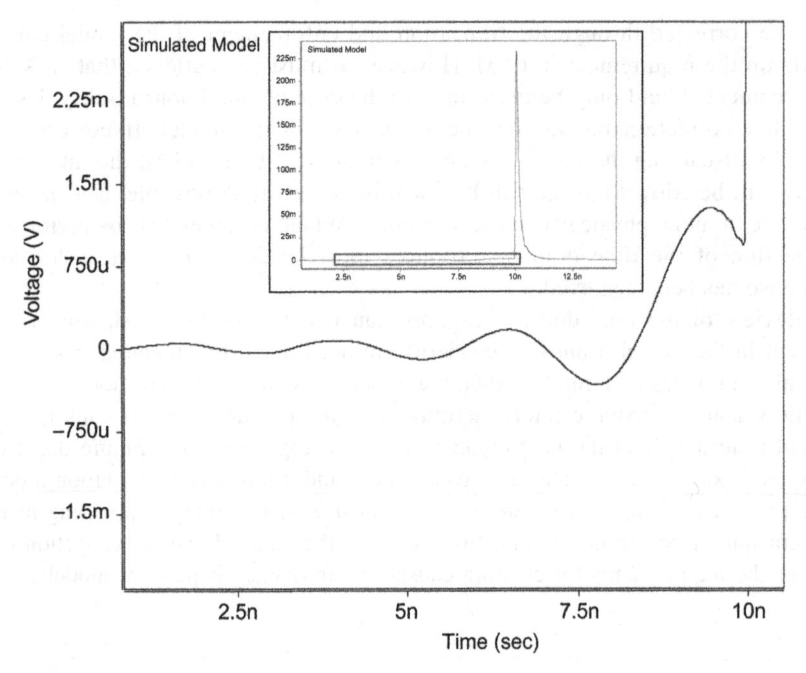

FIGURE 3.2

Noncausal pulse response from modeled interconnect.

CAUSAL FREQUENCY-DEPENDENT MODEL

The R, L, C, and G matrices for a frequency-dependent model must be constructed with adherence to strict rules to maintain causality. The methodology discussed to obtain a frequency-dependent model begins with a single reference frequency from a 2D field solver and measured (or known) real and imaginary permittivity values. Hall et al. [3, p. 2616] states that using the Djordjevic model is sufficient to derive the frequency dependence of the conductance and capacitance from a single frequency result and provides the expressions to do so as:

$$C(\omega) = C_{ref} \frac{\varepsilon'(\omega)}{\varepsilon'_{ref}} \tag{3.5}$$

$$G(\omega) = G_{ref} \frac{\varepsilon'(\omega)}{\varepsilon'_{ref}} \frac{\tan \delta(\omega)}{\tan \delta_{ref}} \frac{\omega}{\omega_{ref}} \tag{3.6}$$

Causal relationships must also be maintained between conductor inductance and resistance. When the solve frequency for inductance is high enough, the resulting inductance can be expected to be primarily the external inductance, which is on the surface of the conductor. The total inductance of the conductor is the sum:

$$L = L_e + L_i \tag{3.7}$$

though the internal inductance is small at high frequency. To properly account for all the inductance, and to maintain causal relationships in the conductor, the total inductance must be related to resistance in (3.8) by a Hilbert transform, such as the example in Hall (2009:649):

$$L(\omega) = L + \frac{Hilbert[R(\omega)]}{\omega} = L_e + L_i \tag{3.8}$$

$$R(\omega) = R_{DC} \tag{3.9}$$

$$R(\omega) = R_S\sqrt{f} \tag{3.10}$$

Hall et al. [3, p. 2614] clarifies that resistance must be calculated without the direct current (DC) term when the frequency leads to a skin depth that is equal to or less than the trace thickness. At and below that frequency, $R(\omega)$ is only defined by R_{DC} and $L(\omega)$ may be approximated as the inductance where the skin depth was equal to the thickness. It is important to use this model, expressing the reduction in internal inductance and corresponding increase in resistance, as frequency increases that is otherwise missing in noncausal models, allowing for an over-prediction in transferred energy. The calculations for (3.10) are valid until copper foil roughness begins to contribute to resistive losses and must be accounted for in the model.

COPPER SURFACE ROUGHNESS

Copper foil surfaces are found to be covered in surface imperfections that are intentionally created by the roughing of the copper to promote adhesion to the dielectric. Traditionally, high-frequency losses have been described by the $\sqrt{f}$ present in skin effect calculations, where the current is no longer uniform throughout the conductor. The descriptions are useful until the current depth is the same height as the microscopic rough features of the conductor surface. These features are shown on a microstrip copper foil under magnification in Figure 3.3, originally published by Pytel et al. [4, p. 1145]. A further magnification up to 100,000× is shown in Figure 3.4 detailing the size of the microscopic features. It

FIGURE 3.3

Microstrip profile from Pytel et al. [4, p. 1145].

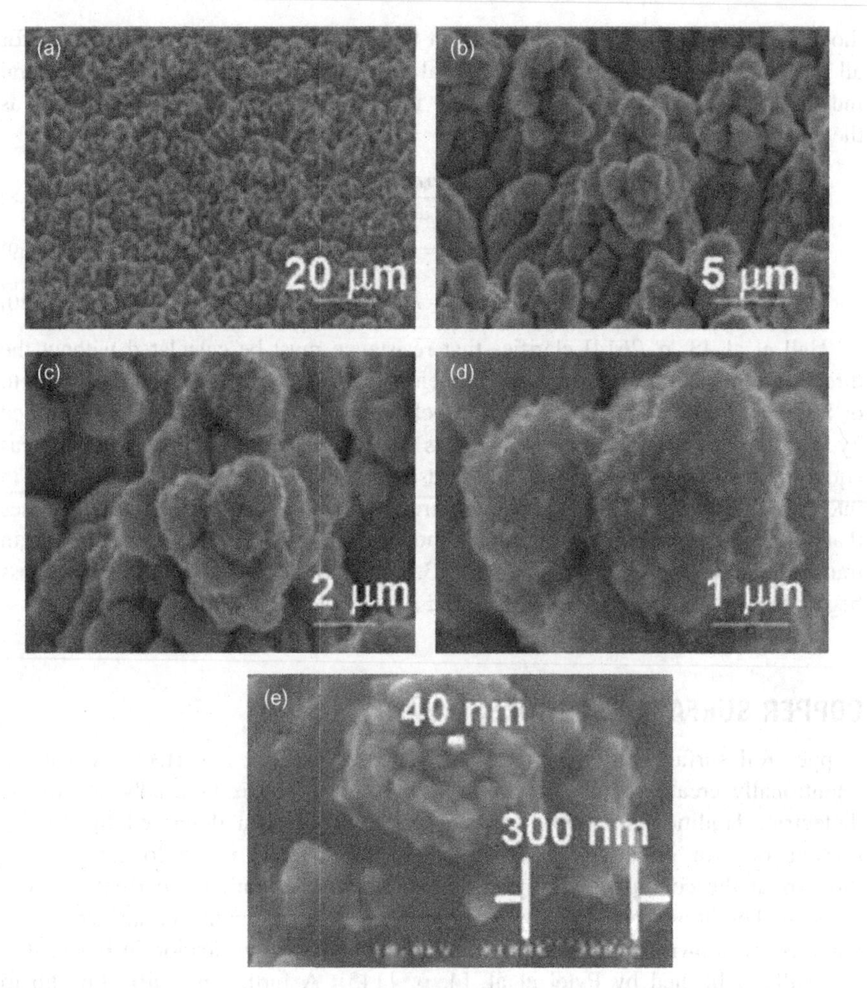

FIGURE 3.4

Copper foil at (a) 1000×, (b) 5,000×, (c) 10,000×, (d) 25,000×, and (e) 100,000× magnification, at a 32-degree offset, from Pytel et al. [4, p. 1145].

is generally around 1 GHz where the skin effect and equation (3.10) no longer accurately describe the conductor loss and a mathematical model is needed to accurately describe the nonideal surface of the conductor.

As design engineers we desire to enter parameters into a simulation that relate directly to the measurable parameters in the manufacturing process. However, detailed parameters describing the surface roughness from suppliers are not common. Research into the surface of the conductors is a new and evolving field that has revealed its complexity, which is difficult to characterize. Surfaces vary from supplier to supplier, over time within a supplier, and also geographically with smoother coppers in the United States.

A mathematical model for surface roughness that can help separate dielectric and conductive losses, which are difficult to isolate in a direct vector network analyzer (VNA) loss measurement, has been well researched. In addition to increased confidence at higher frequencies, the design engineer can now make design tradeoffs in laminate selection independently of the copper foil—thereby having a better understanding of the laminate effect. Two useful methods for modeling include the modified Hammerstad model and the Huray snowball model. These models introduce losses from copper roughness by scaling the resistance (and consequently the inductance) matrix of a frequency-dependent transmission line model by a coefficient, *K*. The resulting high-frequency resistance (skin depth less than thickness) that includes surface roughness losses is calculated by (3.11):

$$R(\omega) = K \times R_S\sqrt{f} \tag{3.11}$$

Modified Hammerstad model

The modified Hammerstad model is an empirical fit, or an approximation, to the work of Morgan Samuel. Samuel's work described the rough copper surface as 2D derivations of Maxwell's equations on repeating equilateral triangles. The original model, which is controlled by one term, roughness height *RMS*, has since been modified to add an additional scaling factor. The new term $(RF-1)$ accounts for variations in the jagged tooth angle while maintaining the original results for equilateral triangles, when $RF=2$. The new modified Hammerstad equation describes *K*, a resistive scaling coefficient, as:

$$K_{rh} = 1 + \left(\frac{2}{\pi}\tan^{-1}\left(1.4\left(\frac{\Delta}{\delta}\right)^2\right)\right)(RF-1) \tag{3.12}$$

where Δ = RMS height, δ = skin depth, RF = roughness adjustment factor [3, p. 6].

Copper surface RMS values may range from 0.3 to 5.8 μm, though the Hammerstad model may not be sufficient for RMS heights above 2.0 μm [3, p. 2618]. It is recommended that insertion loss correlation be performed in addition to requesting RMS values from the supplier when using the model, as it is possible for very rough and rather smooth surfaces to yield the same RMS value. If data are not available from the supplier, an approximation may be taken before correlation is performed with 1.6 μm for microstrip and 0.815 μm for stripline.

Huray model

The Huray snowball model is an analytic method to describe power loss from the scattering of electromagnetic waves across the copper balls, resembling "snowballs" in Figure 3.4, on the conductor surface. Huray (2013:8) describes the resistive scaling coefficient for the model below, where r is radius, δ is skin depth, N is number of spheres, and A_{hex} is tile width:

$$K_{rhu} = 1 + \left(\frac{3}{2}\right)\left(\frac{N4\pi r^2}{A_{hex}}\right) \Big/ \left(1 + \frac{\delta}{r} + \frac{\delta^2}{2r^2}\right) \tag{3.13}$$

Very good correlation can be achieved with the Huray model through 30 GHz [3, p. 2623]. Determining the sphere details is difficult without deep examination. From prior correlations, initial values for sphere radius of 0.5 μm and tile width of 9.4 μm are recommended. The number of spheres, N, may be correlated to various copper roughness profiles.

Table 3.1 expresses recommendations for the number of spheres based on known peak-to-trough values from a supplier. If values are not known, initial values of 50 for smooth, 64 for stripline, and 79 for microstrip are recommended until model correlation is performed. Figure 3.5 plots the scaling coefficients for the modified Hammerstad with 1.3-μm RMS height and Huray snowball with 64 spheres. The values are intentionally selected to provide similar low-frequency performance. It is apparent that there is a divergence in model behavior above 8 GHz.

Insertion loss correlation to measurements is recommended to ensure proper surface roughness modeling where material loss is critical. In many cases, surface

Table 3.1 Recommended Huray Model Inputs

Peak-to-Trough	Number of Spheres, N
1–3 μm	50
4–6 μm	64
7–14 μm	79

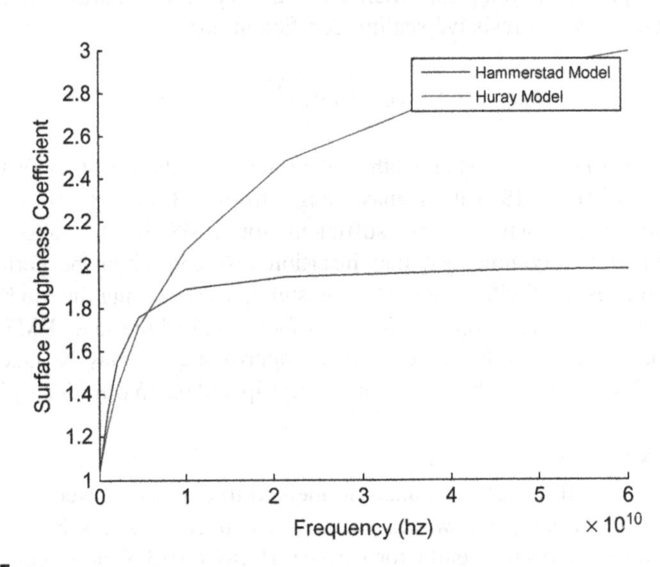

FIGURE 3.5

Surface roughness coefficient comparison.

roughness details are not known and are difficult to acquire. Though costly, cross sections may be performed for improved confidence in transmission line geometries, and dielectric resonators may be used to accurately determine the dissipation factor. At lower operating frequencies, dielectric loss tangent may be used to fine tune insertion loss to meet measurement correlations where precise roughness values are not certain. Note that many factors will contribute to the need to employ a surface roughness model instead of loss tangent adjustment, such as the specific stackup, conductor dominance in insertion loss, and data rate. An example is generated to determine if performance similar to a roughness model can be obtained through adjusting the loss tangent.

A 10-inch asymmetric stripline stackup (h1 = 4 mils, h2 = 13 mils) with a trace width of 5.25 mils is modeled using both techniques. An initial model with no surface roughness coefficient and a loss tangent of 0.022 is generated that yields −0.698 dB/inch at 4 GHz. Then a Huray surface roughness coefficient derived with 64 spheres, 0.5-μm sphere radius, and 9.4-μm tile width is added to the model for a loss of 0.753 dB/inch. An alternative model is generated to achieve the same loss by increasing the loss tangent to 0.030 to achieve 0.753 dB/inch. In Figure 3.6, the differential insertion loss is displayed for the two models

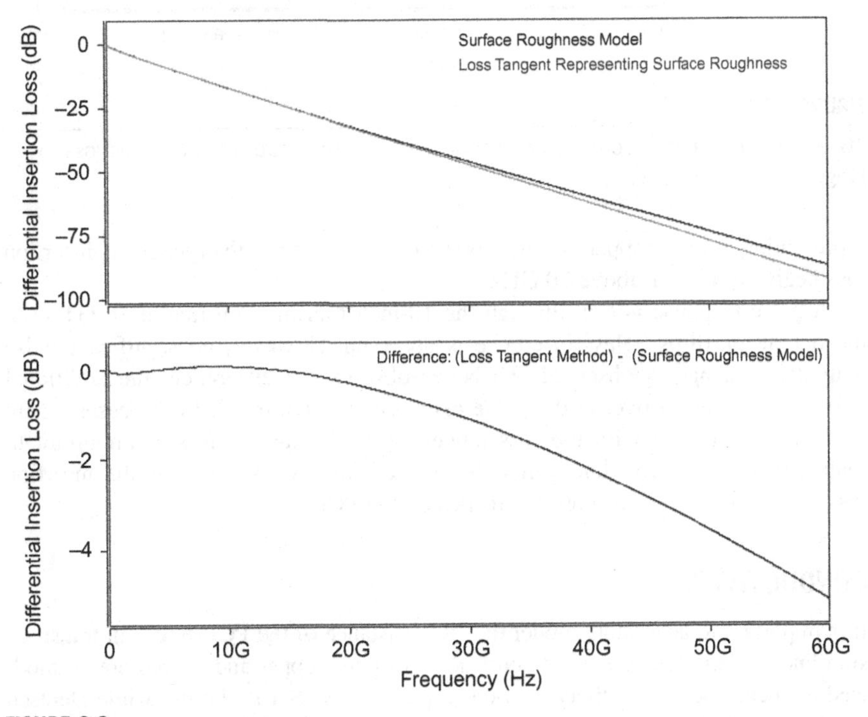

FIGURE 3.6

Insertion loss comparison of surface roughness model and increased loss tangent.

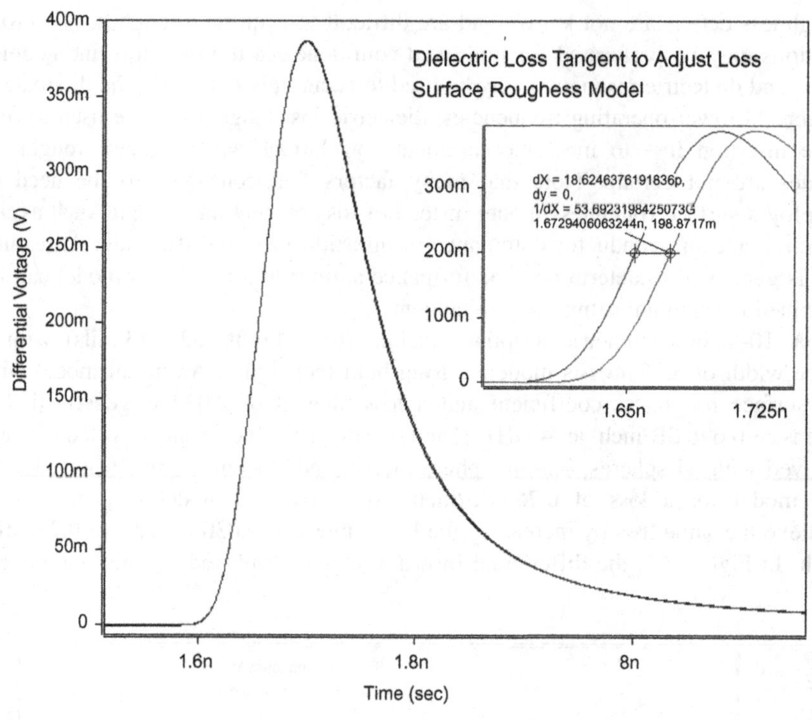

FIGURE 3.7

16-GT/s pulse response comparison of surface roughness model and increased loss tangent.

representing rough copper through 60 GHz. The largest divergence in insertion loss begins to appear above 20 GHz.

A pulse response is sent through the 10-inch transmission line at 16 GT/s. At the output an 18-ps delay difference is observed. However, no significant pulse magnitude change is observed. To better observe the difference, the additional delay is removed to overlie the pulse responses in Figure 3.7. In the cases where equivalent modeling with the loss tangent is achievable, it is still important to realize that surface roughness models are the only means to assess the insertion loss sensitivity to various copper profiles on the PCB.

CONDUCTIVITY

It is important to accurately model the DC resistance of the PCB traces in transmission line models. Traces are not constructed of pure copper and should not be modeled as such. The conductivity for pure copper is 5.88×10^7 ohm/m, while Deutsch et al. [9, p. 282] have completed measurements on six PCB samples from three vendors and finds microstrip trace conductivity ranging from 4.768×10^7 to

4.823×10^7. However, Deutsch do not provide a control for ambient temperature. Loyer et al. [7, p. 17] have determined that the average conductivity values are 4.16×10^7 for microstrip and 4.76×10^7 for stripline at ambient temperature. The average result is from 12 stripline and 10 microstrip measurements from six vendors. Measurements were cross sectioned to yield a precise calculation.

ENVIRONMENTAL IMPACT

The environment's effect on the PCB during operation is a factor that must not be left out of the modeling assumptions. Simulations must be completed in operating conditions with higher temperatures than are present in most validation lab conditions, and this will have an adverse effect on material loss. Environmental boundaries for inlet air conditions are determined by the American Society of Heating, Refrigerating and Air-Conditioning Engineers (ASHRAE). The boundaries are the operating envelope agreed upon by manufacturers. Temperature and moisture boundaries, which vary by class (e.g., controlled data center requirements vs. minimal control on laptops), must be known before modeling. The ASHRAE limits for data center, home office, and portable electronics are available online in the "2011 Thermal Guidelines for Data Processing Environments—Expanded Data Center Classes and Usage Guidance" [11, p. 14]. The resource specifies the allowable ranges as well as the stricter recommended envelope.

Humidity

Higher PCB laminate temperatures will increase the moisture absorption capability of the material, creating another source of PCB loss [7, p. 18]. The absorption of water will increase the dielectric constant (water is 63.78 at 70°C) and loss tangent of the laminate. Not all laminates will respond equally, however, as low-loss materials have observed absorption rates as low as 0.2 percent or less, while standard PCB laminates may be as high as 2 percent [6,10, p. 25]. In a joint paper from Intel and Tyco, "Humidity-Dependent Loss in PCB Substrates," Hamilton et al. [12, p. 30] find that loss increases are possible on PCBs exposed to humidity. The study was completed at 38°C and 95 percent relative humidity (RH) over a 5-month period, creating an extreme humidity case outside of the A1, A2, and A3 ASHRAE class limits. The measurement data demonstrate a small impact from humidity on low-loss Rogers 4350™ (indistinguishable at 4 GHz) and higher loss impacts of approximately 30 percent at 4 GHz on standard loss 1080 and 7628 laminates. In the presented insertion loss data, the percentage increases for microstrip and stripline are approximately equal. The long duration of this study has likely allowed internal layers to become completely saturated. However, it is important to model the PCB in the active environment where it will be used—that is, at high temperature and within appropriate ASHRAE limits. Loyer et al. [7, p. 18] take measurements in worst-case Class 2 ASHRARE conditions, contrasting high-temperature PCBs at low and high humidity. The results indicate that Hamilton's findings may be aggressive, finding high-humidity loss measurements are not significantly higher than their low-humidity counterparts. In retrospect, the sample size of the data is relatively small and a chamber time of 43 days may not have been sufficient to reach equilibrium. Nonetheless, the data are sufficient to indicate that humidity does not appear to have as severe an impact as originally thought.

Conductivity

The resistivity increase of pure copper over temperature is known but is not directly applicable to PCB modeling because the traces are not constructed of pure copper. Loyer et al. [7, p. 17] measures a sample of PCBs from six vendors over an increase in temperature to determine the rate of resistivity increase per degree Celsius. The overall data presented resistivity changes from 0.05 percent to 0.49 percent per degree Celsius. However, on average, the resistivity increase with temperature is approximately 0.35 percent per degree Celsius. This is similar to that of pure copper, which is stated to be 0.4 percent per degree Celsius.

Temperature

Increases in temperature above the ambient are known to increase the dielectric constant by up to 0.24 percent and the dissipation factor by as much as 8.97 percent [8, p. 4]. While this variation has been observed for laminates, it is desired to observe with totality the measurable total insertion loss increase due to dielectric and conductor changes. An assessment of the impact on PCB insertion loss is completed in "Humidity and Temperature Effects on PCB Insertion Loss" [7]. Generally, PCB loss measurements are completed at room temperature near 25°C, whereas operation is much higher. The study tested the impact of coupons measured at 25°C and then increased to 75°C. A summary of the result is shown in Table 3.2. The result is an increase in loss varying from 13 percent to 32 percent measured at 4 GHz. The percentage increase did not correlate with PCB loss, considering three types of materials: low loss, mid loss, and regular FR4. Loss increases in microstrip (17–32 percent) consistently trended higher by

Table 3.2 Worst-Case Temperature Measurements

Nominal Loss (dB/inch at 4 GHz)	Microstrip Worst Case 75°C dB/inch (percentage increase)	Stripline Worst Case 75°C dB/inch (percentage increase)
0.40	0.50 (25%)	0.48 (19%)
0.44	0.54 (23%)	0.52 (17%)
0.5	0.60 (20%)	0.58 (15%)
0.6	0.70 (17%)	0.68 (13%)
0.7	0.93 (32%)	0.90 (29%)
0.8	1.03 (28%)	1.00 (25%)

Note: *Decibels per inch (dB/inch) is measured at 4 GHz.*

3–6 percent than stripline layers (13–29 percent) on the same material. The measurement data show that high operating temperatures do have a severe and adverse effect on PCB loss.

Model and simulation

Below are a set of questions to consider before modeling the PCB environment effect on transmission line models.

- How important is PCB loss for this design, or are the channel lengths moderate?
- What environmental envelope must be supported?
- What is the expected chassis temperature when operating at the boundary?
- What thermal analysis is available for the PCB temperature?

The more important loss is for the design (e.g., for long lengths), the more time that should be spent to properly model the transmission line environment. It is unlikely that the entire PCB dielectric over the length of the interface operates at the same temperature. Areas near voltage regulators and central processing units (CPUs) are expected to have higher temperatures than locations at a distant receiver (RX), as illustrated in Figure 3.8. Modeling a transmission line that traverses an entire PCB at a worst-case high-temperature loss will be a pessimistic modeling assumption. Two options may be presented to the simulation engineer. (1) Segment-based modeling may be used with worst-case losses near hot spots and moderate increases in other regions. This approach requires the buildup and management of several models. (2) A simpler approach is to develop one model to approximate the higher and lower temperatures with an average. In the study reviewed, a recommendation of 10 percent above nominal conditions is made for an averaged loss transmission line model [7, p. 19]. The total loss effect may be achieved through a combination of the known increases in conductivity and supplemental increases in loss tangent.

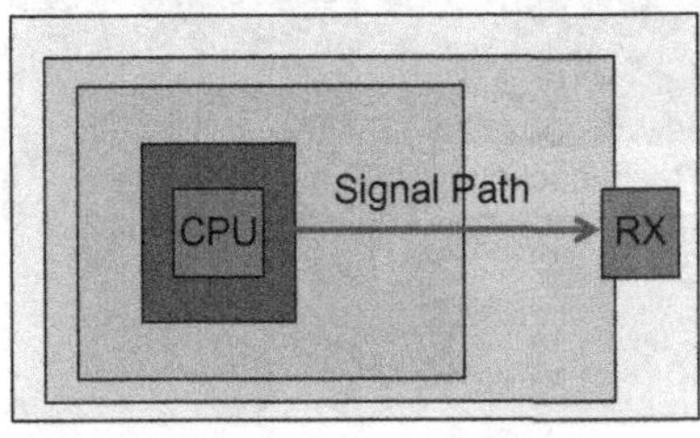

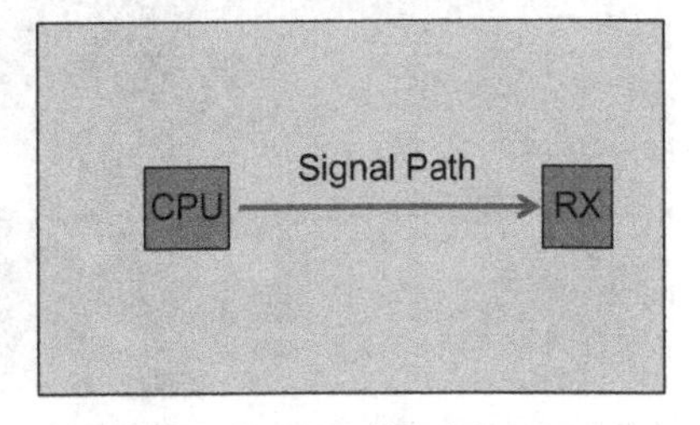

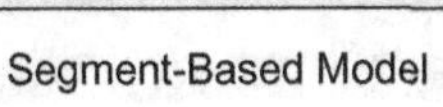

Averaged Model

FIGURE 3.8

Two modeling approaches to PCB temperature gradient.

MODEL GEOMETRIES

This section will dig deeper than the classical stack-up geometries used for transmission line modeling such as trace width, trace thickness, dielectric height, dielectric constant, dielectric loss tangent, trace spacing, and so on. The investigation takes a step further to more accurately represent the transmission line structure as it is manufactured. Cross-section images are provided to aid in this understanding.

Stripline structures

Model input geometries available from board layout or vendor quotes include trace width, inter-pair and intra-pair spacing, trace height, core/prepreg height, dielectric constant, and loss tangent. Trace width etching is not commonly provided in vendor data sheets but is necessary for proper modeling. A common etching of 0.3 mils from each side of the trace is suggested unless otherwise known. Reference plane thickness is available in the stackup and as an input to models but has little sensitivity for high-speed signals. In Chapter 2, the issue of crosstalk uncertainty due to resin concentration in the signal was introduced. The cross section in Figure 3.9 shows the lack of glass in the metal layer, and the proposed model is shown in Figure 3.10. Electrical properties for resin vary based on vendor and material selection and can have a significant effect on crosstalk levels. Vendors should be contacted for details. For the example, a dielectric constant of 2.9 is assumed.

A common stripline structure is modeled to observe the impact of properly modeling the resin concentration in the metal layer. The original model uses a dielectric constant of 4 above, below, and between the conductors. The second

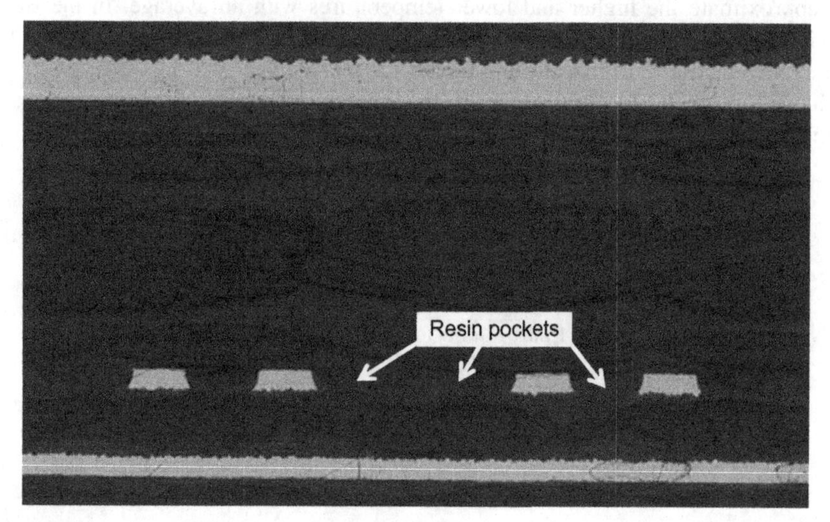

FIGURE 3.9

Asymmetric stripline microscope cross section.

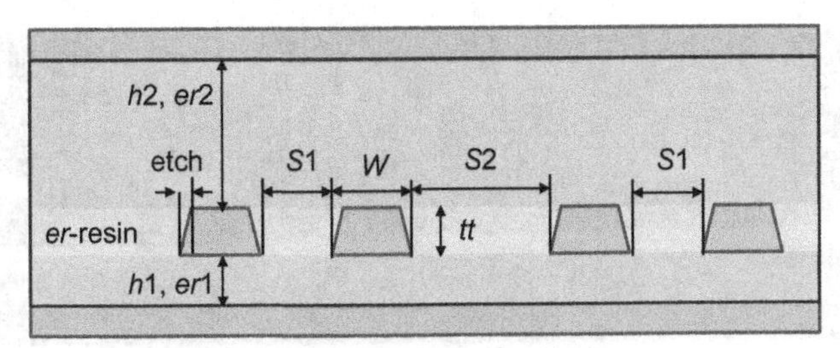

h1 = core thickness, h2 = prepreg thickness,
er1 = core dielectric constant, er2 = prepreg dielectric constant,
tt= trace thickness, S1 = intrapair spacing, S2 = interpair spacing

FIGURE 3.10

Asymmetric stripline model parameters.

Table 3.3 Changes in Model Property With and Without Resin

	Differential Impedance	Differential Insertion Loss	Integrated Crosstalk
No Resin Pocket	85.64 ohms	−0.806 dB/inch at 4.0 GHz	1.70
With Resin Pocket	83.31 ohms	−0.807 dB/inch at 4.0 GHz	6.39

Note: *Integrated crosstalk is the forward wave taken over a 5-inch length at 8 GT/s. Calculation for integrated crosstalk is shown in Chapter 2, "Time Domain."*

model specifies a different resin property between the conductors. Table 3.3 compares the impact on impedance, loss, and crosstalk due to the metal layer resin modeling. A small but not negligible impact is shown on the differential impedance of 2.7 percent, or 2.2 ohms. If modeling the resin concentration is correct, the discrepancies in measurements would have easily been attributable to process variations. The insertion loss is, however, not affected in this model. It is observable that correlation to crosstalk may require resin pocket modeling. Exceptions may exist where the difference between the prepreg dielectric constant and resin dielectric constant are small, such as some low-loss materials. In these conditions correlation might be achievable without the resin pocket model.

Microstrip structures

The following pictures of a microstrip cross section provide a visualization to aid transmission line modeling. The unique features include the mushroom shape due to electrolytic copper and tin plating and the conforming solder mask shape.

Solder mask thickness will vary across the PCB surface. Applications directly on top of the conductor are thin, while applications directly on top of the dielectric tend to be thicker, as shown in Figure 3.11. When applied, the solder mask

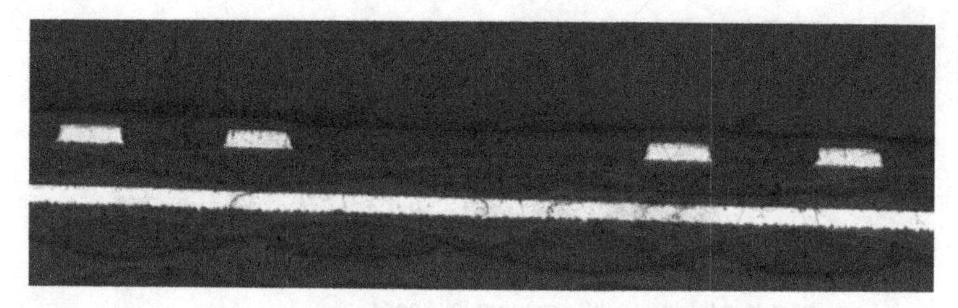

FIGURE 3.11

Microstrip cross section with 1080 glass weave.

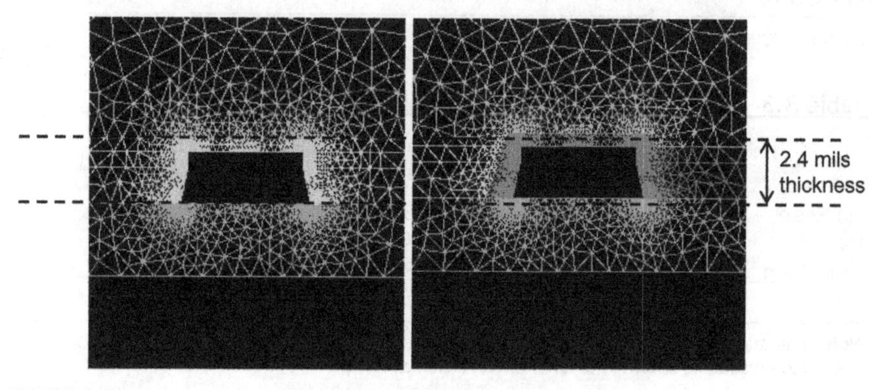

FIGURE 3.12

Blanket (left) and conforming (right) solder mask model.

may pool thicker between differential pair conductors than it would otherwise. The amount of variation in thicknesses may differ from vendor to vender due to solder mask application styles (e.g., horizontal or vertical). Some simulation tools offer model templates for a conforming solder mask model, creating hills and valleys in the applied resin. Other simulation tools provide only a blanket solder mask model. Solder masks that are thinner than the conductor will not be accurately modeled with a blanket model. The effort to model a custom solder mask surface is significant, but is it worth the modeling time?

Conforming and nonconforming microstrip models are created in a 2D field solver for analysis. For the example, the stated solder mask application is 1.9 mils and the conductor thickness is also 1.9 mils. For modeling purposes, it is assumed that 0.5 mils of solder mask will adhere to the top of the conductors. The first model contains the blanket solder mask. In order to achieve the proper thickness on top of the conductor, the solder mask blanket is modeled as a total of 2.4 mils (1.9 conductor + 0.5 adhesion thickness). The second case is the conformal model, accurately able to represent 1.9 mils between conductors and 0.5 mils on top of the conductors. The modeled cross sections are shown in Figure 3.12.

Table 3.4 Comparison of Conformal and Blanket Solder Mask Models

	Differential Impedance	Differential Insertion Loss	Crosstalk (ICN)
Blanket Solder Mask	83.34	−0.812 dB/inch at 4.0 GHz	15.6 mV
Conformal Solder Mask	84.47	−0.789 dB/inch at 4.0 GHz	16.2 mV

Note: *Data for 8 GT/s.*

The simulated results for differential impedance, insertion loss, and crosstalk are reported in Table 3.4. The two modeling approaches are commonly used to represent the same structure. The difference between the results is a 1-ohm differential impedance, 3 percent insertion loss, and 3.8 percent integrated crosstalk noise. It is clear from the model cross section that the model on the right more closely represents the physical reality of the microstrip solder mask.

CORNER MODELS

Manufacturing variation from the nominal impedance is due to inconsistencies during the manufacturing process in the glass-to-resin ratio, lamination, imaging, etching/stripping, solder mask properties, and oxidation. High-speed designs for high-volume manufacturing often target 15 percent stripline and 17.5 percent microstrip control limits on impedance without added cost. In 2013, measured data collected By Intel suggests lower variations of 12 percent and 15 percent, respectively, may be achievable without cost increase. The tolerances may be considered to be 3–4 sigma away from the nominal. A reduction in impedance tolerance improves the defect-per-million performance. Cost and capability to achieve these targets, or better, varies from vendor to vendor. Capability may also be improved when trace width is increased. As nominal trace width is increased, the variation in width remains constant [5, p. 28]. Designs with wider trace width have a reduced percentage change in trace width over HVM. The smaller percentage leads to a tolerance reduction due to the high dependency of impedance on trace width.

Some guidelines for stack-up variations in Table 3.5 can be used for model development and were published by Gary Brist [5, p. 15–59]. The causes shown in the table are exclusive to each prepreg or core and do not provide for any correlation between build-up layers within the stackup. For example, impedances on the top layer may be at the upper tolerance limit, while impedances on the bottom layer may be at the lower tolerance limit. This polarization of the impedance corner creates a worst-case signal integrity situation that must be simulated. Alignment of the full +15 percent with −15 percent within the same design creates the largest electric discontinuity and reflection. Note that the occurrence of the polarized impedance case carries a statistical significance that can be modeled with the details in Table 3.5. Two methods for creating the corner model are discussed: iterative and Monte Carlo.

Table 3.5 Stack-Up Parameter Variations

Parameter	± 3 Sigma	Units	Causes
Trace Width	0.25–1.0	Mils	Imaging, Stripping
Trace Thickness	½ oz: 0.05–0.1 1 oz: 0.1–0.25	Mils	Plating
Dielectric Thickness	0.11–0.33	Mils	Lamination
Dielectric Constant	<0.05		Resin Control
Loss Tangent	Unknown		Resin Control
Solder Mask	Unknown		Application Process

Note: *Summary of parameters referenced in Brist ([5, p. 15–59], [10]).*

Trace spacing is not listed in Table 3.5 because it is a function of, and equal to, the standard deviation in the trace due to stripping. When modeling, maintain the same center-to-center distance for all corner models. Calculate the new edge-to-edge spacing between the signals from the change in trace width. The statistical variation in core and prepreg loss tangent is not listed because it has not been published and depends on resin control (electric properties or prepreg lamination). Similarly, the variability of solder mask over HVM is not known and therefore may be kept constant across microstrip impedance corner models.

Iterative corner model

The first proposed method creates a normal distribution around a nominal 85-ohm stackup. To achieve this, the process involves manual adjustment of the geometries within their 3–4 sigma limits. A 12 percent tolerance around 85 ohms (74.8, 95.2 ohms) will be assumed for the stripline stackup. Recall Chapter 2, in which Figure 2.4 illustrates the most sensitive parameters for differential impedance: trace width and dielectric height. The remaining secondary parameters will incrementally affect the impedance result. The procedure for creating the low impedance corner model is the following:

1. Increase trace width, trace height, and dielectric constant by 3 sigma.
2. Decrease dielectric heights by 3 sigma.
3. Calculate differential impedance.
4. If necessary, increment or decrement geometries to achieve target.

For the test case, an asymmetric stripline stackup with a nominal impedance of 85 ohms is assumed. The stack-up parameters are shown in the nominal column of Table 3.6 and the parameter standard deviations are selected from within the range of parameters shown in Table 3.5. A low- and high-impedance model is achieved with all parameters at 2.7 and 2.3 standard deviations, respectively. Final standard deviations of 5.8 and 6.2 are calculated for the corner models by a root sum square (RSS) of the individual standard deviation of all six iterated

Table 3.6 Methodology Comparisons for 12% Impedance Model

Parameter	Nominal Stackup	Iterative low impedance	Iterative high impedance	Monte Carlo LZ	Monte Carlo HZ
Trace Width	5.25 mils	5.96 mils	4.62 mils	5.71 mils	4.43 mils
Trace Space	6.25 mils	5.54 mils	6.88 mils	5.79 mils	7.07 mils
Trace Thickness	1.9 mils	1.38 mils	1.23 mils	1.30 mils	1.30 mils
Core Thickness	4 mils	3.67 mils	4.29 mils	3.67 mils	4.10 mils
Core Dielectric Constant	4.1	4.24	3.98	4.03	4.09
Prepreg Thickness	13 mils	12.67 mils	13.29 mils	12.99 mils	13.04 mils
Prepreg Dielectric Constant	4.1	4.24	3.98	4.17	4.13
Loss Tangent	0.022	0.022	0.022	0.022	0.022
Differential Impedance	85.10 ohms	75.05 ohms	94.97 ohms	78.99 ohms	93.87 ohms
Insertion Loss (4 GHz)	−0.76 dB/inch	−0.783 dB/inch	−0.746 dB/inch	−0.774 dB/inch	−0.762 dB/inch

Note: *A metal layer dielectric constant of 2.8 is assumed for this example.*

parameters. The process is effective to locate a model at the impedance tolerance, but the standard deviation for the final model is very high, leading to a very low probability corner model.

A few alternative iterative methods are suggested for the quick generation of a corner model with a higher probability of occurrence than the aforementioned process. First, only the parameters with very high sensitivity to impedance may be set to 3–4 sigma from the nominal value. Lower sensitivity parameters may be left at nominal. For the example in Table 3.6, a resulting low-impedance model following this process would increment only the core dielectric height and trace width to 3.5 standard deviations. An impedance of 75.07 ohms is achieved with a combined standard deviation of 4.9. A second proposal is a weighting factor based on impedance sensitivity. Parameters with low sensitivity to impedance are incremented by a reduced amount, perhaps reaching 1 standard deviation or less. The proposal effectively locates a model at the same impedance target with a higher probability of occurrence.

Monte Carlo corner model

The empirical method for selecting impedance corner transmission line models is through a Monte Carlo simulation. The transmission line model is selected by measuring the resulting impedance from a volume of samples. The model is chosen at a desired probability—for example, 3 or 4 standard deviations away from the mean. Multiple models from the sample may be available near the selected probability. At this time, selection based on a secondary metric is possible. Common secondary metrics are crosstalk and insertion loss. Such a selection may lead to "low-impedance high-crosstalk corner" or "high-impedance high-loss corner" models. Note that the ability to select from a secondary population requires a significant number of cases in the Monte Carlo volume (more than 500 for 4 sigma).

The 500-case Monte Carlo simulation is used for the example. Results for impedance and insertion loss are shown in Figure 3.13. Standard deviations were selected from within the range in Table 3.5 to create the random population. The corner model is selected with the most extreme impedance, as the sample size of 500 is not sufficient to fully yield models at the 12 percent target. The high corner is 1.1 ohms below the target, and the low corner is 3.9 ohms higher than the target. The distribution is shown in Figure 3.13 with a normal distribution fit. Indeed, the standard deviation is 2.47 ohms and ±4 sigma is 75.12 and 94.88 ohms, less than a quarter ohm from the 12 percent target. Surely, more simulated models are needed to empirically reach 4 sigma. It can also be observed in Table 3.6 that corner models are achieved with an uncorrelated sensitivity to dielectric constant, indicating its low bearing on impedance for the allowed range.

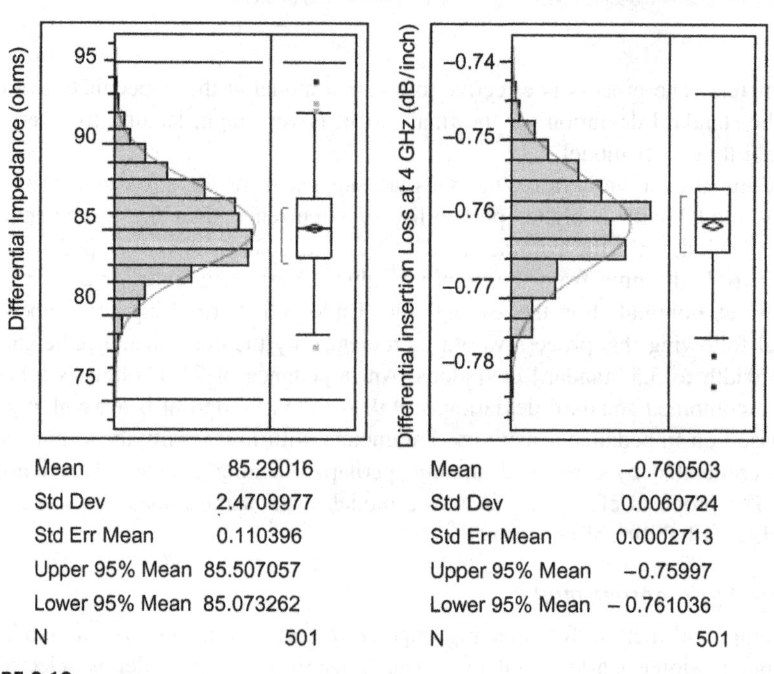

FIGURE 3.13

Simulated Monte Carlo impedance and insertion loss.

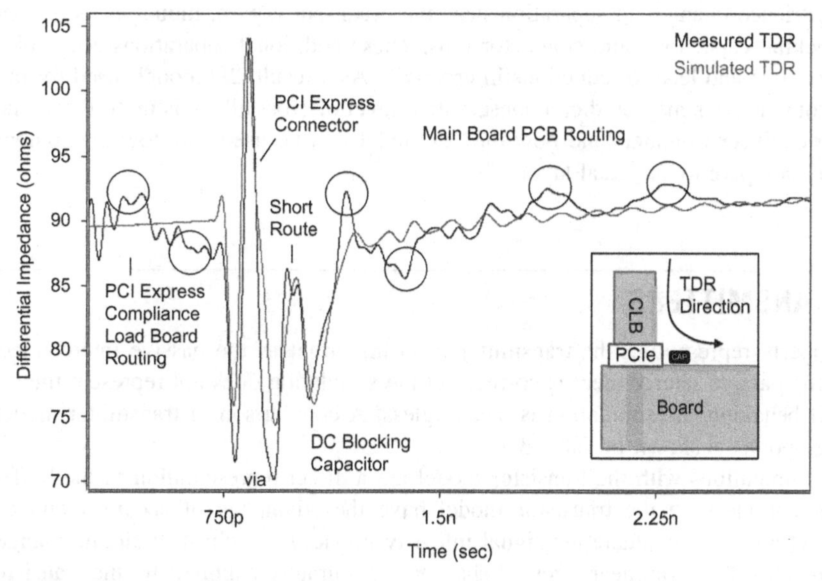

FIGURE 3.14

Transmission line impedance variations shown in TDR.

IDEAL ASSUMPTIONS: HOMOGENEOUS IMPEDANCE

Channel measurements reveal that PCB impedance varies over the length of the route, unlike a homogeneous transmission line model. This inhomogeneity is clear from time domain reflectometry (TDR) results. Variation may be due to the fiber weave effect as described in Chapter 2 or small changes in dielectric height due to copper density. Figure 3.14 compares measured motherboard routing to an idealized model. Impedance appears to vary ± 5 ohms over the duration of the TDR response. It is possible that the variation is greater than observed because energy is lost traveling through the trace. A comparison with a TDR of known-model impedance would confidently determine the actual variation in impedance but will not be shown here. At higher data rates it may become necessary to reflect this physical phenomenon in modeling to accurately predict performance. It is critical to carry out simulation correlation to eye opening for higher speeds to build confidence in model predictability.

IDEAL ASSUMPTIONS: CROSSTALK AGGRESSORS

Pre-layout simulations using models generated by 2D field solvers apply constant coupling levels throughout their use that may not represent the coupled reality of an actual design. Crosstalk measurements on test boards with perfectly straight routing correlate well to 2D models, while measurements on real designs experience changes in separation near the break-out region, mounting holes, DC blocking capacitors, and connector pins. These additional separations accumulate over time and lead to reductions in crosstalk. As a result, 2D models used for pre-layout analysis may predict a conservative level of crosstalk. It is up to the signaling engineer to understand how the 2D model will be used and how the contents may compare to an actual layout.

TRANSMITTERS

Properly representing the transmitter is as important as the passive interconnect. If the passive interconnect is correct but the simulation does not represent the circuit behavior, the simulation is meaningless. A comparison of transmitter modeling options is shown in Table 3.7.

Simulations with the transistor model are a direct representation of the buffer. Simulations with the transistor model have the advantage of accuracy and the direct ability to evaluate the signal integrity impact from physical circuit changes and tradeoffs. Nonlinear circuit behavior is accurately captured by the transistor model. Analysis is computation-intensive, limiting the reasonable simulation length to millions or tens of millions of bits. Since models contain intellectual property (IP) in the form of the exact circuit representation, they generally are not distributable by integrated circuit (IC) vendors.

IBIS MODELS

Many IC vendors provide behavioral models following the format of the I/O Buffer Information Specification (IBIS). Models keep the IP of the vendor safe by representing the circuit behaviorally. Behavioral models do not contain the physical reality of the circuit, but represent the effective electrical circuit response. This is sufficient and accurate if the model properly replicates the

Table 3.7 Transmitter Model Types

Model Type	Advantage	Disadvantage
Transistor Model	Direct representation of circuit	Slow simulation for bit-by-bit Low BER simulation unfeasible Contains IP
IBIS	Can represent V/I nonlinearity Protects IP	Moderate development time
Spice Voltage Source	Fast simulation Use in any simulation tool Protects IP	Excludes nonlinear characteristics Must be correlated to transistor model

output behavior of the transmitter. The output current behavior is dependent on the transitioning voltage. Output voltage varies in amplitude and time, depending on circuit loads that affect rise time, have varying supply voltage, and transmit equalization levels. The current and voltage relationship is described by an I-V curve. The change with time is described by a V-T curve. These tables, plus additional features to represent clamp diodes and die capacitance, represent the transmitter characteristics. Models also represent any nonlinear behavior captured in the I-V and V-T curves. Verification through correlation to the transistor model is necessary to determine modeling accuracy and find the effective capacitance during model development. Note that early revisions of IBIS modeling define a package model in addition to the I/O buffer. Models representing the package by only R, L, and C lumped elements are insufficient and lack the accuracy and mutual coupling (crosstalk) needed for high-speed analysis. Models following the most recent IBIS standards using S-parameter representation of packages are sufficient.

SPICE VOLTAGE SOURCE MODEL

A simple spice schematic may be drawn to represent effective circuit behavior with a voltage source and rise time, resistive termination (R-term), and effective pad capacitance (C-pad) as shown in Figure 3.15. This voltage source schematic is easy to configure and is commonly used in spice and electron design automation tools where IBIS models are not available. This behavioral model can be sufficient for analysis if parameters are set to achieve correlation. In some cases a filter may be applied to achieve proper rise time and wave shaping to more closely represent circuit behavior, such as a Bessel or Gaussian filter. The difference between the wave shape achieved from a linear ramp and a Gaussian filter is shown in Figure 3.16. In the PCIe Base Specification 3.0 (PCIe), Gaussian filters like the circuit shown in Figure 3.17 are recommended for channel compliance simulation.

Linearity test

The behavioral spice model is linear and does not capture nonlinear characteristics of the real circuit. The noise created from nonlinear behavior of the transmitter circuit can be observed in the simulated waveform and can be quantified

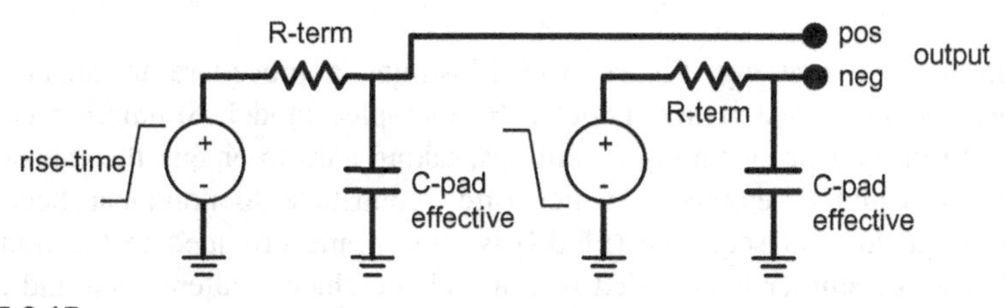

FIGURE 3.15

Behavioral spice transmitter.

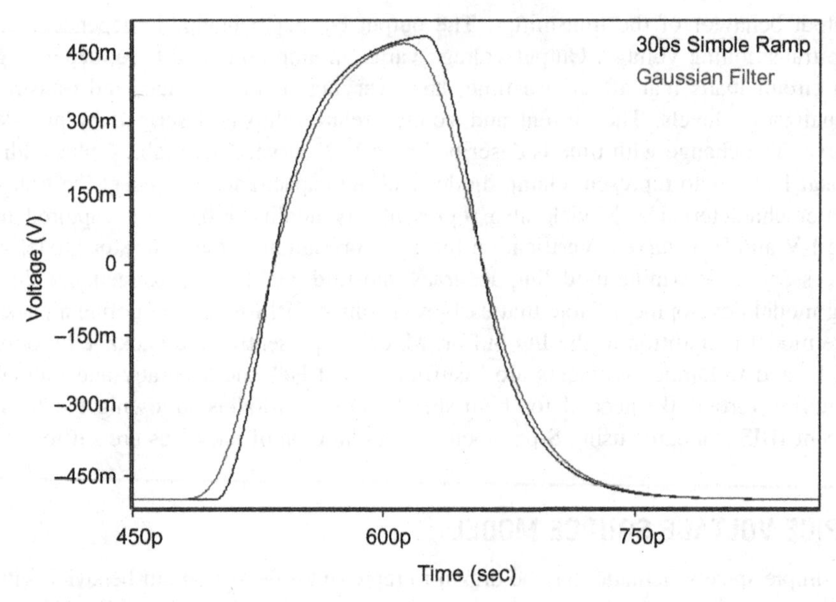

FIGURE 3.16

4 GHz pulse response in spice.

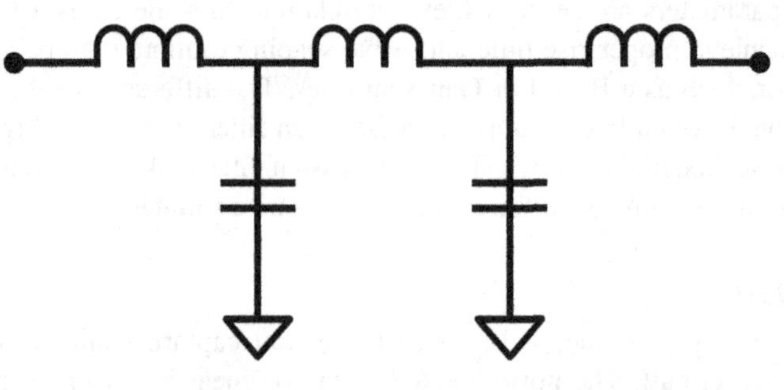

FIGURE 3.17

A Gaussian low-pass filter provides proper output rise time.

through the eye opening. Figure 3.18 illustrates a procedure to quantify the nonlinear behavior that is not modeled by the spice model. Simulation is to be performed in realistic operating conditions, taking care to enable the transmitters pre-emphasis and de-emphasis, which often contribute to nonlinear behaviors. A pseudo random bit sequence (PRBS) is chosen and provided to the transistor model. The transmitter is provided with a realistic channel attenuation and termination. The signal is post-processed to create the composite eye diagram where true eye height and width are measured.

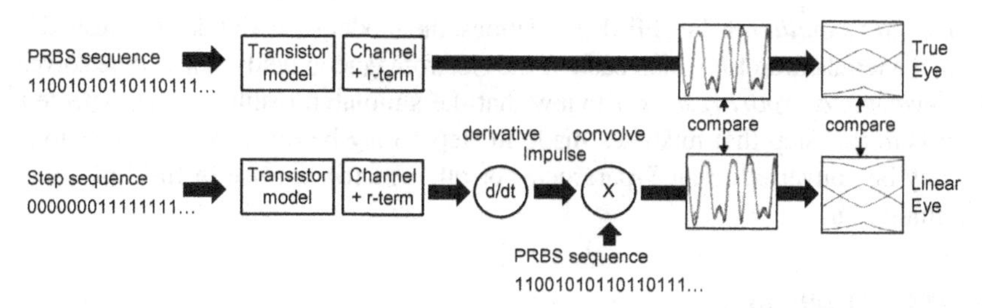

FIGURE 3.18

Spice model linearity test.

The linear model results are processed by convolution to reach the bitstream at the channel output. The resulting bitstream is constructed out of the channel impulse response, which is only valid if the transmitter behaves linearly. The transistor model is driven to create a step response through the channel. The derivative is taken out of the step response to yield the impulse response. The impulse is convolved with the same bit sequence used to create the true eye diagram. The bitstream is post-processed to create the composite eye diagram. Any difference between the eye diagrams is a means to quantify the noise level introduced by nonlinear transmitter behavior. The difference may be used as a guard band for effects that are not simulated and that are not present in full link simulations that operate on the assumption of linearity.

3D MODELING

The use of high-frequency solvers has become increasingly important to simulate the electrical characteristics of interconnects as data rates increase. The guidance in the 3D modeling section is primarily focused on finite element method (FEM) solvers, which have their strength in solving arbitrary 3D objects in a volume. Two examples of commercial tools with FEM are ANSYS HFSS and CST. The section provides guidance with key decisions made when developing 3D models for high-speed modeling and simulation. An overview of the steps of the modeling process is provided.

1. *Creation of the physical model.* The model may be manually drawn or extracted from a computer aided design (CAD) layout. This step includes drawing the structure within a volume, creation of the ports, and defining all metal and dielectric material properties.
2. *Analysis setup.* The boundaries and boundary types for the solution are defined. Wave or lumped ports are created within the 3D model. Simulation range and precision settings are selected.

3. *Execute simulation*. For FEM simulations, the model is subdivided or meshed into a tetrahedron for which each of the electromagnetic fields that will be solved.
4. *Review and export results*. Review that the simulated results are as expected. It is in this step that mistakes made in step 1 may be revealed. The solution may be converted to an S-parameter or other matrix for use in full link simulation.

PORTS/TERMINALS

Ports in 3D modeling tools specify a location for Maxwell's equations to solve for the fields used to excite the model. Selection of a wave or lumped port depends on the terminal location, number of terminals, and terminal separation. Wave ports solve a distributed field in a 2D plane, while lump ports excite a single current source.

Wave ports

Wave ports are the only ports capable of solving the field distributions for multiple propagating modes, transverse electric (TE), transverse magnetic (TM), non-transverse electromagnetic (non-TEM), or multiple modes, and are therefore the preferred port when possible. Closely spaced terminals require the use of a wave port to solve for the coupled interaction between terminals. One wave port may be used for multiple terminals. Only wave ports calculate propagation constants, characteristic impedance at the port location, and compute S-parameters directly.

Wave ports are connected to the modeled structure by a waveguide or transmission line. When modeling in HFSS, the transmission line length should be only long enough for one mode of propagation to exist between the signal and reference at the port. If the length is too short, multiple (higher order) modes may exist at the wave port when the S-parameter matrix is calculated. This leads to erroneous S-parameter results.

In HFSS, de-embedding from the wave port is possible to remove any excess and uniform input waveguide or transmission line. Wave ports are required to exist on an external boundary and should be drawn with a face as small as possible. (It may be possible in some special circumstances to place wave ports internally, though this is not covered here and the reader should check with the particular vendor's documentation.) The characteristic impedance may be used as an indicator to evaluate wave port size in a port-only simulation. If increasing the port size affects the characteristic impedance, the wave port may be too small to calculate contributing fields. A wave port is shown in Figure 3.19. The size of the port should be sufficient to capture all of the TEM energy.

Lumped ports

Lumped ports support only a single mode field. The ports may be placed inside to the model where extension of a uniform waveguide as required by wave ports is not possible.

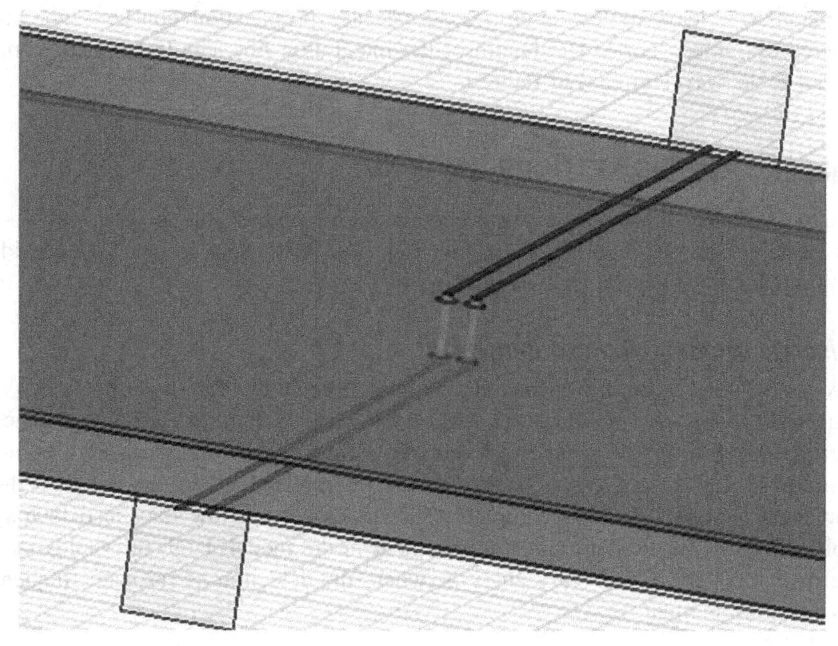

FIGURE 3.19

Properly sized wave port.

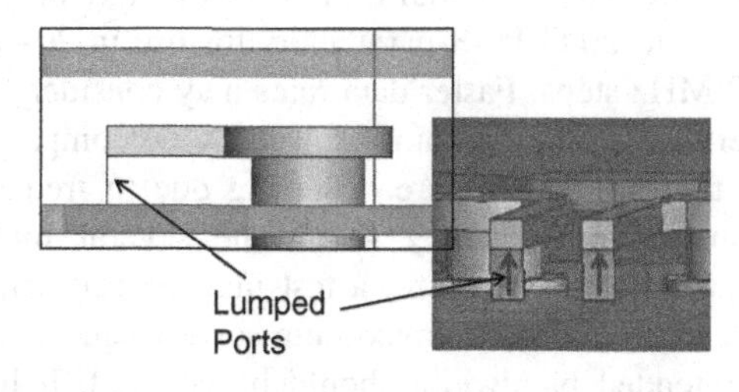

FIGURE 3.20

Lumped port and integration line.

The lumped port requires a user-defined port impedance, which is excited by a fixed current in order to compute the electric field, integrate, and find the voltage. Depending on the tool capability, voltage may be applied and the current solved. An integration line, shown in Figure 3.20, is necessary in all cases to define the direction of the electric field for integration. The integration line is drawn within a 2D sheet that covers the trace width and makes contact with the nearest reference plane. It is assumed that the fields are uniform within the 2D plane.

Lumped ports will not provide the same results as wave ports. S-parameters are converted from Z-parameters using the port impedance supplied by the user.

S-parameters may not be de-embedded because propagation constants are not solved by lumped ports. The length of the integration line should be small compared to the wavelength(s) of interest.

MODEL ANALYSIS SETTINGS

Analysis settings influence accuracy and simulation time. It may be that only one, or perhaps both, is important for a given analysis. This section provides considerations for choosing the right settings.

Discrete or interpolating solutions

Precise models can be achieved in less time with interpolating sweeps. Interpolation sweeps rely on a threshold coefficient (S delta in HFSS) to balance the optimization between precision and solve time. When added discrete points are solved and do not yield changes in the interpolated model greater than the threshold, computation will complete. Thresholds set too low may lead to long solve times without meaningful improvement in the model. High thresholds may be used for a fast preliminary analysis, where the detection of resonant frequencies is not important.

Frequency range and step size

The frequency range and step size necessary depends on the model usage. S-parameters for graphic plots or initial design review may only be sufficient at a 100-MHz step size up to 20 GHz. S-parameters for use in 20-GT/s simulation or less should have 10-MHz steps. Faster data rates may consider a 5-MHz step size. In the case of uncertainty, time domain results may be compared at an increased step size to verify that no changes are occurring due to frequency domain step size. The maximum solve frequency should be several multiples above the Nyquist frequency. Similar to step size, a test may be performed to observe the time domain impact from changes in maximum solve frequency. At least the knee frequency for the intended bandwidth should be covered. It has been observed that the maximum solve frequency may need to be as high as 60 GHz before the time domain response ceases to change at a 10-GT/s data rate. The cost to solve 3D models to 60 GHz is significantly more than the common 20 GHz. It is appropriate to weigh the cost of a longer solve time with incrementally increased data precision. Ultimately, the desired step size and time delay of the user determine how much frequency bandwidth is needed.

A DC point is required to perform inverse fast Fourier transform (IFFT). Tools may offer a direct solve of the DC point or an extrapolation. Extrapolation is sufficient if correct. Visual inspection of the extrapolated slope of the magnitude, phase, and low-frequency group delay can offer an indication of improper extrapolation. The magnitude must fit the higher frequency trend and phase must go to zero at DC. Figure 3.21 shows an example of improper DC extrapolation.

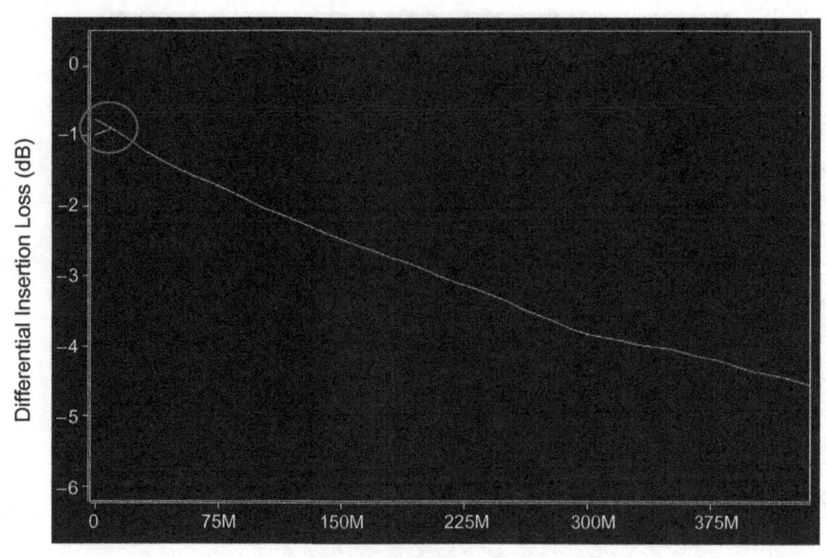

FIGURE 3.21

Proper and improper DC extrapolation.

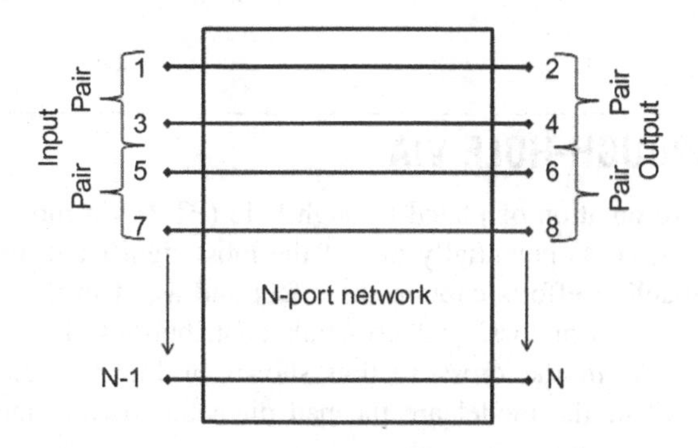

FIGURE 3.22

Odd in and even out port order.

Port order

The most common assignments are odd in and even out, as shown in Figure 3.22. An alternative ordering scheme is sequential order, where 1:N/2 ports are the input and the remainder are the output.

Normalize result to 50 ohms

Ports should be normalized to 50 ohms when generating S-parameters in most circumstances. Ports referenced to something other than 50 ohms should be

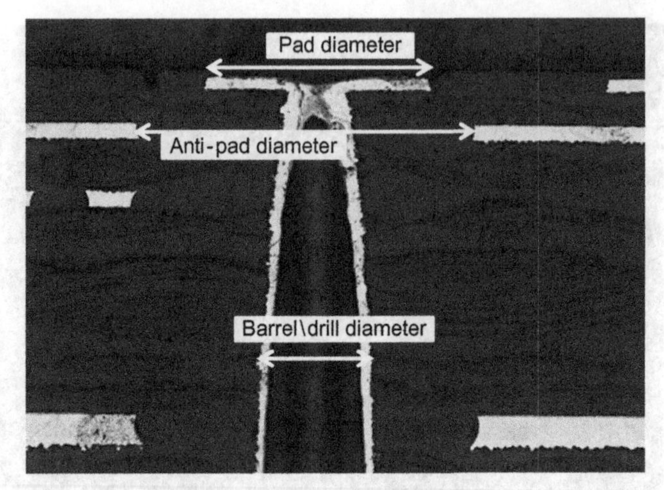

FIGURE 3.23

Microscopic cross section of normal PTH via.

communicated when sharing the models. Simulators following the touchstone format guidelines will automatically detect the port reference impedance in the file header.

PLATED-THROUGH-HOLE VIA

An accurate representation of plated-through-hole (PTH) via models is important in system simulation, as potentially one of the most significant discontinuities in the channel. Modeling efforts capture the effect and assist in the optimization of the insertion loss, return loss, and crosstalk contributions. Features of the via model are captured in the cross section shown in Figure 3.23. The primary geometric features of the model are the pad diameter, barrel diameter, anti-pad diameter, and separation to other vias.

The via models are uniquely made for different categories defined here. Categories are necessary due to unique patterns and drill sizes observed for vias of different purposes that lead to discontinuity and coupling changes. Categories do not correlate to via technology alone. Back-drilled and high-density interconnect (micro, buried, blind, and so on) have unique geometries to be specifically modeled; however, because these via types can be observed in different patterns (BGA or mid-route) they are not considered a separate category for our purposes here. Further, unique categories are not required for interface or data rate. The categories and causes for classification are identified in Table 3.8. In order to complete interconnect simulations, models in most of these categories will be

Table 3.8 Via Modeling Categories

Category	Features	Necessity for Unique Classification
BGA Area	Repeating signal to ground pattern throughout the BGA area. Inner pins transition to PCB layers with the same pattern as the BGA pins.	The BGA pin pattern forces a specific crosstalk pattern. One model generally represents most signals. Exceptions to the pattern, such as pins on the BGA edge, may have higher crosstalk and need to be separately modeled.
Mid-Route Transitional	Mid-route vias have no specific pattern and are placed wherever there is space. Models may represent a typical or worst-case possible pattern.	The lack of pattern allows layout combinations of signal-to-signal and signal-to-ground that can lead to higher crosstalk.
SMT Connector	Via transition patterns occurring very close to SMT pads will be influenced by connector pinout.	Unique patterns driven by the connector pinout may lead to better or worse crosstalk than mid-route transitional vias.
THM or PFT Connectors	Specific patterns and PTH diameters are unique to each connector. Unique vias are needed for every connector.	Changes in discontinuity from barrel diameter and crosstalk from patterns require unique vias for every connector type.

Note: *BGA = ball grid array, THM = through-hole mount, PFT = press fit technology.*

necessary. Once a model is developed, it may be stored in a library for reuse across multiple interfaces and data rates.

MODEL TECHNIQUES

Lumped element models may be used to represent via discontinuities in low-frequency simulation (1 GHz and less). Lumped models relate the physical features of a via to discrete elements. These models are useful for physics understanding, but at higher frequencies the lumped model is not sufficient to capture the performance in high-speed systems—for example, capacitive loading for the via pads and inductance for the through-path in the barrel. Lumped models do not accurately model the Q-factor in the insertion path and they lack via-to-via coupling. Vertical coupling between vias is found to be a significant percentage of system crosstalk that can only be represented in a 3D full-wave simulator. Coupling contributions from vias may be high underneath BGAs, in connector pin fields, and in the open fields. Full-wave 3D-modeling is recommended and the baseline assumption for this section.

PRE-LAYOUT APPROXIMATION

In early and pre-layout analysis stages, it is permissible to use existing via models that are electrically equivalent but may not precisely represent the physical design proposal. Models may be leveraged from either a prior design or CAD extraction. This reuse can provide quick feedback about topology proposals or risk without extensive simulation work. The question is whether physical changes are significant enough to require new model changes when high precision is not required. When substituting a model in simulation, risks from approximations can be reduced by selecting the nearest model with expected worse performance. Some parameters that are found to contribute only small changes in via electrical performance are:

- Layer count
- Dielectric constant
- Loss tangent
- Conductivity

The small significance to dielectric constant and loss tangent is primarily due to the small size of the via structure. Other via parameters remain significant: barrel diameter, via stub length, and pad diameter. Figure 3.24 demonstrates with TDR the small electrical difference due to layer count. In the example, the TDR for a 10-mil via stub is shown for 8-layer and 14-layer stackups.

PRE-LAYOUT MODELING

If design differences are significant, pre-layout modeling will be done according to board layout requirements for signal-to-signal and signal-to-ground separation.

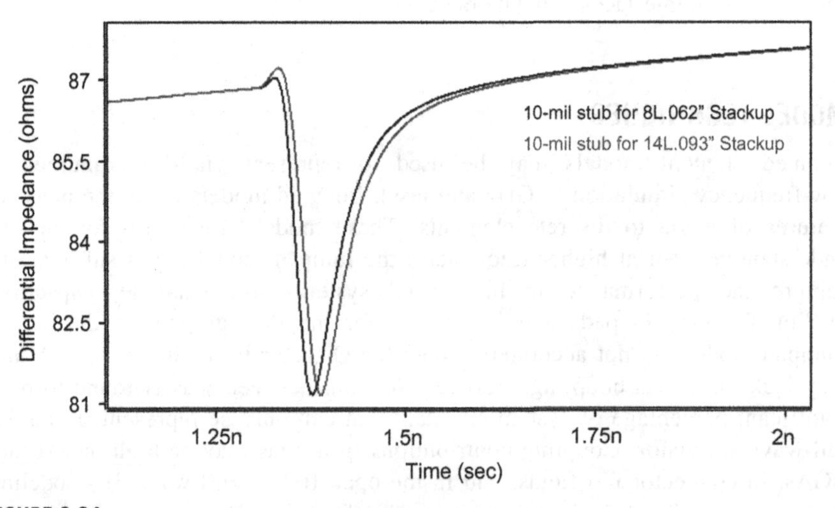

FIGURE 3.24

TDR of vias with different stackup and same stub length.

Models may be drawn manually in 3D FEM solvers. Some software includes via wizards to simplify the modeling process. A simplified process for creating the physical via model is:

1. Build the stackup.
2. Build the signal vias and pads.
3. Create an anti-pad around signal via barrels.
4. Add ground vias and pads.
5. Add signal trace and ports.

The following guidelines should be used when developing the physical model:

1. Do not include pads on unused signal layers. Removal of pads on unused internal layers is a common manufacturing capability with significant signal integrity benefits. The only pads should be on both external layers and any internal layer(s) used for signal transition.
2. Consider an accurate number of signal-to-ground vias and realistic ground via placement as much as possible. An optimistic return path will lead to underestimated losses and crosstalk in the model.
3. Ensure that the transmission line is routed at the geometries (trace width, spacing, etc.) and characteristic impedance of the remainder of the interconnect.
4. Attach transmission lines to lumped ports. Transmission lines connecting the lumped port to the via should be long enough to establish TEM and uniform waves. Creating transmission line segments that are too long interferes with the flexibility to pair the model with short routing or alternate routing impedances. The minimum length is a function of dielectric thickness. Approximately 10 times the distance to the primary reference plane is recommended.
5. Convert power planes to ground planes. This is accomplished by intersecting all power and ground planes with the return path or ground vias in the model.
6. Hollowed cylinders are preferred but not be required. High-frequency current follows on the outer edge of the via barrel, making the hollow feature negligible.

POST-LAYOUT

Board layouts that have interesting via patterns can be extracted directly from CAD for 3D analysis. Due to layout difficulty, such cases may be transmit and receive (TX and RX) vias spaced too closely, a lack of nearby return vias, or trace-to-via coupling. Extraction space should include nearby return-path ground vias. Unused signals within the excitation boundary should be terminated to 50 ohms and not connected to ground (connecting to ground will provide an unrealistic return path). An example extraction space is shown in Figure 3.25.

After extraction, undesired objects should be deleted from the model. Ground planes and dielectric edges should be extended around the via structure to ensure currents are diminished before reaching the radiation boundaries. A distance of 500 mils from any signal via is sufficient.

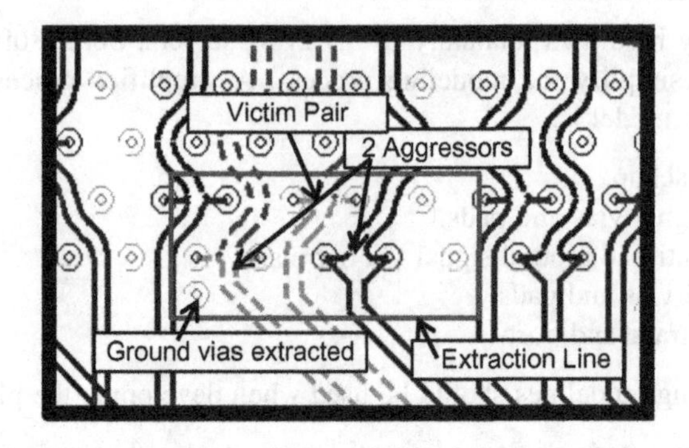

FIGURE 3.25

Post-layout extraction area including ground vias.

CONNECTORS

Connector models are created from 3D full-wave finite element simulators like HFSS or CST. Model collateral is developed by connector suppliers and shared with system design and manufacturers. Collateral may be shared one of three ways: simulated S-parameter file (or spice equivalent), geometry file (*.sat), or completed 3D model file. Table 3.9 defines some considerations when requesting model collateral from vendors.

CONNECTOR VARIABILITY

Changes in a connector's housing dielectric properties or pin shape variations are not part of typical high-speed simulations. It is suspected that variation is small compared to signaling frequencies in that skewed models are not provided from connector vendors. This expectation is confirmed through a set of measurements taken by Intel. Five surface-mount PCIe connectors are mounted to four separate test boards. Insertion loss measurements are taken on five lanes for each connector after de-embedding the test structure. The differential insertion loss profile is shown in Figure 3.26 for all 20 measurements. The insertion loss variation is 0.06 dB at 4 GHz and 0.7 dB at 8 GHz. Intuitively the variation appears negligible. A simulation at 16 GT/s was performed with the measured connector responses to observe the eye height range across the measurements. The result demonstrated negligible changes in eye opening with an eye height variability of ±1 mV and eye width variability of ±0.5 ps. Though only for one vendor, the results indicate that connector variability is not a significant model parameter, at least through a data rate of 16 GT/s.

SIGNAL SELECTION

There are some considerations when choosing the signal pins in a connector for extraction. Some connectors are physically small and may be completely modeled

Table 3.9 Options for Acquiring 3D Connector Models

Model	Advantage	Disadvantage
S-parameter (*.sNp)	Models may be used in simulation immediately.	No flexibility to modify the model such as attached vias, PCB properties, and selected signals. Passivity and causality checks are required. An insufficient step size or max solve frequency can only be changed by interpolation or extrapolation.
Geometry File (*.sat, *.igs)	More easily obtained than 3D model files. Complete model with flexibility to change PCB properties, signal entry layer, and pin selection.	Gaining material properties (dk, tand, and conductivity) are also required. Healing imperfections in imported models is time consuming. PCB modeling is usually done by the user. May require non-disclosure agreement (NDA) with supplier.
3D model (*.hfss, *.cst)	Complete model with flexibility to change PCB properties, signal entry layer, and pin selection.	Sometimes difficult to obtain and may require NDA with supplier. Geometric changes may be difficult if components are united/joined.

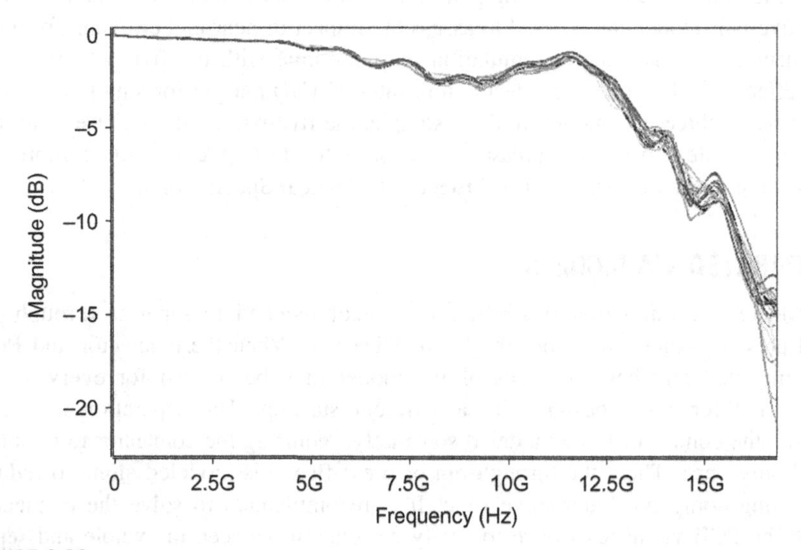

FIGURE 3.26

Measured differential insertion loss for 20 lanes.

(such as mini-SAS) while larger connectors (such as PCIe) will be modeled for only a few selected pins. The first of any adjacent ground pins that affect signal pin impedance should be included in the model. Some questions to determine the signal pin selection are listed here:

- Does the PCB routing layer dictate which pins are coupled?
- Which pins are TX and which are RX? Will the simulation include near-end crosstalk (NEXT)?
- How many differential pairs are simulated in the package and board models? Are there more crosstalk aggressors in the connector than package and PCB models?
- Where is the worst-case crosstalk aggressor?
- Where are the longest (worst-case) signal pins in a right angle connector?

An example of a PCIe connector pin map is shown in Figure 3.27. Two pin extraction choices are highlighted. In the noninterleaved selection, three differential pair signals are selected. This model is useful for simulation decks of only three pairs. Use of the model requires that it be used with other models representing TX signals. It is likely that the simulation deck would be noninterleaved entirely from end to end. The model assumes that the NEXT is negligible or that it is modeled and simulated separately. Due to the connector's symmetry, the model can be used to simulate three TX or three RX pairs.

In the interleaved selection, five differential pairs are modeled. The model may be used for an interleaved simulation deck of TX and RX signals. Alternatively, the model may be used in the noninterleaved three-pair simulation deck in the first example if the two opposing pairs are terminated to 50 ohms. Simulation with the five-pair model can be used to assess the impact of near-end crosstalk from the connector into the system. Simulation may continue with the five-pair model or the effect may be budgeted into the unit interval (UI) margin for simulations with the smaller three-pair model. In this example, the five-pair model represents all the non-negligible crosstalk aggressors necessary for testing connector compliance according to the PCI Express Card Electromechanical Specification.

SEPARATED VIA MODELS

For increased simulation flexibility, it is recommended to separate through-pin and press fit connectors from the board PTH via. When the connector and PCB are modeled together, instances of the model may be needed for every signal layer and for every design with a different stackup. The separation approach allows the connector to be modeled separately, requiring the connector to be modeled only once. Then, the through-pin or press fit via is modeled alone to reduce modeling complexity and solve time. It is recommended to solve the connector with the PCB via at least once to verify correlation between the whole and separated model. Two separation methods are possible, though none is recommended over the other.

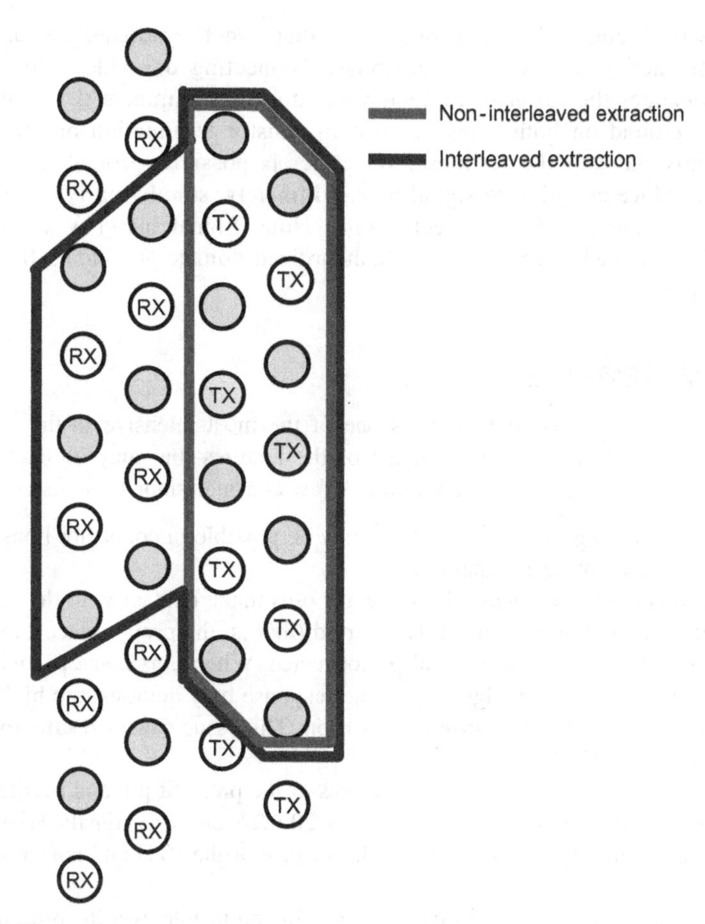

FIGURE 3.27

Selected signals in a PCIe connector.

- The separation plane is the first ground plane in the stackup. The connector model includes the outer dielectric layer and a small height (~4 mils) of the via and surface pad. A circular wave port is placed inside the anti-pad of the via.
- The separation plane insects the pin between the connector housing and PCB surface. Circular wave ports are drawn on the radiation boundary in the air, between the connector and the PCB.

UNCONNECTED PINS

Signal pins reserved for special purposes where connectivity is uncertain must be treated very carefully. Unused signals that are floating pins in the final design will become absorbers of energy that can be radiated out as noise, essentially acting as accidental antennas. Connecting one side of the pin to ground changes the resonant frequency but does not eliminate the problem. If possible, ground on both sides or 50-ohm resistor termination on either side will greatly dampen the response. If neither is possible, ground vias can be added to reduce coupling to signal pins. Ultimately, simulation is necessary to evaluate the impact of unconnected pins. Note: Signal integrity assessments should have already been done for standardized connector pinouts that allow floating pins.

PHYSICAL FEATURES

Development of connector models is one of the most intensive in the interconnect. Some consideration can be taken of the features that may be excluded or features that require additional attention. A few considerations to be mentioned:

- *Reduced meshing.* Reduced meshing may be possible in connector housing areas that are not near signal pins.
- *Edge finger insertion depth.* If connector pins make contact with the edge finger in a depth that is not fully inserted (that is, the proper place), the model will not represent normal performance. When pins make partially inserted contact to the edge finger, the response may demonstrate higher inductance than the fully inserted position. This issue causes results to appear optimistic.
- *Press fit pin and needle eye.* The features of the press fit pin and needle eye are small compared to the rise of today's 20 GT/s and less signals. High-frequency currents can be found on the surface of the PTH and not inside the pin.
- *Through-hole pin.* Similar to the press fit pin and needle eye, the pin internal to the PTH via is not necessary to include in the PTH model. It is found that the protruding pin beyond the board length has a small but incremental effect on electrical performance and should be modeled to the connector part specifications.

DESIGN OPTIMIZATION

Electrical considerations when making a connector choice begin with a frequency domain review of insertion loss, crosstalk, and impedance and return loss profiles. Lower crosstalk levels will improve the available margin, and maximum length can be achieved for a given connector. There are fewer connector vendors optimized for impedance; however, these connectors can yield observable benefit. Impedance optimization is not specified by interface specifications but is often provided in a vendor's electrical report in the form of a TDR plot. Insertion loss and return loss are often controlled parameters by most interface specifications.

It should be noted that insertion and return loss are often comparable between suppliers, and variations often do not translate into significant time domain eye height differences. With that being said, in many instances the via associated with the connector often has more significant effect on loss, crosstalk, and reflection than the connector itself. Therefore, connector via optimization is elevated to a top priority in high-speed design. For more details on important via parameters for optimization, refer to Via Stub Mitigation in Chapter 2.

Voiding edge fingers

The pads or "fingers" on edge mounting cards create a high capacitance due to the large size of the pads with respect to the main routing. The common practice is to remove ground plane immediately beneath the pads. The removal of the ground plane has been missed by some board houses even when it has been removed in the CAD and Gerber drawings. It is important to explicitly include notes for the ground plane removal in the CAD file. Failure to remove the ground plane below the pads has been correlated to bit errors in the laboratory. Figure 3.28 shows a TDR measurement of two edge finger card designs. The card with poor ground plane voiding demonstrates a low-impedance spike at the edge finger location. Note: In order to obtain a realistic TDR response, provide an appropriate rise time and a modest amount of pre-DUT insertion loss (at least 2 dB) to the simulation.

Plug-in cards with only four layers will experience crosstalk increases due to the cross-layer exposure demonstrated in Figure 3.29. Even on cards with six or

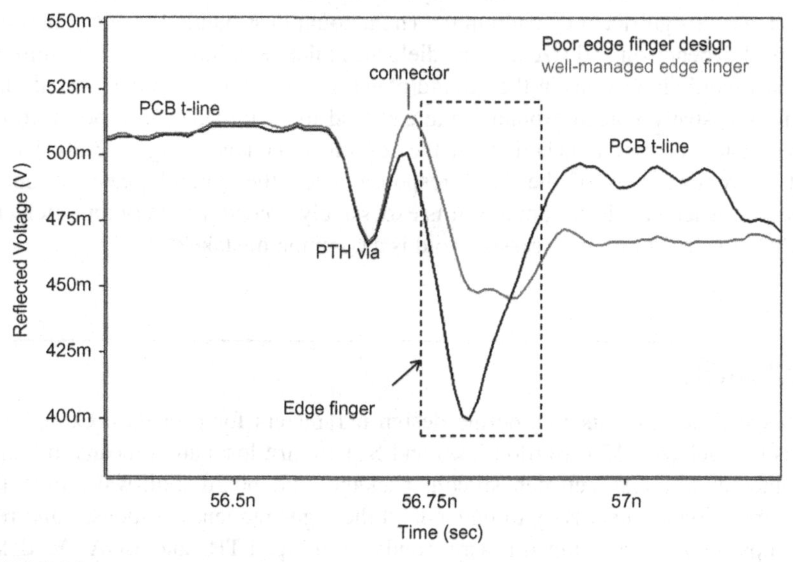

FIGURE 3.28

Measured differential TDR of connector and edge finger.

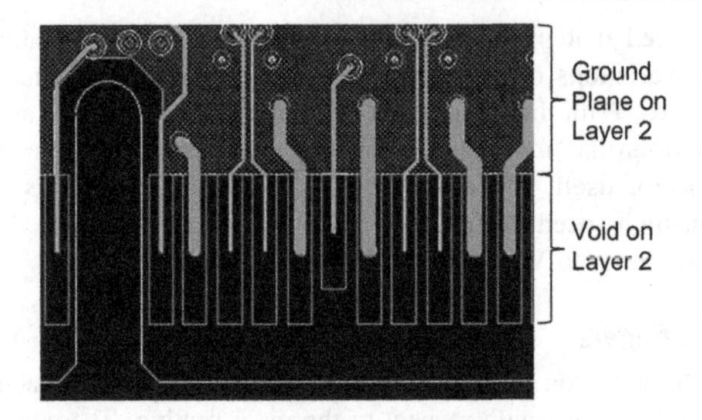

FIGURE 3.29

Removed ground plane below edge fingers.

more layers, it is better to remove the ground planes than to allow them to remain for crosstalk reductions. High layer-count stackups with thin dielectric layers will almost certainly require voiding of several ground planes below the edge fingers. Removing ground planes to lower impedance is the first priority until no benefit is realized; the ground planes may remain for crosstalk shielding.

Voiding SMT connector pads

Ground planes should be removed below the mounting pads for surface mount connectors. Similar to connector edge fingers, the pads create a capacitance increase that creates a significant discontinuity. The amount of ground plane that should be removed depends on pad size and the dielectric thickness. Simulation is required to optimize void dimensions in the ground plane. Failure to remove the ground plane or an excessively voided ground plane can lead to an impedance response to that lower or higher (not matched) than the routing impedance. Figure 3.30 demonstrates two examples of the TDR response when the ground plane voiding is improper, whether it is not done or it is excessively voided. For very thin dielectric heights of near 3 mils, an excessive void is a common mistake.

PACKAGES

High-speed design standards define design parameters for packages. Compliance limits on package S12 (insertion loss) and S11 (return loss) are a means to budget the interconnect loss between silicon, package, and board. Full-wave modeling and simulation are necessary to understand the high-frequency responses and relationships among C4 bump (or wire bonds), routing, PTH, and BGA. Modeling assists in optimization among loss, crosstalk, and reflection.

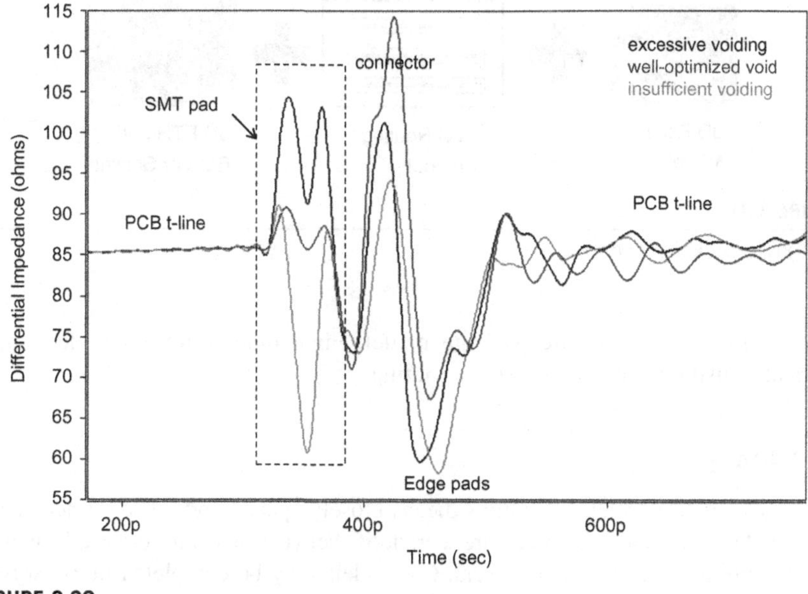

FIGURE 3.30

TDR response due to voiding the ground plane beneath SMT pads.

Complete package models may be created through CAD extraction and simulation in a planar or full-wave simulator. Planar solutions may only be completed post-layout because they are extracted from the design. In instances of large package sizes (greater than 1 inch) or the desire to extract more than a few nets, solve time may increase from hours to days. While this method may be the fastest, it offers the least flexibility in design simulation and optimization. Extraction lacks the capability to complete HVM analysis. Planar extracted models may be the best choice when the package design is not a bottleneck to signal integrity solutions. This may be the case when excess eye margin exists or the package design was previously optimized. Planar extraction is beneficial to find lane-to-lane design issues through the extraction of an entire port. Lane-to-lane comparison of an eye height, insertion loss noise, signal-to-noise ratio, or other metric can quickly locate design issues.

When HVM analysis and design optimization is possible, a composite package model of 2D and 3D components is recommended. The composite model is subdivided at every significant horizontal-to-vertical transition. Horizontal routing is modeled as a 2D scalable frequency dependant transmissionl line model or an s-parameter. This provides flexibility to simulate the impact of short or long routing lengths. Transmission line models may be created at low-, typical-, and high-impedance corners and used to develop composite corner models for HVM analysis. Figure 3.31 illustrates the

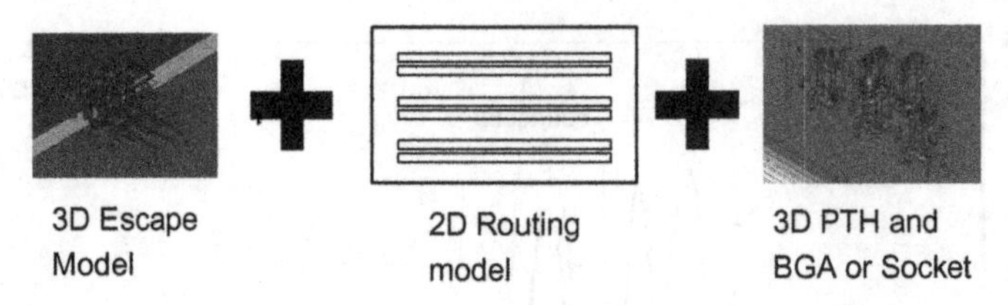

FIGURE 3.31

Composite package model.

components of a composite package model. Final models may include bump, vertical transitions, and short escape routing.

C4 ESCAPE

The small pitch between C4 bumps means closely spaced and narrow traces near the die. Trace geometry choices are a tradeoff between loss and crosstalk, which can be optimized in 2D field solvers. C4 models may be completed in 3D solver and include initial escape routing until it reaches a uniform layout that can be represented by 2D scalable models.

TRANSMISSION LINE

Routing length should be kept as short as possible due to the high resistivity and insertion loss from the small trace width and height (compared to the PCB). As a rule of thumb, every 1 inch of package routing leads to a 2-inch reduction in length on standard loss PCB and 3 inches on low-loss PCB. Modeling with 2D solvers provides the opportunity to make design tradeoffs among impedance, crosstalk, and loss. Spice models or S-parameters used for this optimization may be leveraged directly into full simulation to assess eye height and margin. Figure 3.32 is an example of a package transmission line parametric sweep, observing differential impedance. The sweep covers nominal stack-up parameters, including trace width (TW), spacing (TS), and dielectric height (h1, h2). The impedance is calculated from each RLGC file at 1 GHz, using the formulas for differential characteristic impedance in Chapter 2. For many high-speed I/Os, 85 or 100 ohms is the typical routed impedance. For this example, it is assumed that 80 to 85 ohms is the optimal routing impedance. The optimal impedance is a relationship between the termination resistance, transmission line, and PTH discontinuities. As geometries change to create new impedances, loss and crosstalk are also influenced. Optimization is complex and is best determined through simulated eye height.

A review of additional metrics can give an indication to the optimal dimensions before eye height simulation. In Figure 3.33 the differential insertion loss per inch at 8 GHz is observed. The analysis can be carried further to include

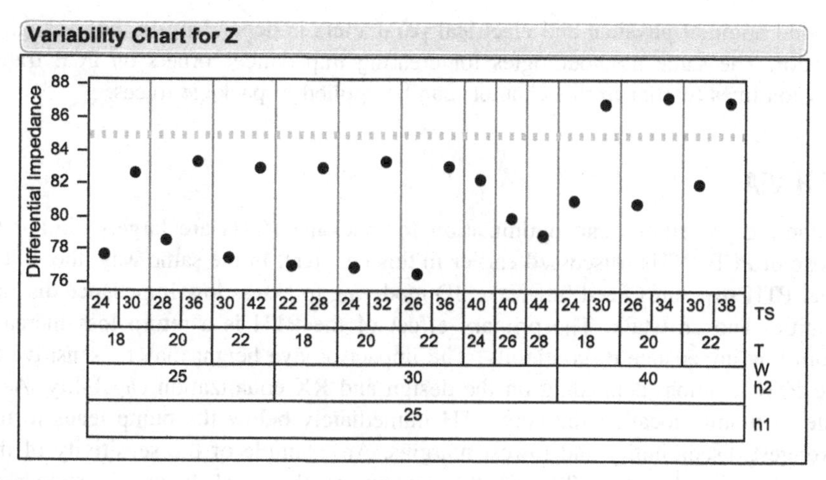

FIGURE 3.32

T-line parametric sweep against differential impedance.

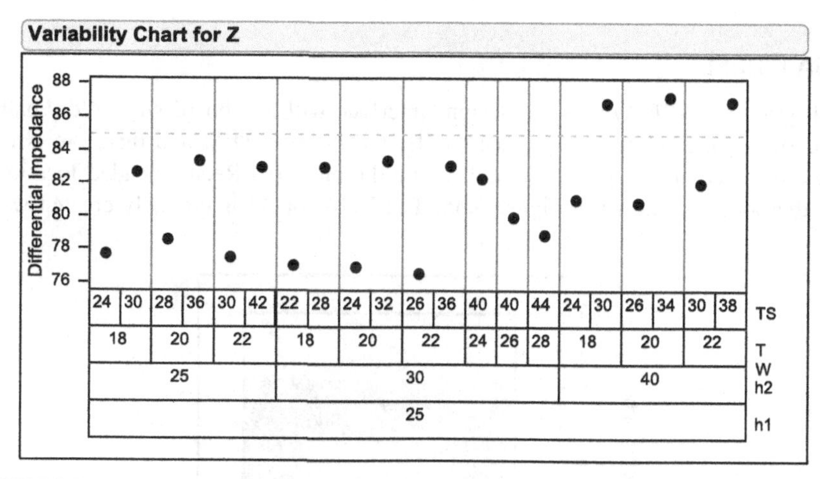

FIGURE 3.33

T-line parametric sweep against differential insertion loss.

coupling and layout pitch. A balance among layout, signal integrity, and other interfaces is needed in selecting the optimal dimensions. Geometries for decreased loss and decreased crosstalk consume layout resources or require thicker dielectrics that can lead to very significant crosstalk impacts to signal-ended interfaces like double data rate memory (DDR).

Impedance must be controlled for high-speed signaling, and it can be expected that today's manufacturing can meet a 15 percent impedance target. Variability around nominal physical and electrical parameters is dependent on the packaging vendor. The same methodologies for creating impedance corners on PCB transmission lines (earlier in this chapter) can be applied to package traces.

PTH VIA

Modeling procedures and optimization for package PTHs are largely similar to those of PCB PTHs (discussed earlier in this chapter). In the same way, these vertical PTH structures need full-wave 3D modeling to assess low-impedance discontinuities and crosstalk. The primary effect of the PTH is a return loss increase from low-impedance discontinuity. The impact on eye height may be sensitive to the PTH location, depending on the design and RX equalization capability. As a rule of thumb, locating the core PTH immediately below the bump leads to the strongest discontinuity and lowest margins. An example of the sensitivity of the PTH location between 20 and 80 percent of the total length is shown in Figure 3.34. In the example the eye height is discovered to be reduced when the core PTH is located precisely in the middle of the design. The severity of the reduced eye opening is dependent on design and EQ capability.

BGA MODEL

Ball grid array (BGA) package designs interface with the board with solder balls. The electrical field at the interface (solder ball) is non-TEM and therefore is not the most accurate location to establish simulation ports. Recommended locations for the ports are shown in Figure 3.35. The BGA model is typically created with

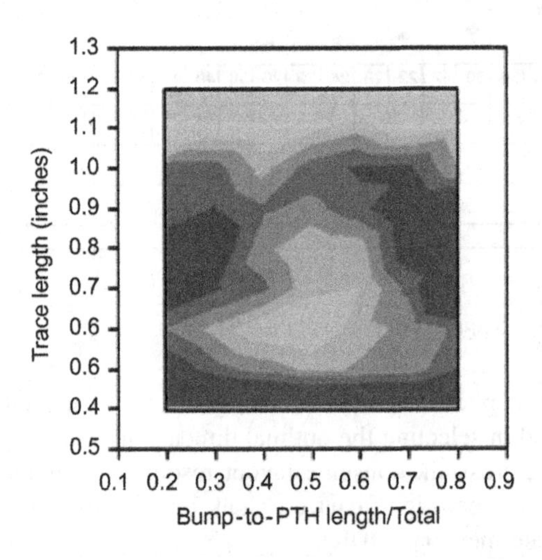

FIGURE 3.34

Eye height in response to core PTH location.

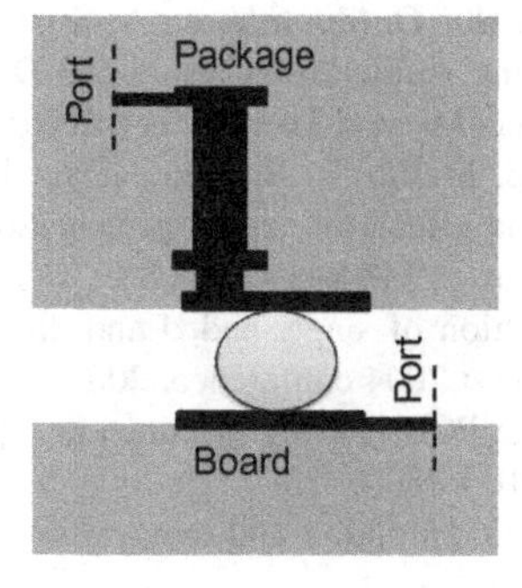

FIGURE 3.35

Port locations for BGA model.

a 3D layout extraction of the actual package design. The 3D extraction must include the vertical path until a TEM field is established, such as a transmission line. Therefore the extraction may include several layers of microvia and core PTH. At the board interface the solder ball, pad, and surface transmission line are added to the extracted package design.

SIGNAL SELECTION FOR 3D PACKAGE STRUCTURES

A review of the signals present in a post-layout design is necessary to determine the appropriate signals for modeling. The review is meant to locate the patterns, exceptions, and potential worst-case signals in a design for modeling. In most cases a common and repeating vertical pattern is the first choice in modeling. Exceptions to the pattern with higher crosstalk may not be the best choice in modeling if the trace length is significantly shorter than other signals. Such signals with shorter lengths may result in a greater eye opening. The number of signals to extract should be equal to or greater than the number of signals present in other available interconnect models (generally at least three differential pairs). Additional signals may be modeled to create multiple crosstalk analysis configurations from the model, where unused signals may be terminated at the time of the full link simulation.

REFERENCES

[1] Djordjevic A, Biljic R, Likar-Smiljanic V, Sarkar T. Wideband frequency-domain characterization for FR-4 and time-domain causality. IEEE Trans Electromagn Compat 2001;43(4):662–7.

[2] Bogatin E, DeGroot D, Huray P, Shlepnev Y. Which one is better? Comparing options to describe frequency dependent losses. DesignCon 2013.

[3] Hall S, Pytel S, Huray P, Hua D, Moonshiram A, Brist G, et al. Multigigahertz causal transmission line modeling methodology using a 3-D hemispherical surface roughness approach. IEEE Trans Microw Theory Tech 2007;55(12):2614–24.
[4] Pytel S, Huray P, Hall S, Mellitz R, Brist G, Meyer H, et al. Analysis of differing copper treatments and the effects on signal propagation. In: Electronic components and technology conference. p. 1144–149, 2008.
[5] Brist G. Design optimization of single-ended and differential inpedence PCB transmission lines. In: PCB West 2004 conference, 2004.
[6] Coonrod J. Understanding PCBs for high-frequency applications. Printed Circuit Des Fab Circuits Assembly 2011;28(10):25.
[7] Loyer J, Kunze R, Brist G. Humidity and temperature effects on PCB insertion loss. DesignCon 2013.
[8] Hinaga S. Thermal effects on PCB laminate material dielectric constant and dissipation factor. IPC Expo/APEX 2010. p. S16-1, 2010.
[9] Deutsch A, Surovic C, Krabbenhoft R, Kopcsay G, Chamberlin B. Prediction of losses caused by roughness of metallization in printed-circuit boards. In: IEEE transactions on advanced packaging. p. 279–87, 2007.
[10] Brist G. Re: dk standard deviations, e-mail to G. Brist, 2014. 29 Aug. [29.08.14].
[11] ASHRAE Technical Committee. Thermal guidelines for data processing environments – expanded data center classes and usage guidance, [Online]. Available: <http://ecoinfo.cnrs.fr/IMG/pdf/ashrae_2011_thermal_guidelines_data_center.pdf>; 2011 [30.08.14].
[12] Hamilton P, Brist G, Barnes Jr. G, Schrader J. Humidity-dependent loss IN PCB substrates. Printed Circuit Des Manuf 2007;24(6):30.

CHAPTER

Link circuits and architecture

4

The greatest ideas are the simplest.
— William Golding

This chapter describes the physical layer link circuits and architecture used in both sides of the channel, commonly referred to as transmitter (TX) and receiver (RX), or transceiver as a whole. These building blocks are mixed-signal circuits in nature. A transmitter converts a digital data stream from the system agent (SA) to high-speed analog signals, whereas a receiver recovers data and the clock, converts analog data back to a digital data stream and then sends it to the SA on the receiver side.

TYPES OF LINK CIRCUIT ARCHITECTURES

There are two commonly used architectures in high-speed serial links: embedded clock architecture, where the clock is recovered from the data stream, and forwarded clock architecture, in which a dedicated clock is sent in parallel with data.

Throughout the chapter, the PCI Express (PCIe) specifications [1] are often mentioned as a practical example to facilitate the discussion. However, the concepts introduced in this chapter can easily be applied to other high-speed I/O systems as well.

EMBEDDED CLOCK ARCHITECTURE

Figure 4.1 shows a generic block diagram of an embedded clock architecture. The input data stream from the system agent to the transmitter has, for example, either 8-bit or 10-bit digital data in parallel, as defined by the PCIe specifications. Parallel input to serial output (PISO) converts parallel data into a serial data stream; feed forward equalizer (FFE) applies equalization to the data stream; and the line driver (LD) drives the data stream into an impedance-matched channel. On the receiver side, a continuous-time linear equalizer (CTLE) serves as a linear amplifier to increase the signal amplitude, as well as shape the waveform by offering a peaking at high frequencies. Automatic gain control (AGC) is then used to adjust the data amplitude to a proper level before entering into a decision feedback equalizer (DFE) block, which performs further equalization and

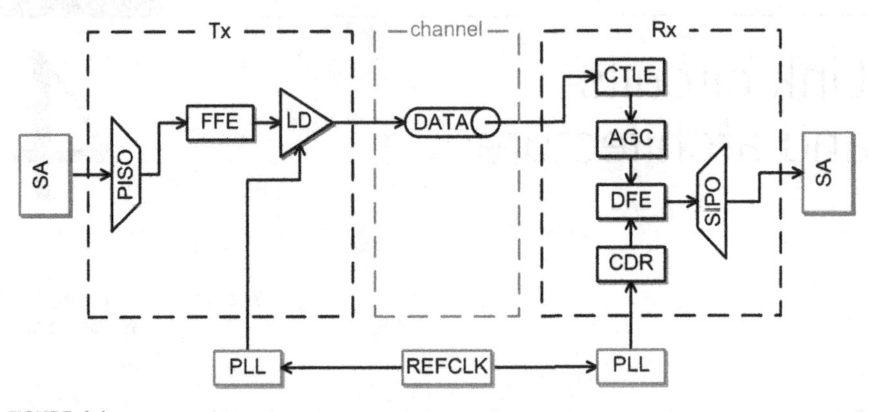

FIGURE 4.1

Block diagram of an embedded clock architecture.

sampling. Serial input to parallel output (SIPO) down-converts data to a parallel bitstream and then feeds data into the SA on the receiver side. Clock and data recovery (CDR) is used to retrieve clock information from the data and use it to sample the data.

In addition, a phase-locked loop (PLL) is required to provide clocks for the transceiver from the reference clock (Refclk). As defined in the PCIe specifications, transmitters and receivers can share one single Refclk as shown in the figure. This is known as common Refclk architecture. They can also use two independent reference clocks, known as separate Refclk architecture. The links that support separate Refclk architecture need to tolerate the parts-per-million (PPM) difference between two reference clocks. PCIe specifications define +/−300 ppm of the nominal frequency of 100 MHz as the maximum tolerance between two reference clocks.

FORWARDED CLOCK ARCHITECTURE

Figure 4.2 shows forwarded clock architecture. Compared to embedded clock architecture, forwarded clock architecture has an additional lane for delivering the clock. Since the same clock used to send data is now available at the receiver end, if the delay between the data and the clock is small enough, the jitter introduced by the clock can be canceled at the receiver. As a result, forwarded clock architecture in general offers better jitter performance than embedded clock architecture. The disadvantage of forwarded clock architecture is that it requires one additional lane—that is, a differential pair of wires—that consumes the real estate of the chip and extra power consumption. Such impact can be minimized by sharing the forwarded clock lane with multiple data lanes. Furthermore, the implementation of the forwarded clock receiver is quite different from the data receiver, which translates to additional resources in design and validation.

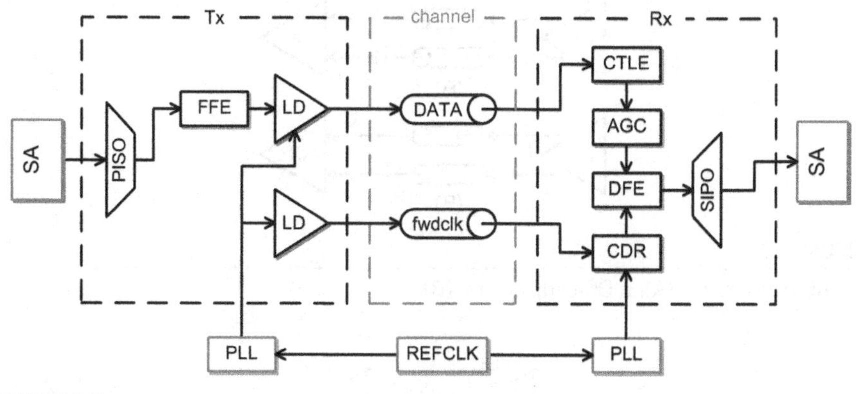

FIGURE 4.2

Block diagram of a forwarded clock architecture.

The rest of the chapter describes individual blocks used in high-speed I/O links. The goal is to describe the function and design considerations of each block, with some examples of the circuit implementation in details.

TERMINATION

A termination refers to the component (usually a resistor) used at both ends of the link to terminate the channel in order to reduce the reflection of the channel. Many high-speed I/O standards, such as PCIe, specify that the termination be 50 ohms. Termination is also used as receiver detection (RXDetect), to indicate whether the receiver is ready to receive data from the transmitter. This section first discusses the considerations of AC coupling versus DC coupling, followed by the types of terminations. Then a typical implementation of the termination circuit is described, and the concept of calibration is introduced. Finally, the implementation of the termination detection circuit is described.

DC AND AC COUPLING

Links between two high-speed I/O blocks can be either DC coupled or AC coupled, as shown in Figure 4.3. Industry standards, such as PCIe, request that links be AC coupled, which has the advantage of isolating common-mode voltage levels between RX and TX. This allows both TX output and RX input common-mode voltage to be set independently, giving maximum flexibility to the design of the TX driver and RX input stages. On the other hand, a DC-coupled link does not require the coupling capacitors on board or on package, which has advantages from the perspectives of component count, real estate, and cost. The DC-coupled link also makes the termination detection circuit relatively easy to design and

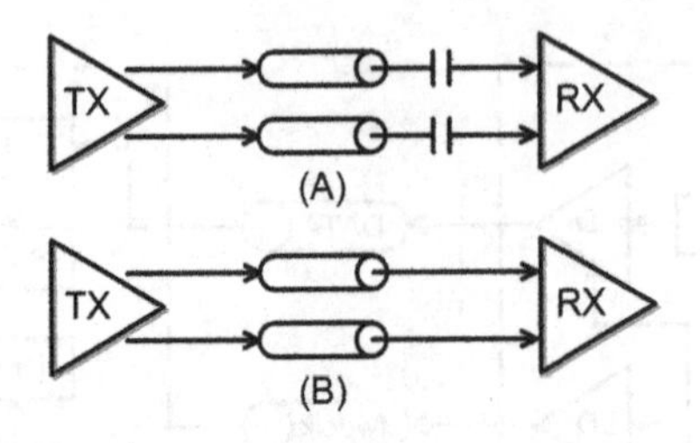

FIGURE 4.3

An AC-coupled link (A); a DC-coupled link (B).

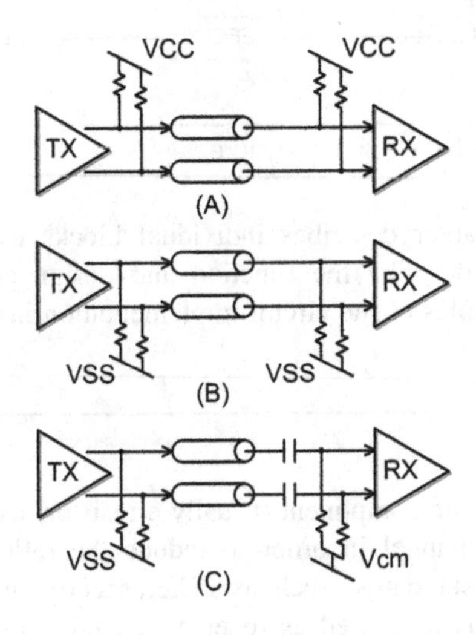

FIGURE 4.4

Terminated to VCC (A); terminated to VSS (B); terminated to Vcm (C).

hence more robust. However, to guarantee the interoperability of a DC-coupled link, the specification of the TX output and RX input common-mode voltage needs to be clearly defined, hence the termination type.

TERMINATION TYPE

A termination can be tied to power supply (VCC), ground (VSS), or some DC voltage in between, as shown in Figure 4.4. For an AC-coupled link, termination type can be independently chosen between RX and TX. In this case one may choose to terminate RX to an input common-mode voltage Vcm, as shown in Figure 4.4(C). Otherwise a DC current path may be formed from the receiver input pins to the power supply used for termination, consuming additional power. For the DC-coupled case, termination type has to be matched between TX and RX, or an undesired DC current path will be formed between VCC and VSS. Some designs have

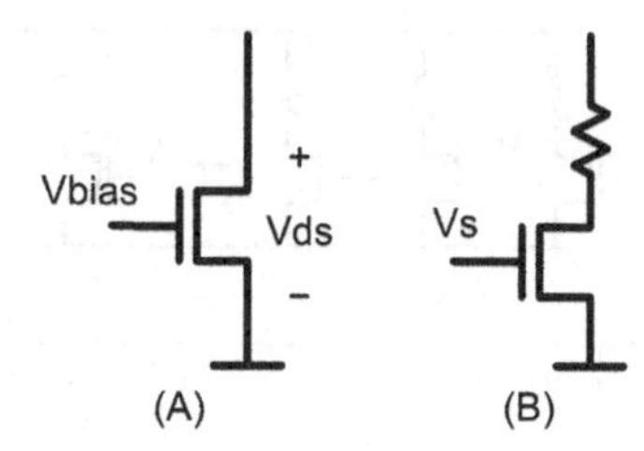

FIGURE 4.5

Active transistor termination (A); passive resistor termination (B).

an on-die AC-coupling capacitor inside the receiver to decouple receiver input pins from the internal receiver input nodes, such as CTLE inputs, which essentially turn a DC-coupled link into an AC-coupled one to accommodate different termination types of remote transmitters, as well as flexibility in the choice of RX input Vcm. However, because of the size limitation of on-die capacitors, the loss of gain due to the AC-coupling capacitor needs to be taken into consideration.

TERMINATION CIRCUITS

Components used for the termination circuit can be either passive resistors or active transistors. As shown in Figure 4.5(A), one can bias a transistor in the linear region to make it behave like a resistor. Such implementation is commonly used as internal loads in many circuits. However, since the requirement of the dynamic range—that is, the voltage swing across a component—is quite high for the termination used in high-speed I/O links, using an active transistor may lead to unacceptable nonlinearity because the effective resistance is a strong function of drain-source voltage (Vds) of the termination transistor. Therefore, common implementation is a resistor in series with a transistor, as shown in Figure 4.5(B), in which the transistor simply behaves as a switch to turn the resistor on or off.

The accuracy requirement for the termination resistor is usually loose. For example, PCIe specifications define the range between 40 ohms and 60 ohms. At high frequencies the link performance is usually bounded by the pad capacitance, where the return loss (RL) is used as the design specification. The termination resistor can be implemented off-chip. However, this increases the component count, and it may not be practical to place the off-chip resistor close enough to the I/O pins. As a result, most termination resistors are implemented on-die. For the process technologies in which precision resistors are available, this can be easily done by a single resistor with a transistor switch as mentioned above. For those technologies where precision resistors are not an option, the termination circuits are implemented as a set of programmable resistors with some kind of trimming or calibration capability. It consists of a group of resistor legs in parallel, each of which can be turned on or off independently such that the total resistance is close enough to the target value, say, 50 ohms. Figure 4.6 shows

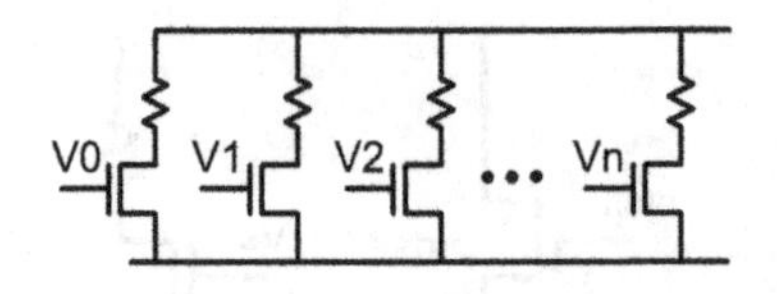

FIGURE 4.6

Programmable resistor termination circuit.

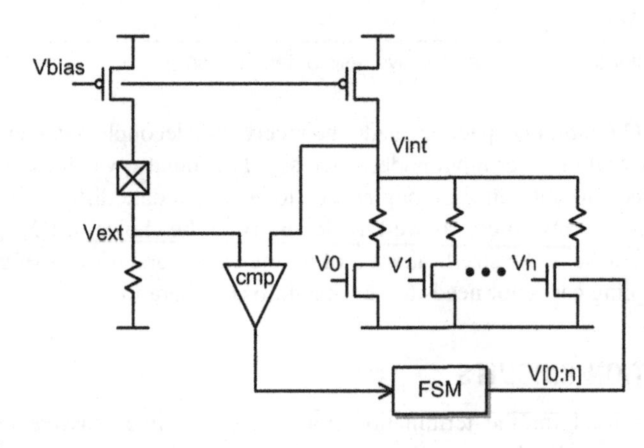

FIGURE 4.7

Termination calibration circuit.

such an implementation. A calibration circuit is required to maintain the accuracy of the resistance value across the variations of silicon process skew, VCC, and temperature (PVT).

TERMINATION CALIBRATION CIRCUITS

Figure 4.7 shows one possible implementation for the termination calibration circuits. A pair of bias transistors is used to provide identical bias current, one of which is connected to an external precision resistor, and the other to an identical copy of the termination circuits at TX output and RX input. A low-offset comparator is used to compare two voltages across the resistors, and a digital finite state machine (FSM) is used to adjust the programmable resistor set until two voltages are the same. Since the bias current is also the same, the on-die programmable resistor will have the same value as the off-die precision resistor. Then the digital code can be sent to all the lanes that share the same termination circuits. If the process variation and temperature gradient are small enough across the die, all termination circuits on the same die can share the same digital code. In this case only one termination calibration circuit is needed for the entire high-speed I/O module. Considering the resistance variation across PVT, calibration circuits can be activated once during the power-on to compensate for the process variation. VCC variation is usually not a concern for the passive resistor termination circuit. If temperature drift also needs to be compensated, the calibration circuits can be reactivated periodically while the link is in operation. In such a case, the implementation of thermometer coding is required to eliminate undesired glitches when turning on and off the resistor legs.

TERMINATION DETECTION CIRCUITS

As mentioned previously, termination can also be used as an indication of the existence or readiness of receiver circuits. Many high-speed I/O standards define the specifications for receiver input on resistance (Zrx-on) and receiver input off resistance (Zrx-off). For example, PCIe specifications require Zrx-on to be between 40 and 60 ohms and Zrx-off to be more than 10 kilohms. Then the detection circuit can be implemented in the transmitter to distinguish between 60 ohms and 10 kilohms in order to tell whether the receiver is ready to start or not. Depending on whether it is a DC-coupled or an AC-coupled link, the implementation of detection circuits may be different.

For the DC-coupled case, the detection circuit can be as simple as the one shown in Figure 4.8. A pull-high resistor (if termination type is VSS) is implemented at TX output. During the detection phase, the TX driver is in high-impedance mode, and, therefore, a resistor divider is formed between the pull-high resistor at TX and the termination at RX. Then a comparator can be used to compare TX output voltage against a known reference voltage Vref to detect the RX termination. If RX termination exists ($\sim$50 ohms), TX output voltage is lower than Vref; if RX termination is off (>10 kilohms), TX output voltage is higher than Vref.

For the AC-coupled case, since no DC path is available to form a resistor divider, one way to detect RX termination is by the difference of the time constant between the on and off states of RX termination. Curve (a) in Figure 4.9 shows the case where the RX termination is on ($\sim$50 ohms). When the pull-high resistor attempts to pull TX output voltage high, RX termination will form a low-impedance path at the RX side of the AC-coupling capacitor such that TX output voltage is slowly charging up. If RX termination is off (>10 kilohms), the

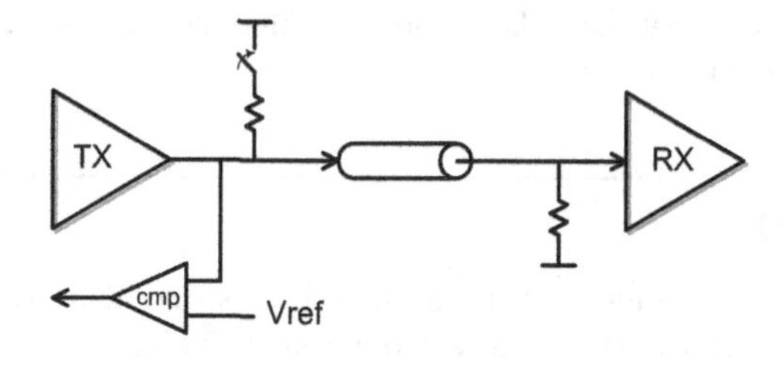

FIGURE 4.8

Termination detection circuits.

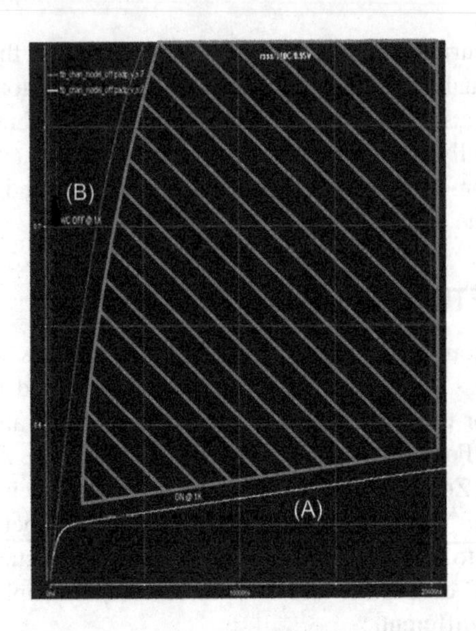

FIGURE 4.9

TX output voltage in RX detection in an AC-coupled link.

RX input is high impedance such that the AC-coupling capacitor is dangling as if no AC-coupling capacitor existed. Therefore, when TX is trying to pull the output voltage high, it can be charged up within a much shorter time, as indicated by curve (b) in Figure 4.9.

The time constant can be estimated as:

$$\tau = RC \tag{4.1}$$

where R is the RX termination resistance, and C is the AC-coupling capacitance. Since both curves are moving as a function of time, the detection circuit needs to set the comparator voltage Vref in between two curves and the output is only valid during the sampling window in between, as illustrated by the shaded area in Figure 4.9.

Termination detection circuits can also be implemented by sending a pulse to effectively form a resistor divider across the AC-coupling capacitor. PCIe specifications limit the height of the pulse to no more than 600 mV in order not to damage the receiver input circuits.

TRANSMITTER

The purpose of a transmitter is to send a high-speed data stream through a channel to a receiver by delivering a large enough voltage swing at the pins of the transmitter. Figure 4.10 shows a typical transmitter implementation. Three major functions are performed in this transmitter: parallel-to-serial conversion,

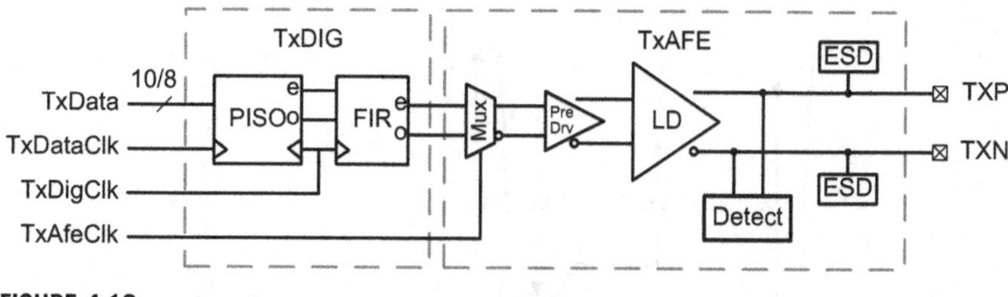

FIGURE 4.10

Block diagram of a transmitter.

feed-forward equalization (FFE), and line driving. Usually the input of the transmitter is a parallel bus of relatively low-frequency data from the SA. For example, PCIe specifications define a 10-bit-wide parallel bus for Gen1 (2.5 Gb/s) and Gen2 (5 Gb/s), and an 8-bit-wide bus for Gen3 (8 Gb/s). PISO converts 10- or 8-bit data to serial data as input to the pre-driver. The pre-driver is used to precondition data to the level suitable for the line driver to deliver large enough swing to TX output pins. The FFE can be realized anywhere along the data path. In this particular implementation, it was chosen to divide the data serialization into two stages: PISO up-converts 10- or 8-bit data to even and odd paths at the half rate, and then a 2:1 mux mixes them to a full-rate data stream. Such implementation allows the finite impulse response (FIR) perform the FFE at half rate instead of full rate. Because the transmitter largely consists of digital logic gates, it is usually implemented by a synthesized logic block except for the high-speed data path, as shown in Figure 4.10. How to partition between the digital synthesized logic block and the custom-drawn logic block is usually determined by the process technologies used in the design, as well as the preference of the designers. The advantage of the synthesized logic blocks is that the design is relatively easy to port from one technology to the other. On the other hand, custom-drawn logic blocks can be optimized for a specific technology and hence potentially consume less power.

A few other blocks are shown in the block diagram. ESD (electrical static discharge) diodes are attached at output pads to protect devices connected to the pads. The major concern of ESD in high-speed I/O design is the additional parasitic capacitance at output pads, which has a direct impact on the transmitter performance. The parasitic capacitance from ESD and other circuitry associated with the TX pads has become a limiting factor in the maximum operating frequency a transmitter can achieve. The detection circuit mentioned in the previous section is also tied to output pads, as Figure 4.10 shows.

TRANSMITTER EQUALIZATION

As described in the previous chapter, FFE in a transmitter is used to cancel inter-symbol interference (ISI). For low-loss channels, a 2-tap FFE (one main cursor and one post-cursor) has been found sufficient. For higher loss channels, a

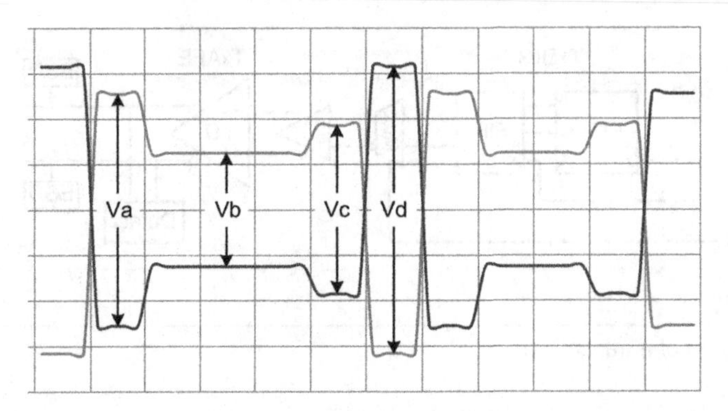

FIGURE 4.11

Definition of TX voltage levels and equalization ratios.

3-tap FFE (one main cursor, one post-cursor, and one pre-cursor) is often required. PCIe specifications require that a transmitter supports 2-tap FFE, also known as de-emphasis, for Gen1 and Gen2 and 3-tap FFE for Gen3.

A 3-tap FFE is typically expressed by:

$$X_n = c_{-1}a_{n+1} + c_0 a_n + c_{+1}a_{n-1} \tag{4.2}$$

where c_i is the tap coefficient and a_i is the data stream. Data a_{n+1} is commonly referred to as the pre-cursor, a_n the cursor, and a_{n-1} the post-cursor. Therefore, the tap coefficients associated with them are called pre-cursor tap coefficient (c_{-1}), main cursor tap coefficient (c_0), and post-cursor tap coefficient (c_{+1}), respectively. Data stream a_i can be either +1 or −1, whereas c_0 is usually a positive number; c_{-1} and c_{+1} have the opposite sign of c_0 (hence usually negative), and $|c_{-1}| + |c_0| + |c_{+1}| =$ maximum swing.

Since a_i is either +1 or −1, there are only eight possible TX output levels, as illustrated in Figure 4.11, where *Vd* is referred to as the maximum swing (or the full scale, FS), which occurs when the data pattern is (−1, +1, −1) or (+1, −1, +1), and *Vb* is the DC pattern (or the low frequency pattern, LF), which occurs when the data pattern is (+1, +1, +1) or (−1, −1, −1). *Va* is the post-cursor amplitude, which occurs when the data pattern is (−1, +1, +1) or (+1, −1, −1), and *Vc* is the pre-cursor amplitude, which occurs when the data pattern is (+1, +1, −1) or (−1, −1, +1).

PCIe specifications define:

$$\text{de-emphasis} = 20\log_{10}\left(\frac{Vb}{Va}\right)$$

$$\text{pre-shoot} = 20\log_{10}\left(\frac{Vc}{Vb}\right) \tag{4.3}$$

$$\text{and boost} = 20\log_{10}\left(\frac{Vd}{Vb}\right)$$

PS DE Boost		C+1								
		0/24	1/24	2/24	3/24	4/24	5/24	6/24	7/24	8/24
C−1	0/24	0.0 0.0 0.0	0.0 −0.8 0.8	0.0 −1.6 1.6	0.0 −2.5 2.5	0.0 −3.5 3.5	0.0 −4.7 4.7	0.0 −6.0 6.0	0.0 −7.6 7.6	0.0 −9.5 9.5
	1/24	0.8 0.0 0.8	0.8 −0.8 1.6	0.9 −1.7 2.5	1.0 −2.8 3.5	1.2 −3.9 4.7	1.3 −5.3 6.0	1.6 −6.8 7.6	1.9 −8.8 9.5	
	2/24	1.6 0.0 1.6	1.7 −0.9 2.5	1.9 −1.9 3.5	2.2 −3.1 4.7	2.5 −4.4 6.0	2.9 −6.0 7.6	3.5 −8.0 9.5		
	3/24	2.5 0.0 2.5	2.8 −1.0 3.5	3.1 −2.2 4.7	3.5 −3.5 6.0	4.1 −5.1 7.6	4.9 −7.0 9.5			
	4/24	3.5 0.0 3.5	3.9 −1.2 4.7	4.4 −2.5 6.0	5.1 −4.1 7.6	6.0 −6.0 9.5				
	5/24	4.7 0.0 4.7	5.3 −1.3 6.0	6.0 −2.9 7.6	7.0 −4.9 9.5					
	6/24	6.0 0.0 6.0	6.8 −1.6 7.6	8.0 −3.5 9.5						

Full swing limit

FIGURE 4.12

TXEQ coefficient space triangular matrix example.

Figure 4.12 shows the relationship between the normalized tap coefficients—that is, when $|c_{-1}| + |c_0| + |c_{+1}| = 1$—and levels of de-emphasis, pre-shoot, and boost. Columns correspond to a specific post-cursor tap coefficient (c_{+1}), while rows correspond to a pre-cursor tap coefficient (c_{-1}). Three numbers inside each tile are de-emphasis, pre-shoot, and boost, as denoted in the upper left corner of the figure. As the dashed line indicates, tiles along the diagonal line have the same boost—that is, *Vb* is kept constant—while de-emphasis and pre-shoot vary. PCIe specifications bound the maximum boost to be 9.5 dB because beyond that point, *Vb* will be too small to meet the minimum low-frequency swing requirement of 250 mV. The grayed tiles are 10 presets defined by PCIe specifications to allow a transmitter to perform a specific FFE function in order to maximize the link performance for a given channel and receiver design.

TRANSMITTER DATA PATH

This section describes the implementation of PISO, FIR, 2:1 mux, and the pre-driver. One possible implementation of PISO and TX FIR for a half-rate architecture is shown in Figure 4.13. Ten-bit parallel data are aligned along the even and odd paths, and a 2-UI (unit interval) clock is used to clock the FFs (flip-flops). Odd_sel and even_sel are the data pattern $[a_{n+1}, a_n, a_{n-1}]$. Since there are only eight possible outcomes (L0, L1, L2, L3, −L0, −L1, −L2, and −L3), they can be pre-computed based on tap coefficients, and then an 8-input mux can be used to select one of the pre-computed values as the equalized TX data (fir_odd and fir_even). In this particular case, six bits are used to represent the equalized data. Depending on the technology used, this block can be either a synthesized logic block or a custom-drawn high-speed digital block.

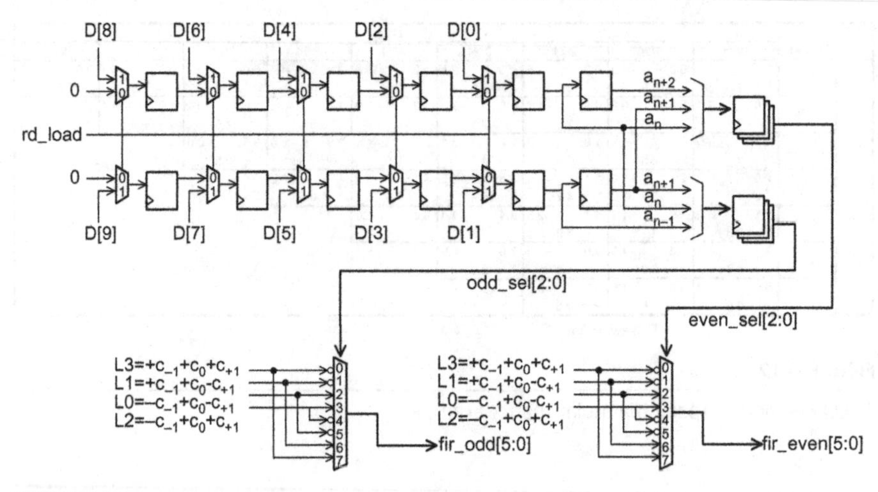

FIGURE 4.13

The implementation of PISO and FIR.

The equalized half-rate TX data then goes through another level of 2:1 mux to generate the full-rate data stream. Figure 4.14 shows a typical implementation of a 2:1 mux and its timing diagram. When clock phase 1 is high, data out equals data phase 1 and vice versa. Since both edges of the clocks are used to generate the final data output, the duty cycle of the clocks is critical in this design. Usually a duty-cycle correction (DCC) circuit is implemented along the clock path to guarantee a good duty cycle.

The output of the 2:1 mux is usually CMOS level, but it may not be strong enough to drive the relatively large input load of the line driver. Therefore, a pre-driver is included along the data path as a buffer to provide the driving strength. The pre-driver can be as simple as a CMOS inverter. The driving strength can be made programmable to fine-tune the rise time and fall time, as required by some high-speed I/O specifications. If the driving strength of PMOS and NMOS can be programmed independently, one may want to make an asymmetrical rise time and fall time to pre-condition the input waveform into the line driver. This is particularly useful for the current-mode types of line drivers. Figure 4.15 shows one typical CMOS pre-driver design.

CURRENT-MODE DRIVER

There are mainly two types of line drivers: current-mode and voltage-mode. Figure 4.16 shows one implementation of a current-mode driver, which is essentially a group of differential pairs arranged in a binary weighted form and controlled by the equalized TX data (six bits in this case). The equalization—that is, the adjustment of the output swing—is done by directing current to TX rterm

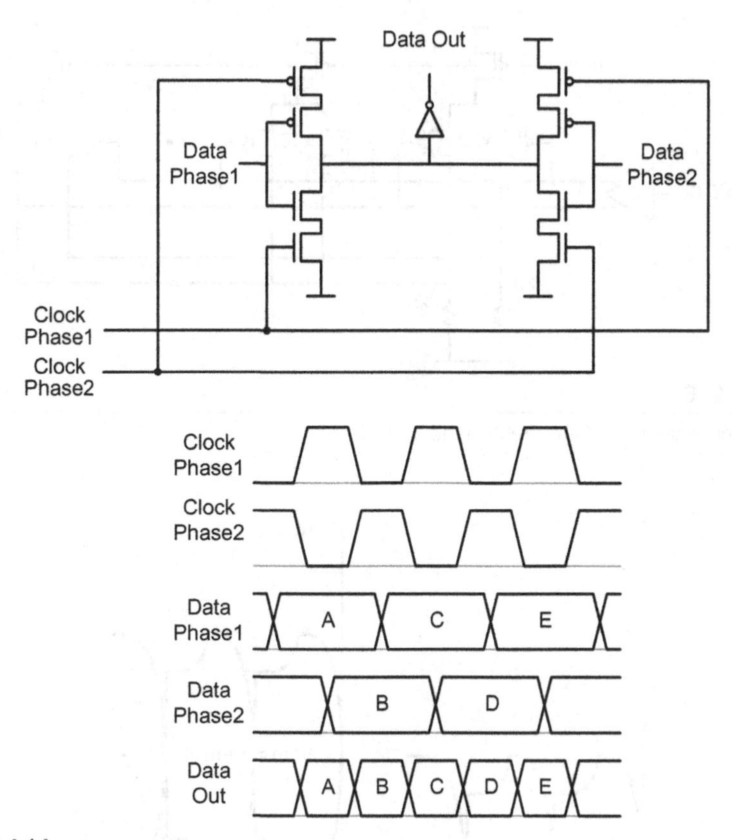

FIGURE 4.14

The implementation of a 2:1 mux and its timing diagram.

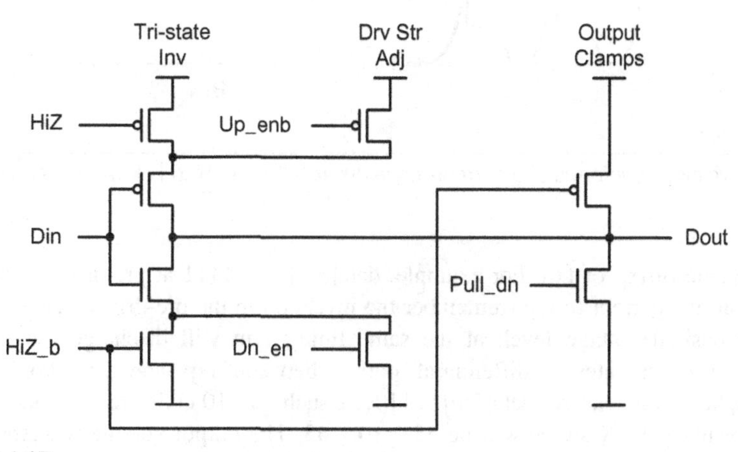

FIGURE 4.15

The implementation of the pre-driver.

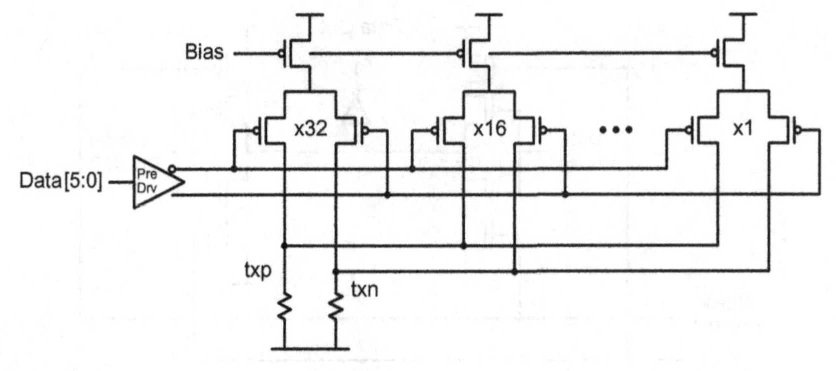

FIGURE 4.16

The implementation of a current-mode driver.

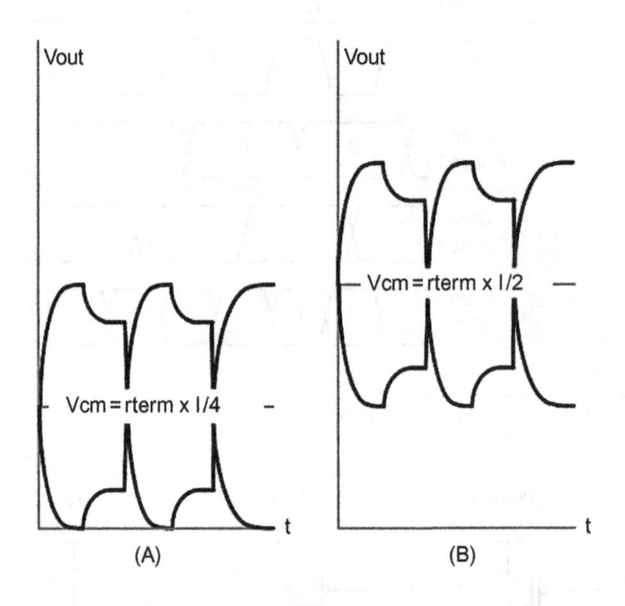

FIGURE 4.17

Output voltage waveforms of a current-mode driver: a DC-coupled link (A); an AC-coupled link (B).

on the side of txp or txn. For example, data[5:0] = 111111 at pre-driver input will direct all the current to txp (remember the inversion in the pre-driver), charging txp to the desired voltage level; at the same time rterm will discharge txn to VSS and hence generates a differential output between txp and txn. For 3.5-dB de-emphasis, one can set data[5:0] = 110101 such that 10 cells are switched to txn. As a result, the TX swing will be 53 − 10 = 43. The output voltage waveform can be seen in Figure 4.17(A) for a DC-coupled link. For an AC-coupled link such as

PCIe, due to the AC-coupling cap between TX and RX, the common-mode voltage of the TX output will be shifted up by the amount of (rterm × I/4), as illustrated in Figure 4.17(B). This is because the direct current is fully consumed by the TX rterm, whereas the alternating current still splits between RX and TX terminations.

VOLTAGE-MODE DRIVER

The other type of line driver is the voltage-mode driver. Figure 4.18 shows the conceptual diagram of these two types of drivers. Assuming $R_{tx} = R_{rx} = R$, then to achieve the same output swing V_{rx} between these two types of drivers, the current-mode driver requires:

$$I_{tx_i} = 2 \times \frac{V_{rx}}{R} \tag{4.4}$$

while the voltage-mode driver only needs:

$$I_{tx_v} = \frac{V_{rx}}{R} \tag{4.5}$$

Therefore, theoretically voltage-mode driver can achieve 50 percent power saving compared to the current-mode driver.

In the voltage-mode driver, the output swing V_{rx} is directly proportional to the source voltage on the transmitter side:

$$V_{rx} = \frac{V_{tx}}{2} \tag{4.6}$$

which sets the maximum output swing for the voltage-mode driver. Fortunately, if a 1-V power supply is used, the maximum peak-to-peak differential swing is also 1 V, which falls into the requirement of many high-speed I/O links such as PCIe. In fact, VCC may be one of the most stable voltage references on die to sink current due to its low impedance. In voltage-mode driver design, careful attention needs to be paid to the supply ripples introduced by the sinking and sourcing current of the driver so that the supply-induced jitter can be estimated as part of jitter budget.

Similar to the current-mode driver, one of the requirements of the voltage-mode driver is equalization. In addition, it is desirable to keep the output

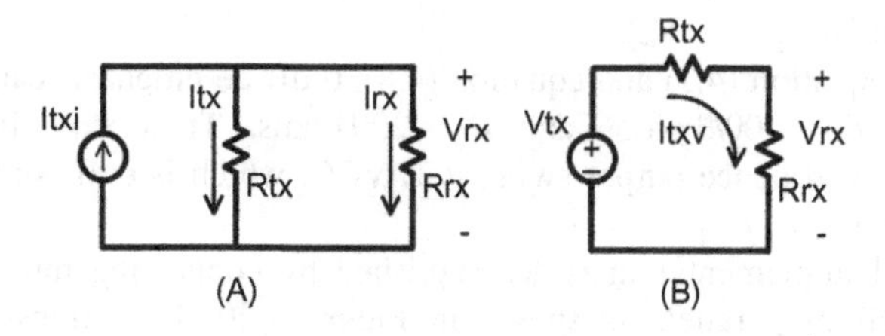

FIGURE 4.18

Conceptual diagram of line drivers: current-mode driver (A); voltage-mode driver (B).

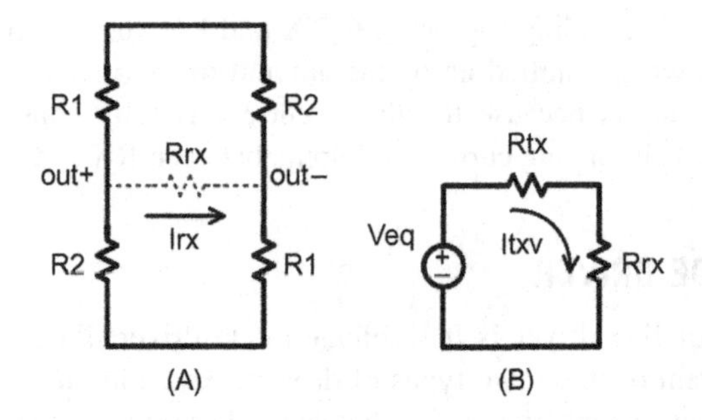

FIGURE 4.19

The topology of a source-series terminated (SST) driver (A); single-ended equivalent circuit (B).

impedance R_{tx} constant while varying TX swing to perform the equalization. One driver topology to fulfill this goal is the source-series terminated (SST) driver [2]. Figure 4.19 shows such a topology. A pseudo-differential output stage is formed by two pairs of resistors, R_1 and R_2. By properly adjusting the ratio of R_1 and R_2, current can be steered from out+ to out− or vice versa, while keeping the termination constant. For example, if R_1 is 50 ohms and R_2 is near infinity, Figure 4.19(A) is the differential equivalent of the circuit in Figure 4.18(B). In this condition the current will flow from out+ to RX and then back to out−, and the termination is kept at 50 ohms. In the case R_2 is 50 ohms and R_1 is infinity, current will flow from out− to RX and then back to out+ and the termination is still maintained at 50 ohm. In both cases the differential output zero-to-peak voltage is ½ VCC.

The impedance looking into both sides of voltage-mode driver is:

$$R_{tx} = 2(R_1 || R_2) \tag{4.7}$$

Assuming $R_{tx} = R_{rx}$, where R_{rx} is the receiver termination across out+ and out−, then the equivalent circuit of Figure 4.19(A) can be shown as Figure 4.19(B), where:

$$V_{eq} = \frac{VCC(R_2 - R_1)}{(R_2 + R_1)} \tag{4.8}$$

and the output swing is ½ V_{eq}.

Based on equation (4.7) and equation (4.8), 6-dB de-emphasis can be achieved by choosing $R_1 = 200/3$ ohms and $R_2 = 200$ ohms. Then $R_{tx} = 100$ ohms and $V_{eq} = ½$ VCC, and hence output swing is ¼ VCC, which is 6 dB smaller than the maximum swing.

The actual implementation is accomplished by connecting multiple slices of the driver cell in parallel, as shown in Figure 4.20. Four transistor switches can be controlled by data and equalization logic to perform FFE while maintaining the proper termination.

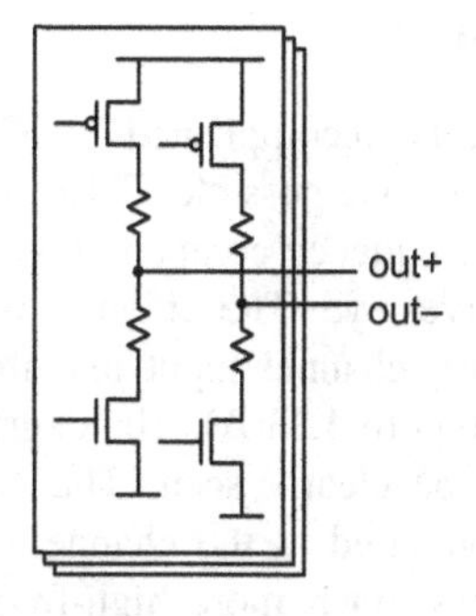

FIGURE 4.20

The implementation of an SST driver.

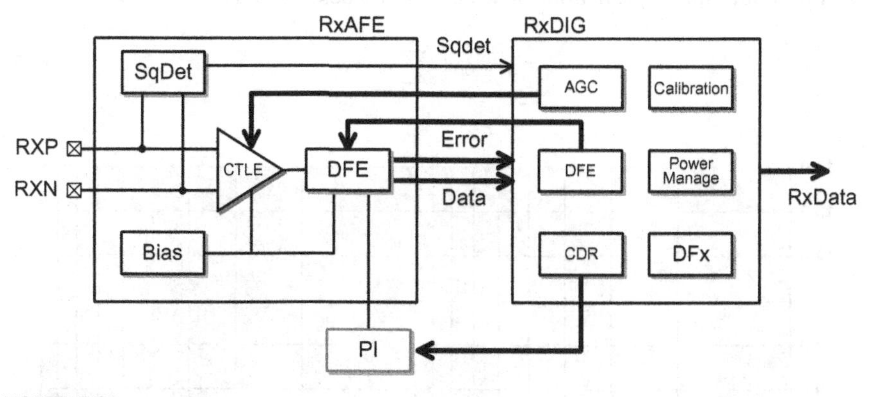

FIGURE 4.21

Block diagram of a receiver.

RECEIVER

The main function of a receiver is to perform equalization and recover data and the clock. Figure 4.21 shows a typical receiver block diagram. A significant portion of the receiver is digital, including control loops (AGC, DFE, CDR), calibration loops, power management, and DFx. The analog portion mainly consists of CTLE and DFE. Squelch detector is an envelope detector to indicate whether incoming data are present or not. Bias provides current and reference voltages for the analog blocks. In this particular implementation, AGC is built inside CTLE. A phase interpolator (PI), which is a separate block from RX, provides the recovered clock to DFE. For an embedded clock architecture, PI takes I/Q clocks directly from PLL; for a forwarded clock architecture, it takes I/Q clocks from the forwarded clock lane.

RECEIVER EQUALIZATION

Receiver equalization is mainly accomplished by CTLE and DFE. Figure 4.22 shows the frequency response of one possible CTLE implementation. Providing frequency peaking near Nyquist frequency compensates for the loss due to the channel and hence opens up the receiver eye. The effect is illustrated in Figure 4.23. The equalized waveform at a lossy channel input and the resulting waveform at the channel output are shown in Figure 4.23(A). The degradation of the signal integrity due to the channel loss can be clearly seen. The output waveform of CTLE is shown in Figure 4.23(B). Compared to the channel output in Figure 4.23(A), the CTLE output waveform shows much more high-frequency content. Those points barely or not even crossing zero are now crossing zero with certain margin. As a result, the receiver eye is opened up for better data sampling.

Figure 4.24 shows a conceptual block diagram of DFE (4 tap is used in this example), the transfer function for which can be described as:

$$y_n = g \times x_n + \sum_{i=1}^{4} c_i \times d_{n-i} \tag{4.9}$$

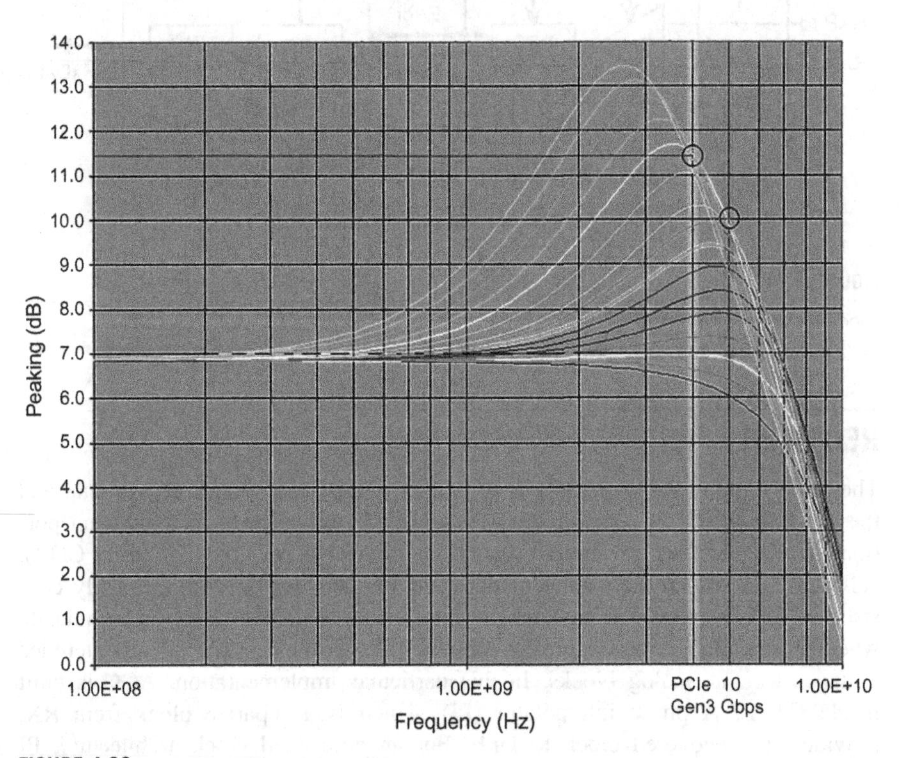

FIGURE 4.22

Frequency response of one CTLE design.

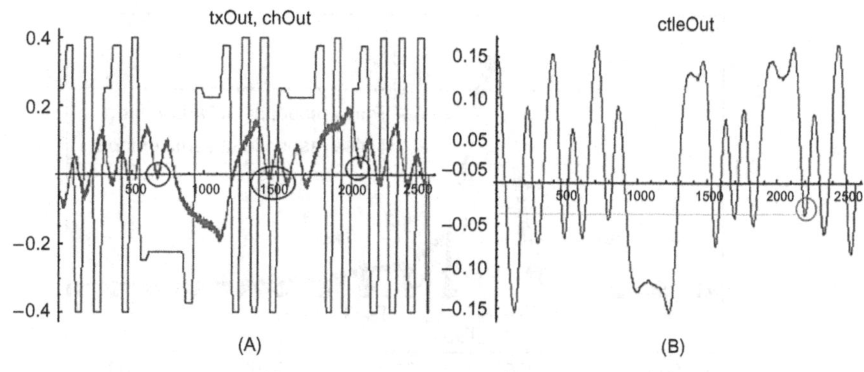

FIGURE 4.23

Waveforms of a lossy channel input and output (A); output waveform of CTLE (B).

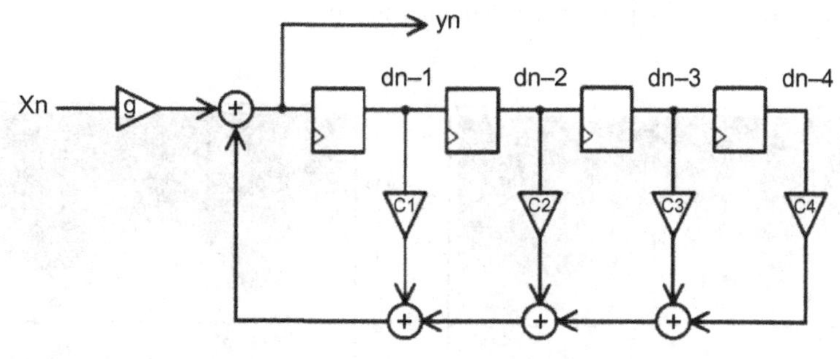

FIGURE 4.24

A conceptual block diagram of DFE.

where x_n is the nth sample of the incoming data, g is the combined gain of CTLE and DFE summer, d_{n-i} ($i = 1, 2, 3, 4$) is the detected previous data or decision with value of $+1$ or -1, c_i is the ith DFE coefficient ($i = 1, 2, 3, 4$), and y_n is DFE summer output, to be sampled by the data sampler. The purpose of DFE is to cancel the post-cursor ISI from the present bit x_n by using previously received bits d_{n-i} ($i = 1, 2, 3, 4$), with a certain set of coefficients c_i computed by the DFE algorithm. This is illustrated in Figure 4.25, where the pulse responses of DFE input (x_n) and DFE summer output (y_n) are both shown. As can be seen from the figure, DFE Tap1 is canceling the first post-cursor ISI by pulling the waveform down, such that the sampled value is near zero. Similarly Tap2 to Tap4 are canceling the second to fourth post-cursors, respectively. In this particular example, all four taps are subtracting the waveforms. Generally speaking, other than Tap1, the rest of the taps can correct the residual ISI in either positive or negative direction.

The effect of residual ISI, or the effectiveness of DFE, can be seen in reconstructed receiver eye diagrams. Figure 4.26 shows the eye diagrams when DFE is off (left) and DFE is on (right) of a particular implementation [3].

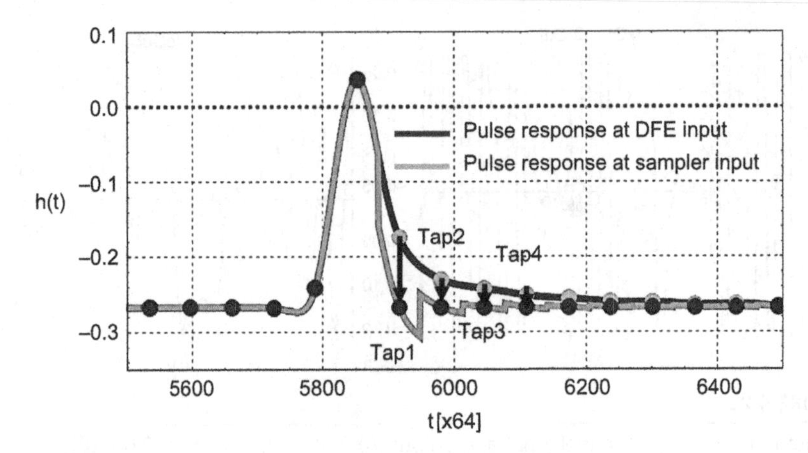

FIGURE 4.25

Pulse responses of DFE input and DFE summer output.

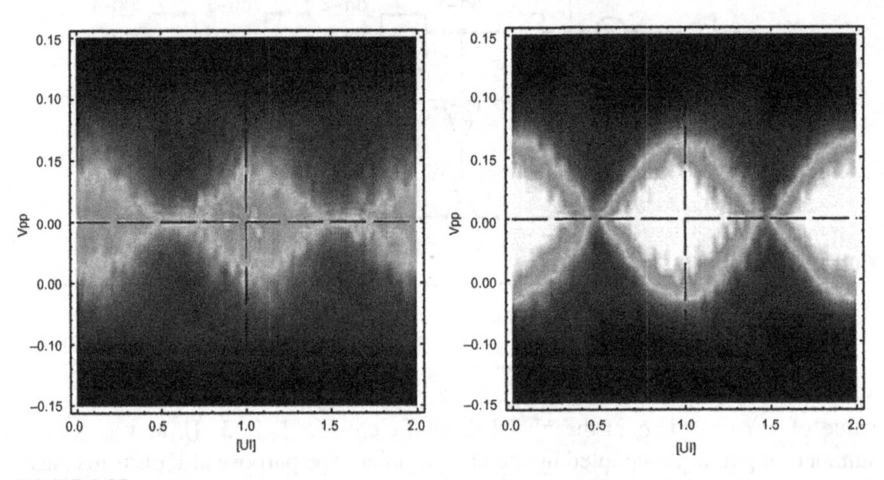

FIGURE 4.26

Receiver eye diagrams when DFE is off (left) and on (right) [3].

RECEIVER DATA PATH

Figure 4.27 shows further details along the receiver data path with the implementation of a 4-tap DFE [3]. The CTLE adjusts signal level by the variable gain amplifier (VGA) and performs equalization by the frequency peaking. DFE is formed by a summer, a data sampler, and current DACs (I-DACs). The data sampler samples the output of the summer and feeds the sampled data (+1 or −1) into I-DACs. The coefficients of DFE ($c_1 \sim c_4$) are computed by a digital DFE loop algorithm and apply to I-DACs such that the output of I-DACs is the representation of ISI of the previous four bits, which is then used to cancel their residual effects to

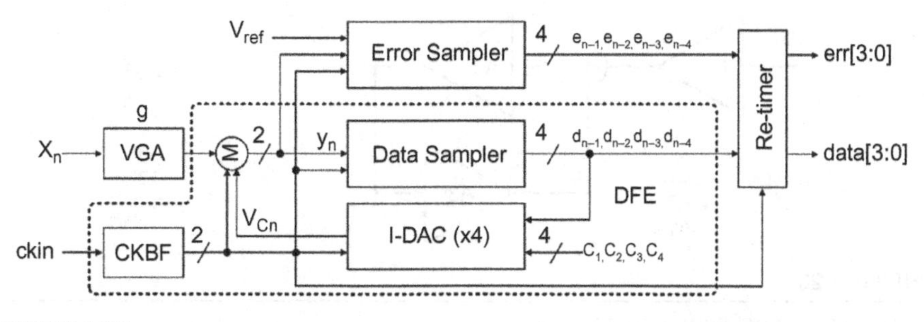

FIGURE 4.27

Block diagram of the receiver data path [3].

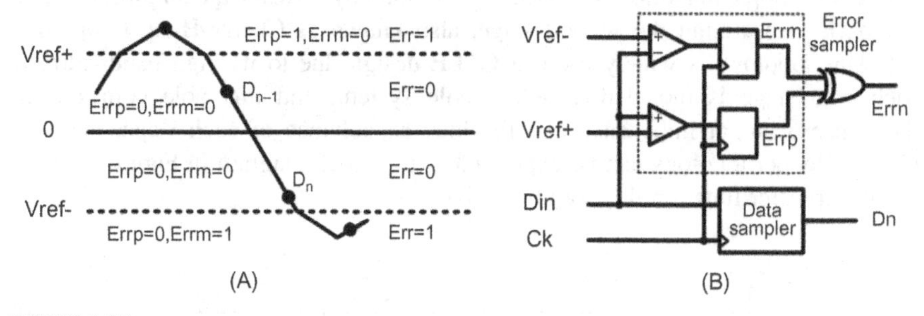

FIGURE 4.28

Operation of data samplers and error samplers (A); The equivalent logic block diagram (B).

the current bit. The summer sums (or subtracts) the outputs of CTLE and I-DACs to achieve such a function. A re-timer is used to align the output of the data sampler before sending data to the receiver digital engine and the system agent. A clock buffer is needed to provide a proper clock to all blocks other than CTLE.

An additional block in this particular implementation is the error sampler [4]. Unlike the data sampler, which compares the sampled data against zero, the error sampler compares data against a nonzero reference to generate the sign of the error with respect to the reference. The sampled error, together with the sampled data, is then used by the receiver digital engine for CDR, AGC, and DFE loops. This idea is illustrated in Figure 4.28(A). The data sampler outputs $D_n = 1$ when incoming data are above differential zero, and $D_n = 0$ when data are below. Two error samplers are implemented, one for V_{ref+} and the other V_{ref-}. The sampled error $E_{rrp} = 1$ when data are above V_{ref+}, and 0 when below; $E_{rrm} = 1$ when data are below V_{ref-} and 0 when above. As a result, $Err_n = 1$ when data are outside of the range between V_{ref+} and V_{ref-}, and 0 when within. As shown in the figure, the quadrant of the sampling point can be determined by reading D_n and Err_n. Figure 4.28(B) shows the equivalent logic operation conceptually.

The next few sections describe receiver blocks in further detail with circuit implementations.

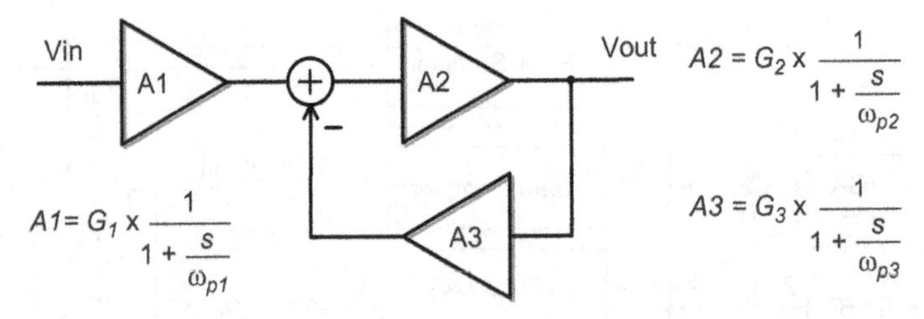

FIGURE 4.29

Block diagram of Cherry-Hooper topology.

CONTINUOUS-TIME LINEAR EQUALIZER

One circuit implementation of CTLE can be done by a two-stage amplifier with a feedback path around the second stage, also known as Cherry-Hooper topology [5]. This topology is widely used in CTLE design due to its high bandwidth. If each gain stage is modeled as a one-pole system, and this pole comes from the nonzero output impedance and the load capacitance of each stage, then the Cherry-Hooper topology can be expressed as the block diagram in Figure 4.29.

The transfer function then can be derived as:

$$\frac{V_{out}}{V_{in}} = G_1 G_2 \frac{\left(1 + \frac{s}{\omega_{p3}}\right)}{\left(1 + \frac{s}{\omega_{p1}}\right)\left[G_2 G_3 + \left(1 + \frac{s}{\omega_{p2}}\right)\left(1 + \frac{s}{\omega_{p3}}\right)\right]} \tag{4.10}$$

Assuming ω_{p1} and $\omega_{p2} \gg \omega_{p3}$, the first-order high-pass filter (HPF) can be given as:

$$\frac{V_{out}}{V_{in}} \approx \frac{G_1 G_2}{1 + G_3 G_2} \cdot \frac{\left(1 + \frac{s}{\omega_{p3}}\right)}{\left(1 + \frac{s}{(1 + G_3 G_2)\omega_{p3}}\right)} \tag{4.11}$$

The detailed circuit implementation is shown in Figure 4.30 [4]. The AGC function is built in as part of the CTLE by two input differential pairs, and their tail-current sources are continuously adjusted by the digital AGC loop. The first stage gain is equal to the difference of g_m of two input differential pairs ($g_{m1} - g_{m2}$). The peaking or the boost level can be archived by programming the variable resistor along the feedback path. The advantage of this implementation is to allow the independent control of the DC gain and frequency peaking and hence reduce undesired interactions between the AGC loop and the receiver equalization.

DECISION FEEDBACK EQUALIZER

Figure 4.31 shows one circuit implementation of DFE with a current integrating summer [3]. This particular implementation has even path (data sampler 0) and

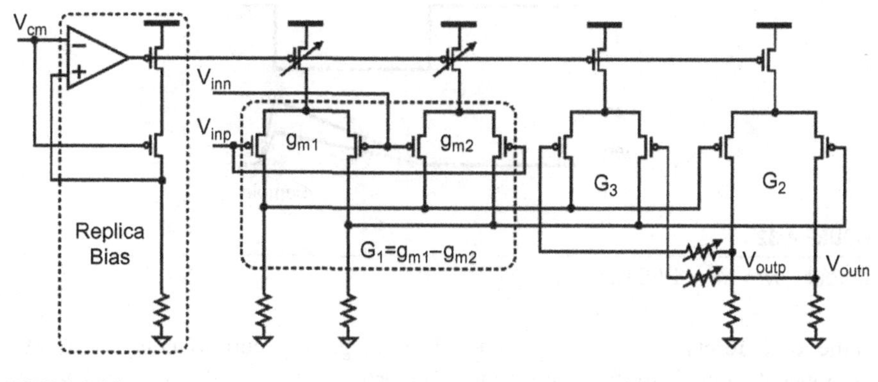

FIGURE 4.30

Circuit implementation of CTLE [4].

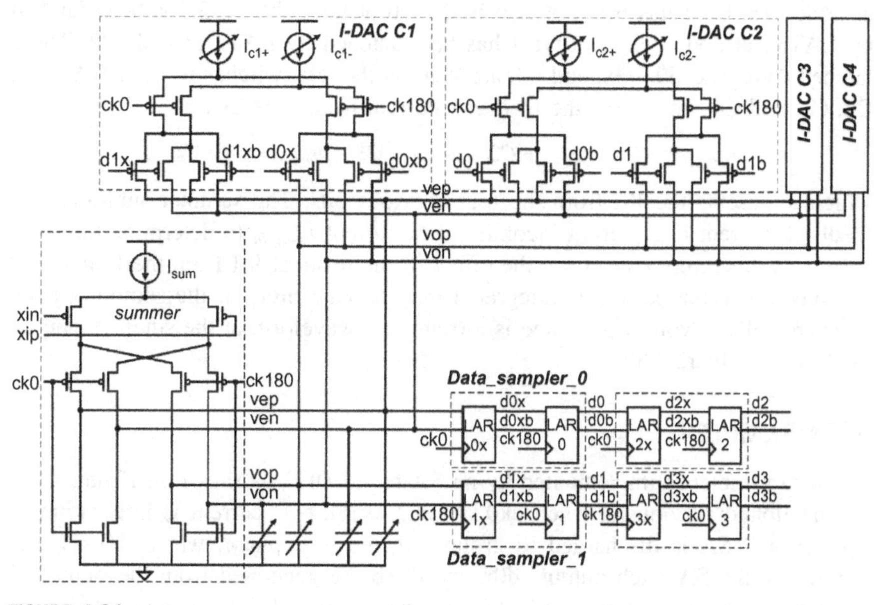

FIGURE 4.31

Circuit implementation of DFE with a current integrating summer [3].

odd path (data sampler 1) for the half-rate operation, which has the advantage that when the even path is evaluating, the odd path is resetting, and the I-DAC current can be reused between even and odd cycles.

The operation of the DFE can be explained as follows: in the negative cycle of ck0—that is, ck0 = 0 and ck180 = 1—data bit 0 is valid at the summer input. The summer then integrates the current to the capacitors on vep/ven, and at the

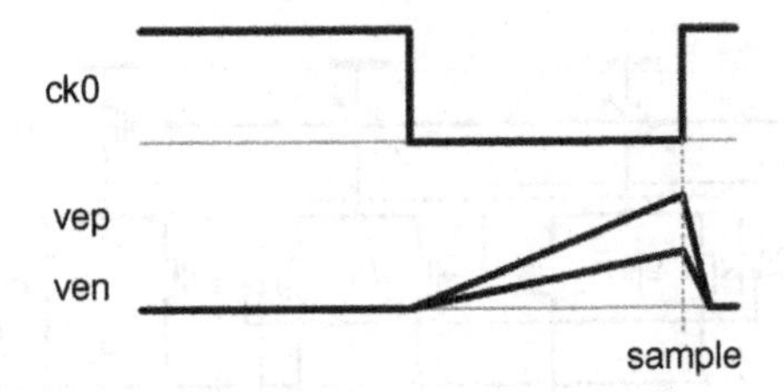

FIGURE 4.32

Output waveform of DFE summer.

same time resets vop/von to ground. The integrating current consists of two components: one from the input of the summer:

$$I_{sum\_in} = g_m \times (v(x_{in}) - v(x_{ip})) \tag{4.12}$$

and the other from I-DACs. During this phase, bit 1 has been latched by LAT1x at node d1x, bit 2 has been latched by LAT0 at node d0, bit 3 has been latched by LAT3x at node d3x, and bit 4 has been latched by LAT2 at node d2. These latched data, d1x, d0, d3x, and d2, are valid at the left switches of the I-DAC C1, C2, C3, and C4. Therefore, the feedback current from I-DACs is:

$$I_{sum\_fb} = C1 \times d_{bit-1} + C2 \times d_{bit-2} + C3 \times d_{bit-3} + C4 \times d_{bit-4} \tag{4.13}$$

which represents the ISI from the four previous bits. The summer subtracts this feedback current I_{sum_fb} from incoming data current I_{sum_in} to determine the present bit and therefore cancel out the effect of the residual ISI from the four previous bits. Since the current is integrated into the capacitors of the summing node vep/ven and vop/von, the voltage is a triangular waveform at the summer output, as shown in Figure 4.32.

DATA SAMPLER

The data sampler is implemented by an SA-based DFF, as shown in Figure 4.33. As mentioned previously, when ck0 = 0 and ck180 = 1, current is integrating to vep/ven and SA is discharged to ground. In the next phase, when ck0 = 1 and ck180 = 0, the SA latch outputs d0x and d0xb are generated from the output of the cross-coupled inverter pair in the SA, while d0 and d0b are generated from the SR latch.

ERROR SAMPLER

Similar to the data sampler, the error sampler also uses an SA-based DFF, as shown in Figure 4.34, except that one extra differential pair is added to introduce the reference voltage. The operation is the same as the data sampler as described in the previous section. For the half-rate architecture there are a total of four error samplers required in the actual implementation.

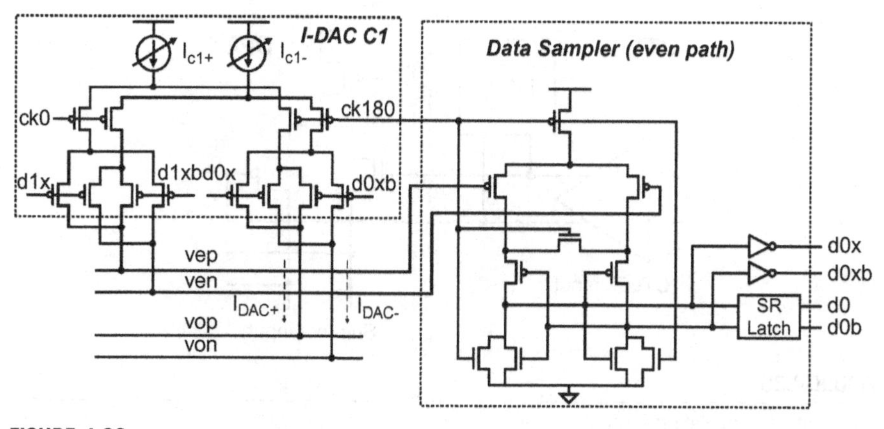

FIGURE 4.33

Circuit implementation of data sampler [3].

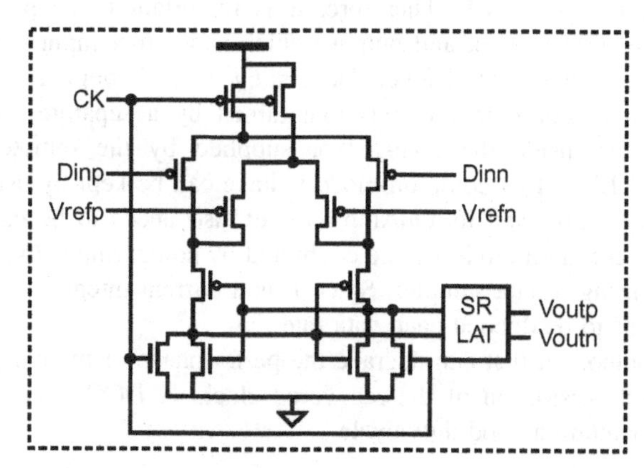

FIGURE 4.34

Circuit implementation of error sampler [4].

RECEIVER CALIBRATION

Because of the device mismatch, offset can be introduced along the receiver data path and hence reduces the opening of the receiver eye. Therefore, offset calibration (or cancelation) is required for key blocks along the receiver path such as CTLE, DFE summer, data samplers, and error samplers. For example, as shown in Figure 4.35, an I-DAC can be added at the CTLE output and controlled by the receiver digital engine. By analyzing the sampled data and errors, differential current can be injected into CTLE output to compensate for the offset present at the output node.

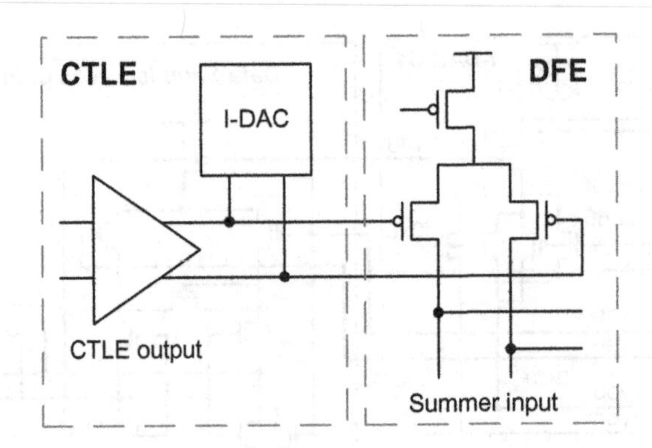

FIGURE 4.35

CTLE offset cancelation circuit.

Not only will the offset degrade the receiver performance, so will the device performance across the PVT. Therefore, it is important to keep the common-mode voltage of CTLE input and output (which is summer input), as well as the summer output, at the desired level. Depending on AC-coupled or DC-coupled links, CTLE input common mode is maintained by a separate common-mode voltage generator inside the receiver or supplied by the remote transmitter, respectively. CTLE output common-mode voltage can be kept by using a replica-bias (see Figure 4.30), and its OpAmp's offset also needs to be calibrated. The summer output common mode can be calibrated by comparing it to a known voltage reference using a data sampler. Since it is a current-integrating summer, the calibration needs to be done at each data rate.

Another component that can degrade the performance of the half-rate receiver is the duty cycle distortion of the recovered clock. A DCC circuit needs to be employed to maintain a good duty cycle.

RECEIVER ADAPTATION

While the calibration is performed only once during the initial linkup or data rate change, some loops are required to update continuously (hence the term adaptation) to maintain the receiver performance. Those loops are AGC, DFE, and CDR loops. AGC and DFE loops are used to open up the vertical eye of the receiver, while the CDR loop is to position the sampling point properly such that the horizontal margin is maximized.

The AGC loop can be adapted by monitoring the sampled errors. As shown in Figure 4.36, the loop will decrease the AGC gain when both $e_{i-1} = 1$ and $e_i = 1$ and increase it when both $e_{i-1} = 0$ and $e_i = 0$. The adaptation gain can be set to a larger value during the initial acquisition phase and reduced to a smaller value during the tracking phase. When sampled data reach the desired V_{ref} value, the distribution should exhibit a bimodal behavior, which is also shown in Figure 4.36.

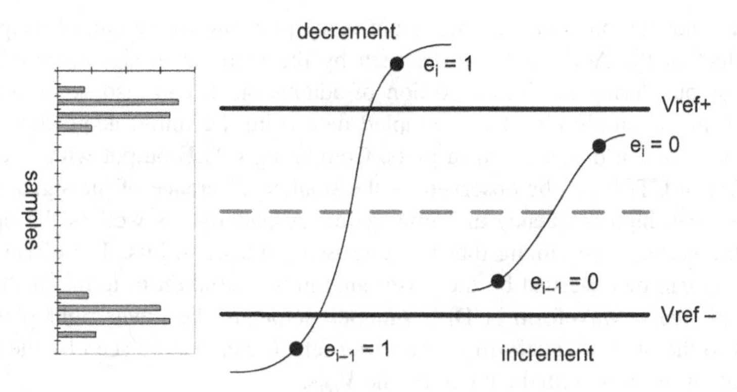

FIGURE 4.36

The operation of an AGC loop.

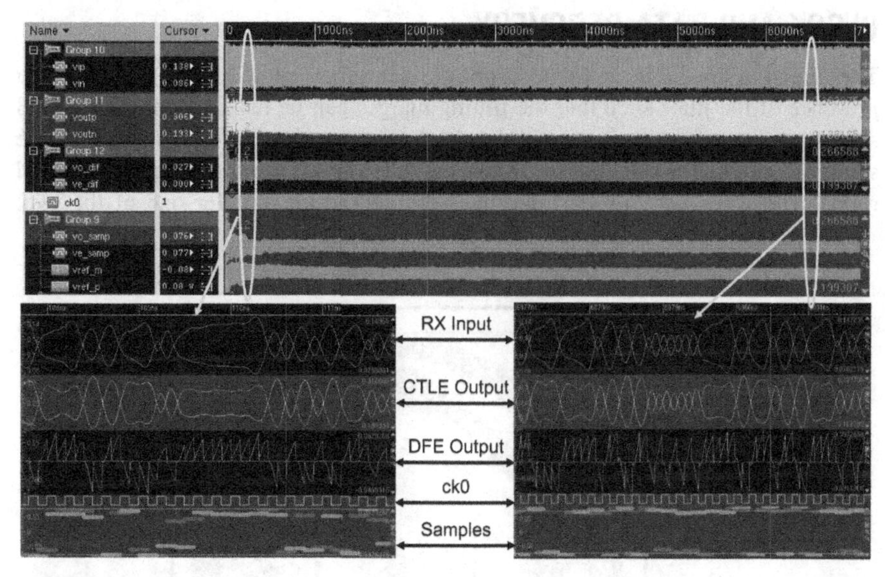

FIGURE 4.37

Simulated waveforms of the receiver data path.

Similarly, a DFE loop can also be adapted using the sampled errors. The goal for an AGC loop is to center the distribution along V_{ref+} and V_{ref-}, while the goal of a DFE loop is to tighten the distribution of the sampled errors by adjusting DFE coefficients. As a result, the vertical eye of the receiver opens up as mentioned in the previous section.

Figure 4.37 shows simulated waveforms of the receiver data path, where the plot on the top is the overall waveform, the lower left is the zoom-in plot during the initial adaptation, and the lower right is the plot toward the end of adaptation. The effect of the AGC loop can be seen by the increase in signal amplitude at CTLE output during the initial portion of adaptation. It can also be seen by the increase in the amplitude of the sampled data from the initial acquisition to the end of acquisition in the zoom-in plots. Comparing CTLE output with RX input, the effect of CTLE can be observed by the smaller difference of the signal amplitude between high-frequency and low-frequency patterns, as well as the opening up of the vertical eye among those zero-crossing transition bits. The effect of the DFE loop can be observed by the lesser amount of variation in terms of the peak of the triangular waveform at DFE summer output in the lower right plot, with respect to the output waveform in the lower left. It can also be seen by the tighter distribution of the sampled data along the V_{ref}s.

The CDR loop is discussed in the next section.

CLOCK AND DATA RECOVERY

The function of CDR is to position the sampling edge of the clock in the presence of the jitter such that the timing margin can be maximized. Figure 4.38 shows one possible receiver eye together with the distribution of edges of the sampling clock. Two double errors on the side are the timing margins (left and right). The worst-case timing margin is defined as the smaller one of the two.

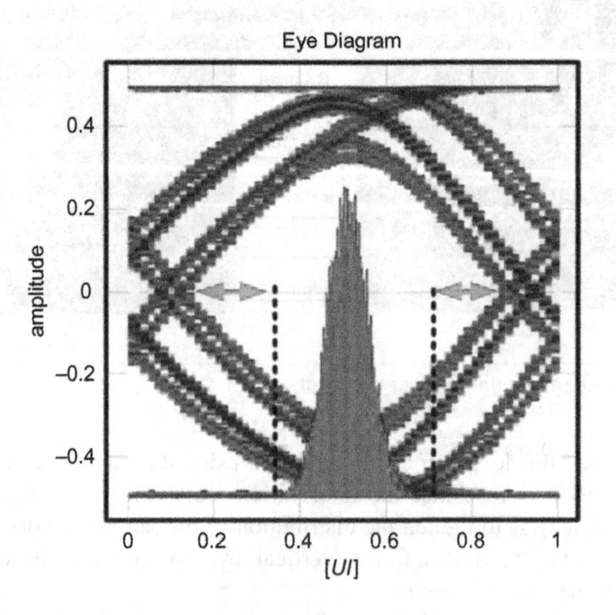

FIGURE 4.38

Receiver eye with the distribution of the sampling clock edges.

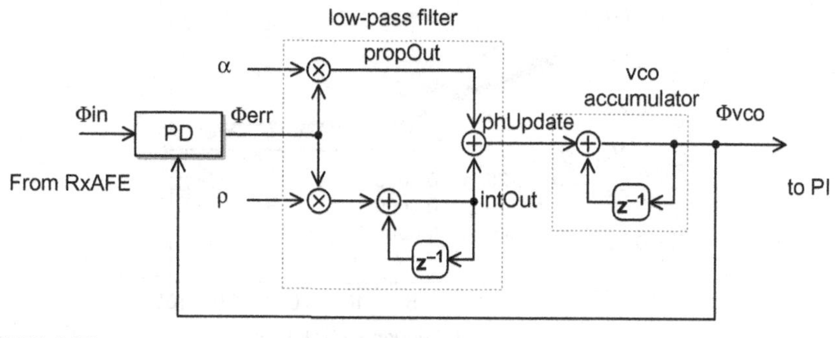

FIGURE 4.39

Block diagram of a second-order CDR loop.

The goal of a CDR is to position the mean as well as reduce the standard deviation of the sampling edge distribution such that the worst-case timing margin is maximized.

CLOCK AND DATA RECOVERY LOOP

A typical digital second-order CDR loop, which incorporates both the proportional path and integral path, is shown in Figure 4.39. It consists of a phase detector, a low-pass filter, and a VCO accumulator. The phase detector compares the incoming phase Φ_{in} with the phase interpolator phase index Φ_{vco} to generate the phase error Φ_{err}. Phase error then is multiplied by the proportional gain α and the integral gain ρ, respectively, to generate the phase update, phUpdate, by summing them together as the input to the VCO accumulator, which accumulates the phase updates and sends the phase index Φ_{vco} to the phase interpolator (PI).

The behavior of the CDR mainly depends on how the phase detector is designed and the choice of the loop gain α and ρ. The proportional part of the CDR is to track the phase difference between the incoming clock and the local clock, whereas the integral part is to track the frequency error of the clocks. In the case of the forwarded clock system, in which only phase delay is present in the system, the integral gain ρ can be set to zero.

Figure 4.40 shows the effect of the proportional gain, illustrated in the plot commonly known as "jitter tolerance." The incoming data to the receiver were modulated by a sinusoidal jitter with its modulation frequency shown on the x-axis and the peak-to-peak modulation amplitude shown on the y-axis. If the modulation frequency is low enough for the CDR loop to track, the loop with larger proportional gain will tolerate bigger jitter amplitude as shown in the left part of the plot. On the other hand, when the modulation frequency is higher than the tracking capability of the loop, the loop no longer responds to it, and, as a result, the sinusoidal jitter will be present at the receiver and reduce the margin of the receiver eye. This explains why the curves become flat when the jitter

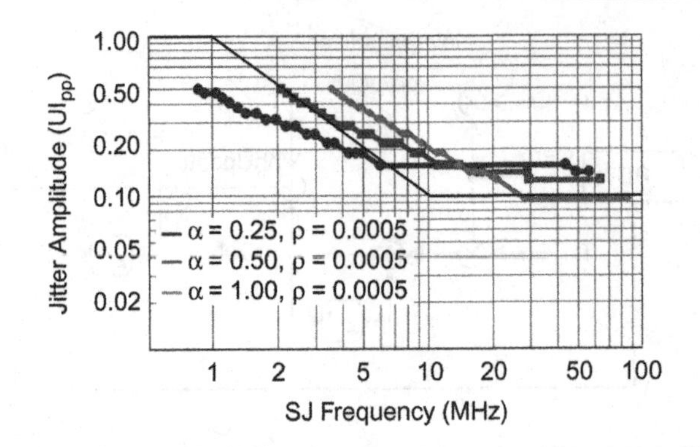

FIGURE 4.40

Jitter tolerance of a receiver.

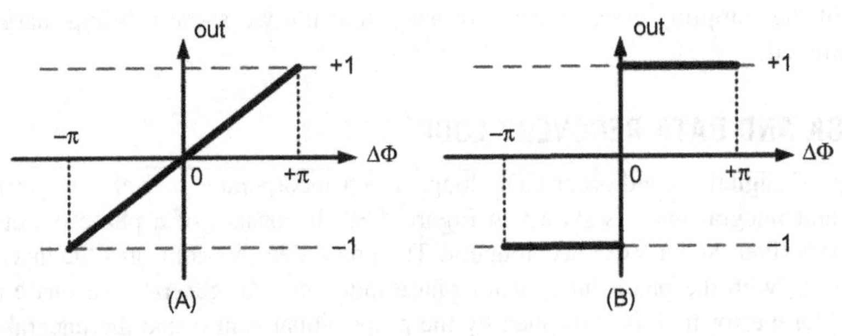

FIGURE 4.41

Two classes of phase detectors: (A) linear phase detector; (B) binary phase detector.

frequency goes beyond a certain frequency. The right part of the plot also shows that the higher the proportional gain, the less tolerable of high-frequency jitter is. This is because the higher proportional gain makes the clock edges easier to move and hence creates more inherent jitter in the system. Most standards define a mask as a specification to bound the CDR loop behavior. As an example, the mask of PCIe specifications is also shown in Figure 4.40. The design choice is to pick the proportional gain just large enough to meet the low-frequency portion of the mask while keeping the tolerance as high as possible for the high-frequency portion of the mask. In this case, one should pick $\alpha = 0.5$ to meet the requirement of PCIe specifications.

PHASE DETECTORS

There are two classes of phase detectors: a linear phase detector, in which the transfer function is somewhat linear between the output and the input phase error, as shown in Figure 4.41(A), and a binary phase detector (or bang-bang phase

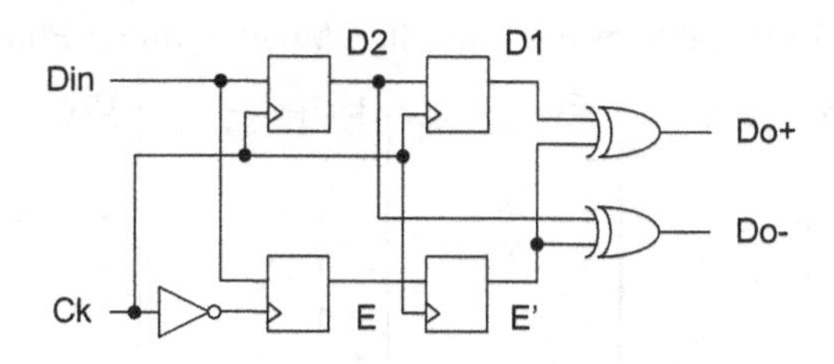

FIGURE 4.42

Block diagram of an Alexander phase detector.

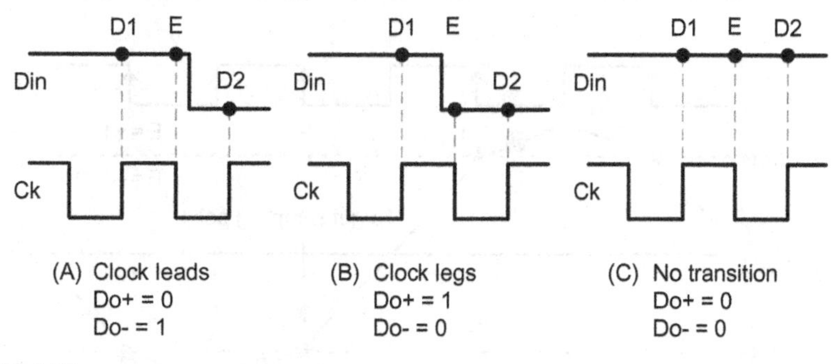

FIGURE 4.43

Operation of the Alexander phase detector.

detector), in which only the sign of the phase error is extracted, as shown in Figure 4.41(B). The rest of the section describes two binary phase detectors, since they are more commonly used among the digital CDR circuits.

Figure 4.42 shows an Alexander phase detector, also known as a 2× sampling phase detector. It uses both edges of a Nyquist-rate clock to sample two consecutive data bits and the data edge in between. The early and late information can be generated based on the sampled results. This is illustrated in Figure 4.43, where (A) shows the clock edge is early with respect to the data edge, (B) shows the clock edge is late with respect to the data edge, and (C) shows no edge transition and hence no update is generated. The goal of the CDR loop is to align the clock edge to the data edge such that the other clock edge can be positioned in the middle of the data to maximize the timing margin.

One limitation of this phase detector is that it requires a Nyquist-rate clock with a good duty cycle since it uses both edges of the clock, which translates to higher power consumption and a more stringent requirement of the clock in high data-rate systems.

Recall the DFE loop described in the previous section where both data and errors were generated for AGC and DFE loops. The same information can be used to generate the phase error using a sign-sign Mueller-Muller phase detector [4].

Table 4.1 The Truth Table of the Sign-Sign Mueller-Muller Phase Detector

D_n	D_{n-1}	E_n	E_{n-1}	Dt_n	Ph_{err}
1	1	1	−1	+	Early
−1	−1	1	−1	+	Early
1	−1	1	−1	−	Late
−1	1	1	−1	−	Late
1	1	−1	1	−	Late
−1	−1	−1	1	−	Late
1	−1	−1	1	+	Early
−1	1	−1	1	+	Early

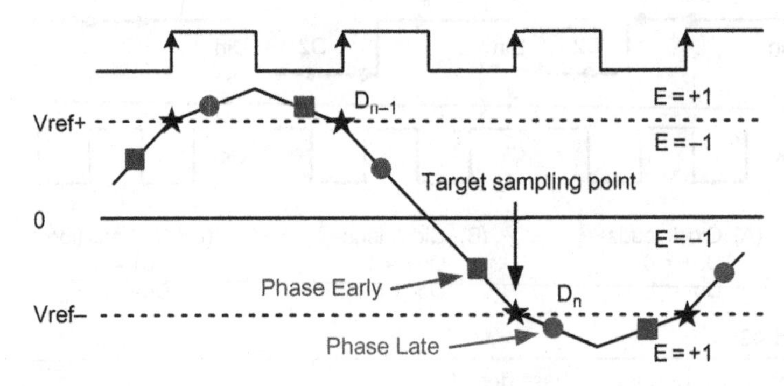

FIGURE 4.44

The operation of the sign-sign Mueller-Muller phase detector.

This type of phase detector is also called a baud-rate phase detector because 2× sampling is no longer required as is the case with the Alexander phase detector, and, therefore, it opens up the possibility of using the half-rate clock for power saving purpose.

Using both current and previous data bits and error bits, the phase error Dt_n can be described by the following equation:

$$Dt_n = D_n \times D_{n-1}(E_n - E_{n-1}) \tag{4.14}$$

where D_n and D_{n-1} are the current and previous sampled data, and E_n and E_{n-1} are the current and previous sampled errors, respectively. The truth table of equation (4.14) is shown in Table 4.1.

The operation is illustrated in the timing diagram shown in Figure 4.44. In the cases where the sign of the previous error versus the current error changes, depending on the previous and current data, phase error information can be extracted. If the sign of errors does not change from previous one to the current one, then no phase information is updated.

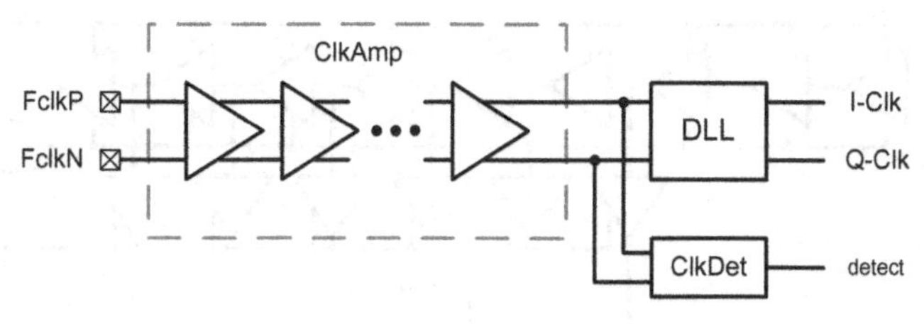

FIGURE 4.45

Block diagram of a forwarded clock receiver.

FORWARDED CLOCK RECEIVER

In a half-rate forwarded clock system, the forwarded clock lane may reuse the same transmitter from the data lane, and the clock can be transmitted by sending out a 1010 data pattern. On the other hand, since the receiver is not required to recover data, it can be as simple as a clock amplifier. In the case where I and Q clocks are needed for the phase interpolator in data lanes, a delay-locked loop (DLL) is often used.

Figure 4.45 shows a block diagram of a forwarded clock receiver. The clock amplifier consists of multiple CML stages to amplify the incoming clock with small amplitude. Due to the large output amplitude and the low jitter requirement, it often requires a $10\times$ gain at the clock frequencies. In addition to the clock amplifier and DLL, a clock/frequency detector is added to indicate whether the clock is present or not.

DELAY-LOCKED LOOP

Figure 4.46 shows the block diagram of a DLL. The voltage-controlled delay cells generate multiple phases of the incoming clock, where I and Q clocks can be tapped off and buffered before sending to PI. A linear phase detector uses I and Q clocks to compute the quadrature phase error, together with a charge pump (CP) to generate a control voltage for the delay cells such that the quadrature phase relationship can be maintained.

DESIGN FOR TEST/MANUFACTURE

Due to the complexity of modern integrated circuit design and the accessibility of the I/O pins, it is necessary to have additional circuitry built in as part of the design, both analog and digital, to measure parameters of the circuits and monitor the performance of the link. This is particularly true for post-silicon validation and debugging in the lab. In the high-volume manufacturing environment, those monitoring data are often used to screen parts and analyze yield. The circuit design for such a purpose is commonly referred to as design for test (DFT) and design for manufacture (DFM), or simply DFx.

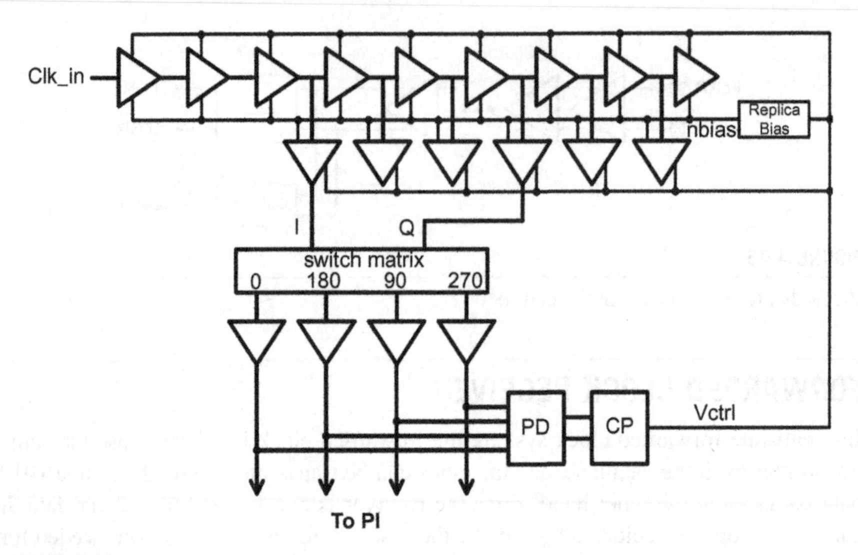

FIGURE 4.46

Block diagram of a DLL.

ANALOG DFX FEATURES

Amonv and Amoni are two separate single-ended analog probe points to monitor internal DC voltage and current, respectively. A lookup table is usually associated with them to allow access to multiple internal nodes one at a time. The outputs are usually brought out to two separate pins, which can be connected to digital multimeters, current meters, or scopes for monitoring.

Some designs may employ an internal analog-to-digital converter (A2D) to convert Amonv into digital data and read them out through a digital bus. This feature is particularly convenient when the Amonv pin is not accessible in some system boards.

High-speed clock buffer (HSCLK) is a pair of differential pins capable of delivering high-speed clocks, which is necessary to characterize PLL performance. With proper design of the clock distribution network, HSCLK can also be used to monitor the TX clock, PI clock, or the recovered RX clock. Although the TX clock can be easily measured by connecting TX pins to the scope, in the case where TX pins are not available, HSCLK may become handy to monitor the TX clock.

DIGITAL DFX FEATURES

Dmon is a single-bit digital probe point to monitor internal digital signals that can be easily connected to a scope. Dmon is mainly used to capture a slow-changing signal as an indicator of a state, and, therefore, the bandwidth requirement is low.

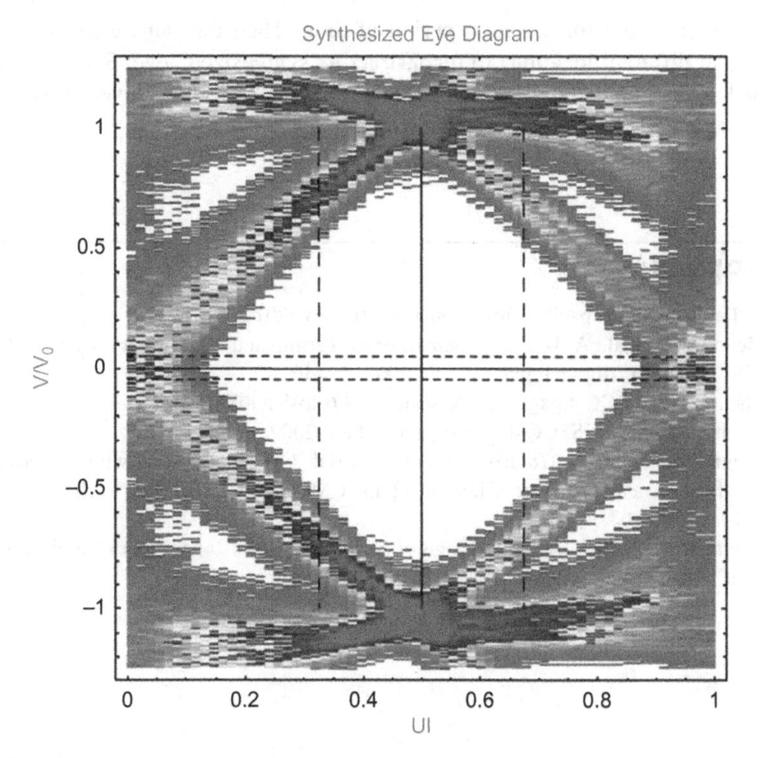

FIGURE 4.47

Example of a synthesized eye.

One very powerful usage of Dmon is the "event" trigger. For example, if the digital signal indicating the transmitter transitioning from the active state to the idle state can be made available to Dmon, it can be used to trigger the scope to capture the waveform around the transition period.

The most useful digital DFx feature is the parallel digital bus, which can be used to read the data stored in the registers. Some buses can be designed to capture the steady state of the register value; others can be designed to capture the time-varying signals updated periodically by a clock. For example, the former can be used to read the calibration value; the latter can be used to monitor the adaptation loop by a logic analyzer. In the case where no logic analyzer is available, it may be useful to have a built-in digital-to-analog converter (D2A) to convert the time-varying signals to an analog port, which then can be captured by a scope.

Many creative and yet complicated DFx features can be developed to help monitor the health of the link. One such example is the eye monitor. The basic concept is that for a given timing (horizontal) and voltage (vertical) location of a receiver eye, the statistics of the receiver data can be obtained by collecting the data sampler output for a certain period of time. Then the same exercises can be repeated at different locations to reconstruct a "synthesized" eye. Such an eye can be used to estimate the timing and voltage margin of a receiver. Figure 4.47 shows one example of a synthesized eye.

REFERENCES

[1] PCI express base specifications revision 3.0. Nov. 2010.

[2] Menolfi C, et al. A 16 Gb/s Source-Series Terminated transmitter in 65 nm CMOS SOI. In: ISSCC dig tech papers. Feb. 2007. p. 446–47.

[3] Chen L, Zhang X, Spagna F. A scalable 3.6-to-5.2 mW 5-to-10 Gb/s 4-tap DFE in 32 nm CMOS. In: ISSCC dig tech papers. Feb. 2009. p. 180–81.

[4] Spagna F, et al. A 78 mW 11.8 Gb/s serial link transceiver with adaptive RX equalization and baud-rate CDR in 32 nm CMOS. In: ISSCC dig tech papers. Feb. 2010. p. 366–67.

[5] Cherry EM, Hooper DE. The design of wide-band transistor feedback amplifiers. Proc IEEE 1963;51:375–89.

CHAPTER

Measurement and data acquisition techniques

5

Designs without validation can be very dangerous.
—Anonymous

Tools and methods used to validate that the design meets the high-speed system design specifications are examined in this chapter. It provides information, such as the minimum bandwidth and accuracy requirements of test equipment, as well as the effects that test equipment has on high-speed measurements. Detailed calibration procedures to remove unwanted equipment effects are discussed. The chapter also shows how to set up a PCI Express (PCIe) spec compliance validation test bench designed for highly efficient data acquisition and analysis.

DIGITAL OSCILLOSCOPE MEASUREMENT

A digital oscilloscope like the one shown in Figure 5.1 is an indispensable tool for design verification and debugging in high-speed digital system design. As the development cycle in today's consumer and enterprise electronic product market shortens to an unprecedented degree, engineers need advanced tools to tackle their measurement challenges quickly and precisely. Digital oscilloscopes are the key to detecting signal fidelity issues in the early stages of a project and accelerating design validation and standards compliance tests. We discuss the most important aspects of the modern digital oscilloscope in this section to help readers get a truly complete scope solution and a clean, reliable, and trustworthy signal.

REAL-TIME AND EQUIVALENT-TIME SAMPLING SCOPES

Current high-speed digital design specifications, such as the PCI Express Base Specification (PCI-SIG Nov 10) and Intel® QuickPath Interconnect Electrical Specification (Intel QPI Electrical v1.1), define that valid specification compliance measurements can be done by either real-time (RT) or equivalent-time (ET) scopes that have sufficient analog bandwidth, faster equivalent rise time compared to the incoming signal, adequate buffer memory, and low noise floor.

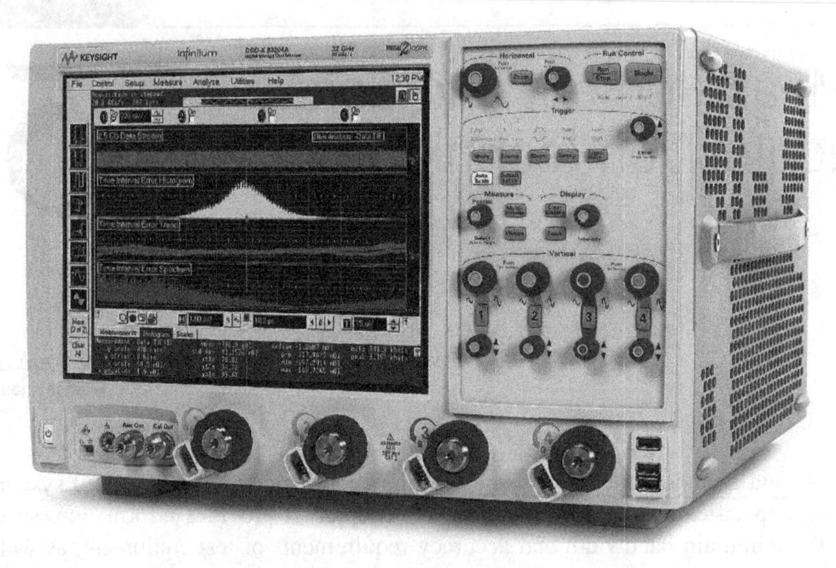

FIGURE 5.1

A Keysight digital scope, from the Keysight website.

RT and ET have their own benefits and disadvantages. For example, RT is capable of capturing unit interval (UI)-to-UI jitter or N-UI jitter, of which ET is not capable. However, ET provides a benefit over RT for sampling signals with fast rise times (<30 ps) [1]. Selecting the right oscilloscope for design validation and debugging is not an easy process. It is helpful to understand the differences between RT and ET scopes. Two very good references [1,2] address this topic and give very thorough comparisons of the two.

BANDWIDTH

Bandwidth is an oscilloscope specification that defines the frequency at which the input signal is attenuated by 3 dB. It is the specification that should be considered first when it comes to selecting an oscilloscope, because it determines frequency contents of a signal that a scope is able to measure accurately. A scope with a finite bandwidth behaves like a low-pass filter that truncates high-frequency components of a signal. This in turn distorts the rising and falling edge of a high-speed signal, because most of the high-frequency energy is contained in these edges. It can cause jitter to increase and the eye diagrams to close. As the data rate keeps climbing, the fundamental frequency of today's high-speed digital and analog signals has reached the multi-gigahertz realm. For example, the 8-Gbps PCIe Gen3 signal would have a fundamental frequency of 4 GHz and harmonics at 8, 12, 16, and 20 GHz, going to infinity. As a result, it is even more critical and difficult to choose a scope with the right bandwidth to satisfy the wide

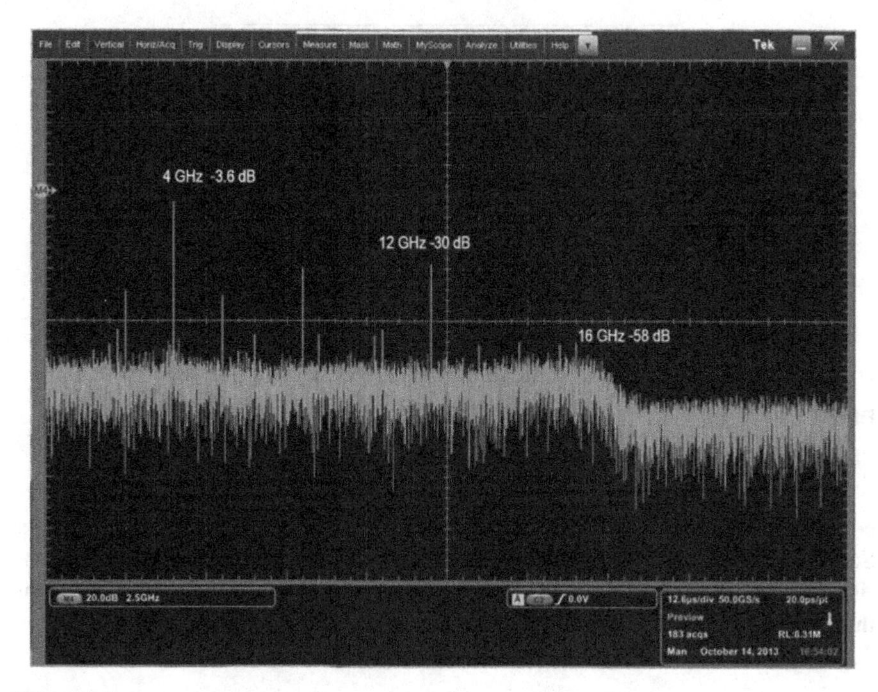

FIGURE 5.2

Frequency spectrum of a PCIe Gen3 PRBS23 pattern.

measurement frequency requirement, as well as to keep the price in budget, because in the marketplace the price of a scope is proportional to its bandwidth.

Complex high-speed signals contain frequency contents up to infinity. It is impractical to capture them all. There is always a tradeoff between cost, noise, and wide bandwidth. The best compromise is to understand your signal and to use a scope with a bandwidth that is wide enough to measure the signal accurately, as well as to keep the noise and cost within budget. Take the PCIe Gen3 PRBS23 signal, for example. The measured frequency component of the signal has a spectrum centered at 12.5 GHz and a span of 25 GHz, which is shown in Figure 5.2. The 4-GHz fundamental harmonic is −3.6 dB, and spurs at the second and third harmonics are −30 dB less than the first harmonic. The fourth harmonic is only −58 dB, and the fifth harmonic at 20 GHz is lost in the scope's noise floor. Despite the different bandwidths, a 12-GHz and a 20-GHz real-time oscilloscope will capture the same harmonic content of the PCIe Gen3 PRBS23 signal.

The rise time of the same signal is measured as evidence that coveting a much higher bandwidth, such as, for example, the fifth harmonic rule that scope manufacturers promote, is unnecessary for this particular measurement. The PCIe Gen3 signal with a rise time of about 55 ps is measured with scopes of different bandwidths. Due to the losses and dispersion of PCB material, transmission line,

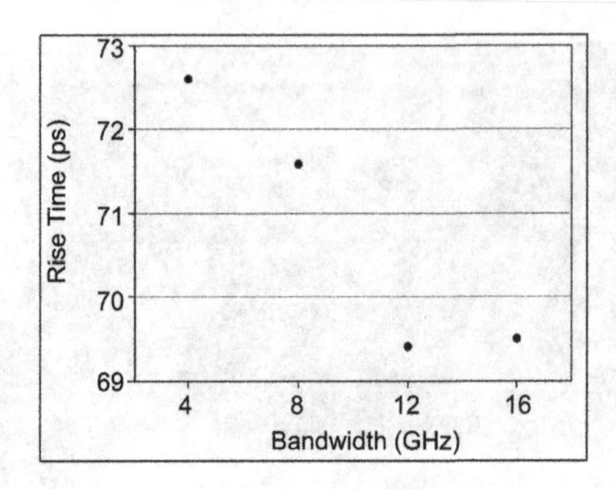

FIGURE 5.3

Rising edge time comparison among different bandwidths.

connectors, and coaxial cables, the rise time is slowed by approximately 15 ps. Scopes with a bandwidth covering only the first and second harmonic degrades the rise time by about 2 to 3 ps. However, scopes with a bandwidth higher than the third harmonic measure the rise time almost identically, as shown in Figure 5.3.

SCOPE DIGITAL FILTER APPLICATIONS

With the advancement of the scopes' A/D converter and high-performance computing, digital signal processing (DSP) is becoming more and more popular because of its flexibility, endurance, lack of exposure to information loss due to power loss, and reconfigurability. Contemporary scopes normally integrate math functions to implement filters using DSP. This section provides details concerning procedures for creating and applying customized DSP filters for high-speed digital and analog system characterization and validation. Note that, although digital filters are generally categorized to infinite impulse response (IIR) and finite impulse response (FIR) filters, only FIR filters are covered here. A brief introduction to FIR filters is to use a set of time domain coefficients, which can be seen as a digital signal processing window, to perform convolution with captured data samples. As a result, specific required shaping of the signal, such as low-pass filtering, can be achieved. Theory of FIR design and digital filters is beyond the scope of this book.

One important application of FIR filters in high-speed system design validation is to compensate for losses introduced by measurement setups and fixtures. An example is a PCIe Gen3 transmitter base specification measurement; the experiment setup is shown in Figure 5.4. The specification defines that the probe point of interest is at the pin of the transmitter at the very left in the figure. However, it is impossible to probe the transmitter pins directly and still be able to automatically select a certain PCIe lane from all 44 PCIe lanes of a recent Xeon processor. One workaround is to route all PCIe lanes through transmission line breakouts to coaxial cable connectors

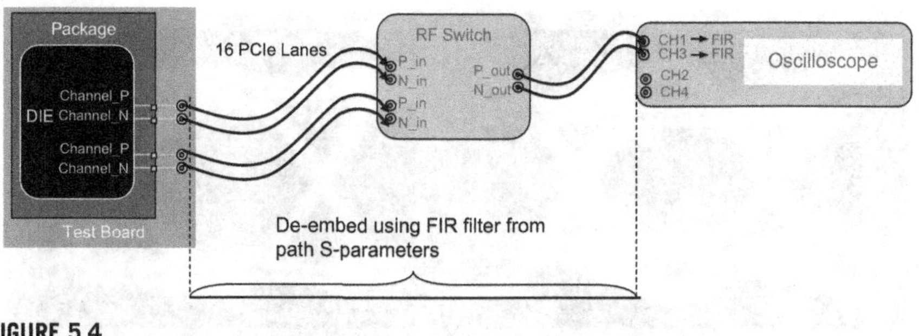

FIGURE 5.4

PCIe Gen3 base specification measurement setup.

on the test board to an external high-frequency compatible switch box, the output of which is connected to a scope. As a result, all PCIe lanes are physically connected in this case, and the test can be controlled by scripts to sweep through all 44 PCIe lanes without human intervention. In order to recover the signal at the transmitter pin, effects from experiment setups, such as transmission line breakouts on the test board, connectors, cables, and switches, need to be removed from the measured waveform. The procedures for creating such an FIR filter, getting the FIR coefficients, and de-embedding the test fixture effects are as follows:

- Measure 4-port differential S-parameters of the whole test fixture path; vector network analyzer (VNA) measurement details are covered in the next section.
- Complete the measurement fixture magnitude frequency spectrum by replicating the differential S21 data measured in the first step in reverse order and append to the end of the first half, H_f.
- Design a desired filter frequency response, H_d.
- Compute the de-embedding FIR filter, $H_{de} = H_d/H_f$.
- Perform inverse fast Fourier transform (FFT) to H_{de} to obtain the desired impulse response of the de-embedding filter.
- Sample the impulse response to obtain the FIR filter coefficients.

A PCIe Gen3 eye diagram comparison between waveforms with and without applying an FIR filter to de-embed the test fixtures is shown in Figure 5.5. Significant differences between eye height and width results due to test fixture losses, discontinuities, and inter-symbol interference can be compensated for by applying the customized FIR filter. It is critical to properly remove the unwanted test fixture effects, because it can easily change the test results from passing specifications to a false failure.

The above procedure is generic to create any arbitrary frequency sampling filter design. However, newer scopes are capable of making the procedures simpler for users by taking the S-parameters and converting them to an FIR filter using scope-integrated hardware. They also provide an arbitrary filter library for use; a low-pass filter library is shown in Figure 5.6.

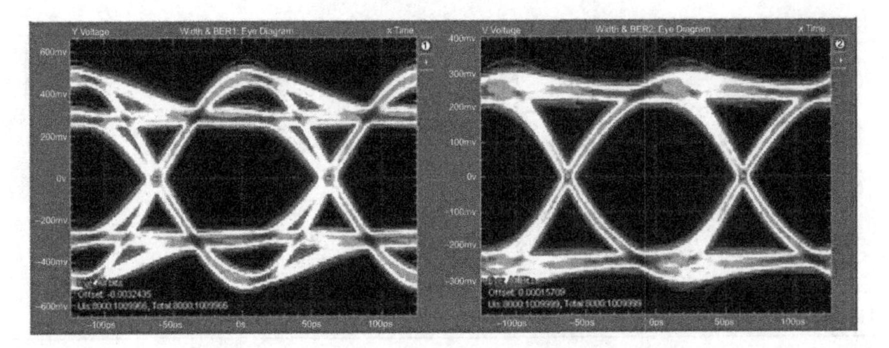

FIGURE 5.5

PCIe Gen3 eye diagram comparison between waveforms with and without applying an FIR filter.

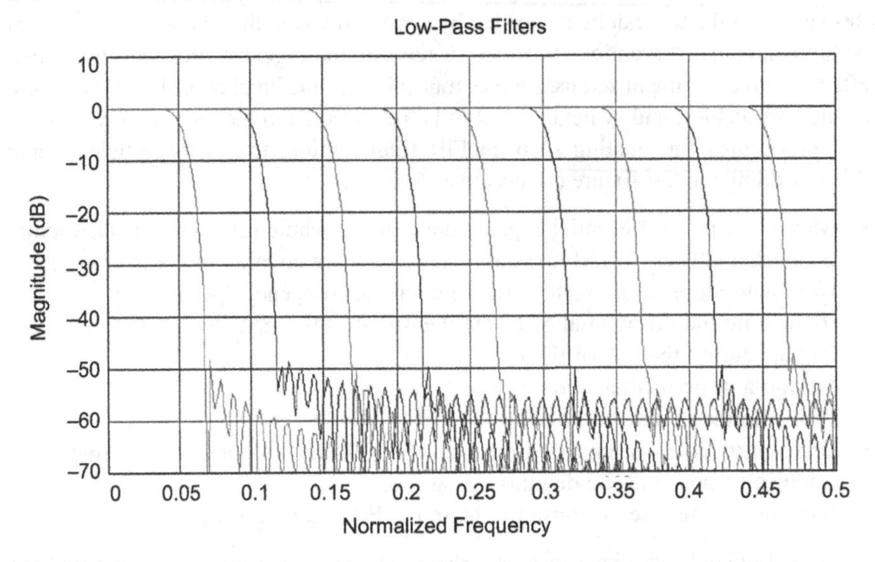

FIGURE 5.6

A low-pass filter frequency response from the Tektronix scope arbitrary filter library.

TDR MEASUREMENTS

Time domain reflectometry (TDR) is used in high-speed system analysis as a nondestructive and broadband method for the characterization of high-frequency properties of platform motherboard traces, connectors, and device packages. By sending a very fast step or impulse function and measuring the reflected "echo" signal and/or the transmitted signal in the time domain, TDR provides characteristic impedance of

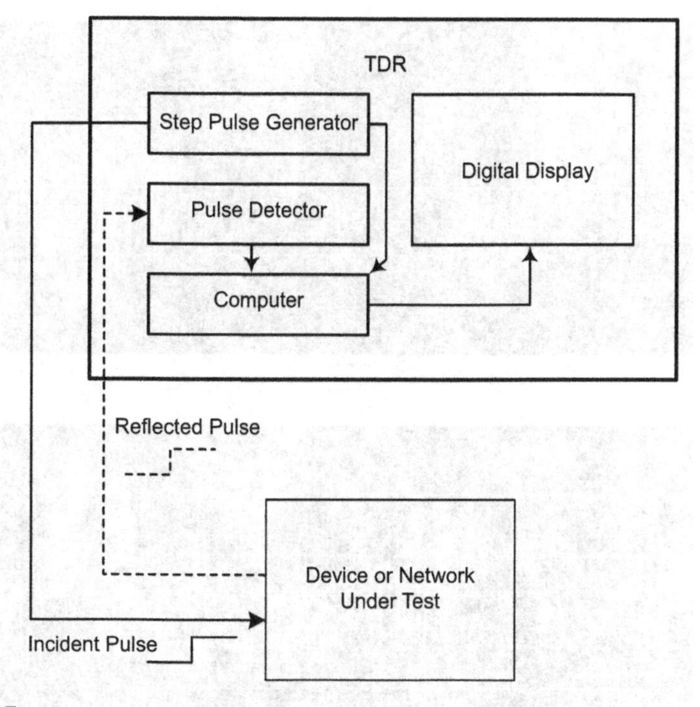

FIGURE 5.7

Schematic of a time domain reflectometer.

the transmission lines and the device package under test directly on its display. TDR is also useful for determining the system discontinuity locations and their electrical natures by looking into the reflected waveform characteristics of polarity, magnitude, delay, exponential rise/decay, ringing, and so on.

A TDR block diagram is shown in Figure 5.7. All the information mentioned above for system characterization can be read on the TDR's oscilloscope display and saved in text format. While a thorough TDR theory is beyond the scope of this book, this chapter attempts a concise presentation of a couple of TDR applications related to high-speed system electrical validation in practical test situations.

DE-SKEW DIFFERENTIAL PAIRS WITH TDR

One application of TDR is measurement channel de-skew. De-skewing is required when measuring a differential signal, for example, a PCIe differential pair. A differential signal or differential pair is comprised of voltage on a positive conductor, V_{D+}, and a negative conductor, V_{D-}. DC common-mode voltage is defined as the average or mean voltage present on the same differential pair ($V_{CM} = [V_{D+} + V_{D-}]/2$) over a large number of UIs [3]. In contrast, AC common

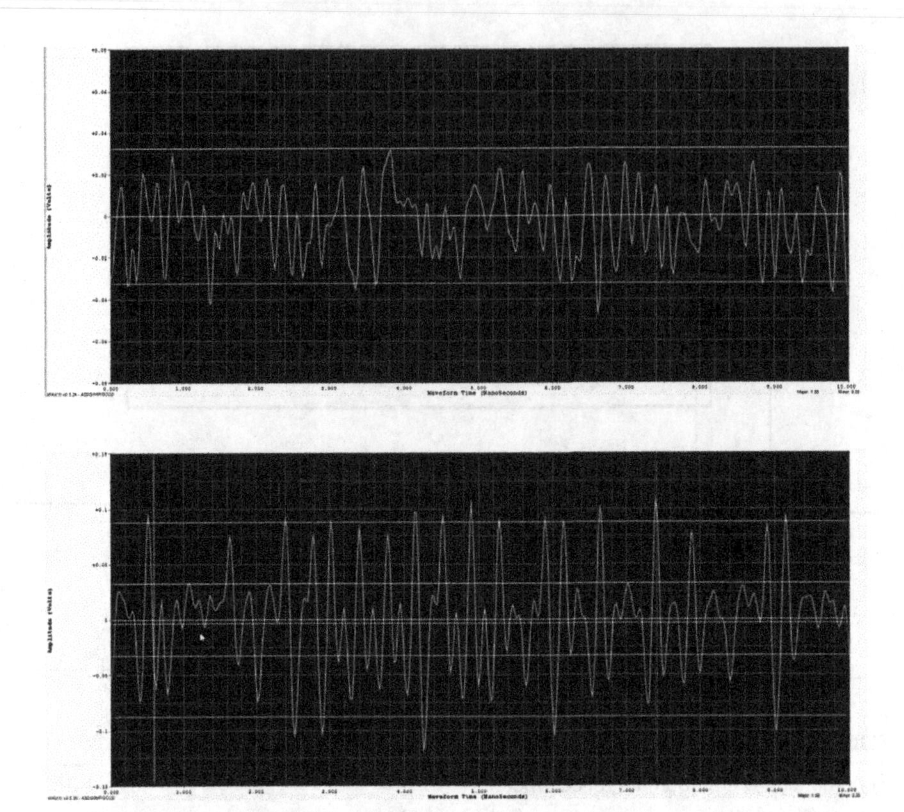

FIGURE 5.8

Channel skew effects on AC common-mode noise.

mode noise is the instantaneous average value of the differential signal and is most sensitive to the phase difference between V_{D+} and V_{D-}. AC common-mode noises with good and bad channel de-skewing are compared in Figure 5.8. The scales for both measurements on the y-axis and in the voltage are the same. It is obvious that peak-to-peak AC common-mode noise can be doubled when the phase mismatch caused by the two individual channels and probes used in the measurement is not properly removed.

A de-skewing setup with TDR is shown in Figure 5.9. As shown in the figure, each single-ended channel is comprised of a breakout channel on the test board, connectors, RF cables, RF switches, and Bias Tees. The total delay between V_{D+} and V_{D-} is in the range of 10 ps. The short rise time of TDR makes it a perfect candidate to capture time delay in this range. The de-skewing procedures are as follows:

- Remove the package from the PCB socket to create an open end.
- Connect the V_{D+} and V_{D-} channels to TDR's two pulse sources.

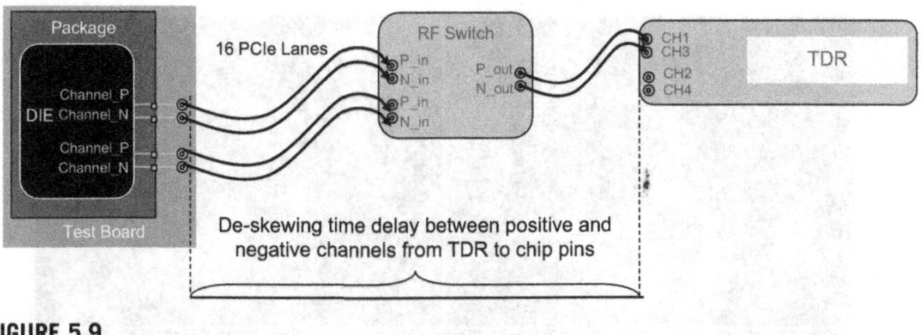

FIGURE 5.9

Schematic of de-skewing a differential pair using TDR.

- Generate the step function with the same polarity to the two channels and record the time delay T_{d1} on the TDR display.
- Swap the V_{D+} and V_{D-} channel leads and measure the time delay T_{d2} from TDR.
- Compute the final time delay between the V_{D+} and V_{D-} channels, $\Delta T = (T_{d1} - T_{d2})/2$.

By using the above procedure, the intrinsic TDR analog front-end time delay is removed and pure time delay between the positive and negative channels is obtained. A typical TDR time delay measurement result on the TDR screen is shown in Figure 5.10.

CHANNEL CHARACTERIZATION WITH TDR

The cross section of a high-speed system built on a ten-layer PCB is shown in Figure 5.11. Two chips, chip A and chip B, are connected through a chip package, micro-via, microstrip line, capacitor, LAI, through via, and stripline. The accuracy of the channel model is critical for system performance, including eye mask and prediction. It becomes more and more difficult to build the model as the system complexity keeps increasing. Model validation also attracts a lot of attention. Due to the nondestructive and broadband nature of TDR and its capability to directly display the electrical nature and location of system discontinuity, it is widely used for system modeling and correlation.

The system model and measurement correlation results are shown in Figure 5.12. The TDR reveals the capacitive, inductive, and/or combined effects of discontinuity locations as well as transmission line characteristic impedance. The measurement provides indispensable information for model correction—for example, the overestimation of the motherboard socket inductance and the underestimation of the microstrip line impedance. Detailed TDR responses to a series of electrical situations can be found in [4].

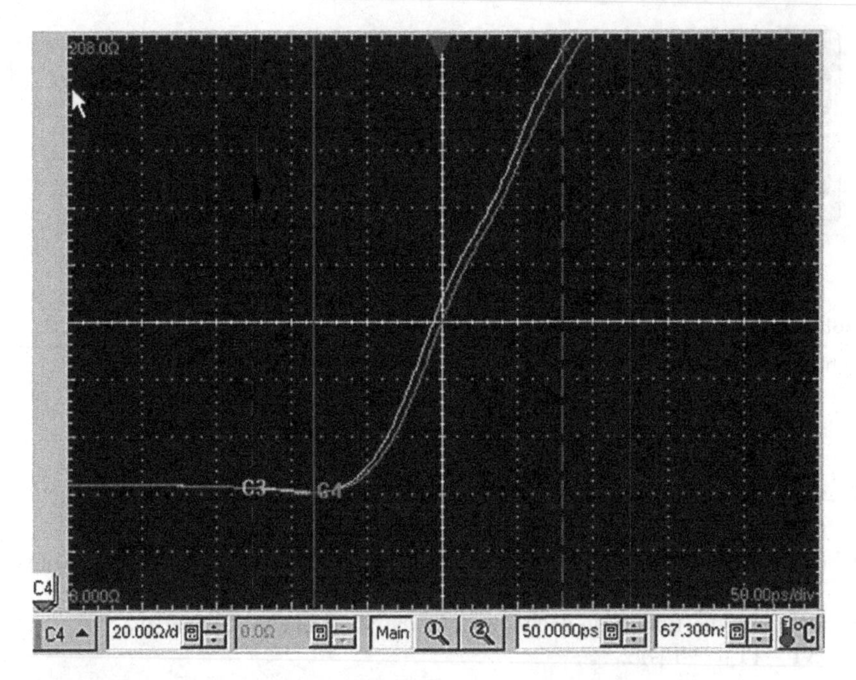

FIGURE 5.10

A typical TDR time delay measurement result shown on the TDR screen.

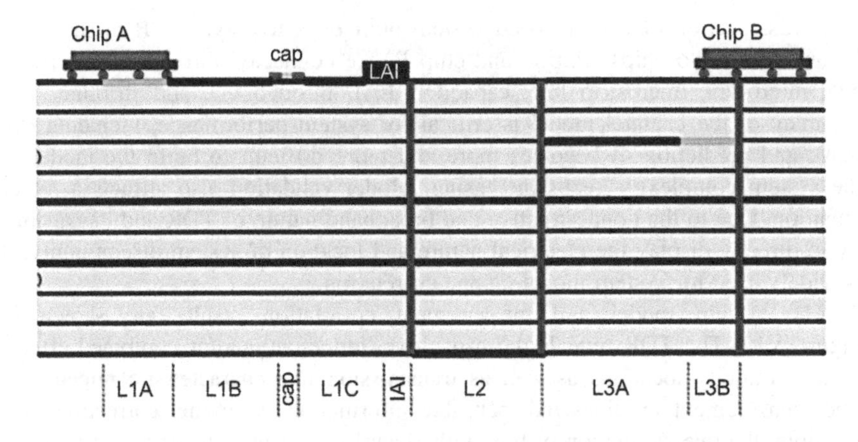

FIGURE 5.11

Cross section of a high-speed system.

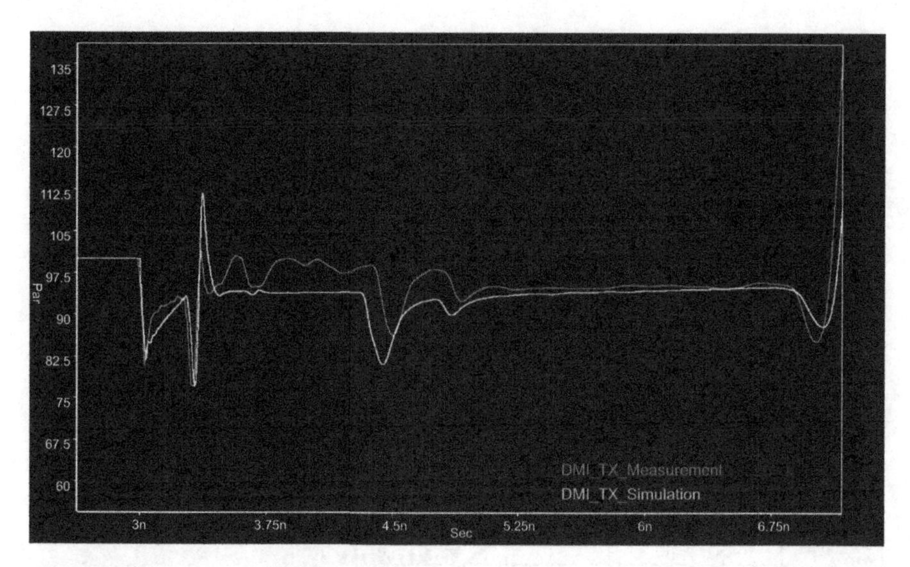

FIGURE 5.12

Time domain to frequency domain conversion software GUI.

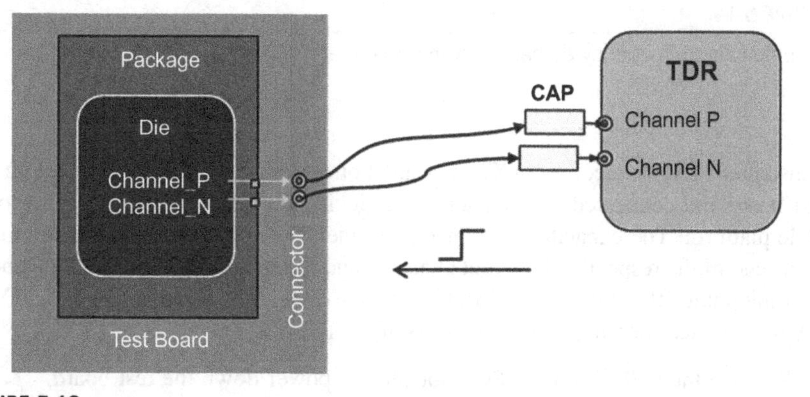

FIGURE 5.13

Schematic of return loss measurement using TDR.

RETURN LOSS MEASUREMENT WITH TDR

Another important TDR application is the return loss (RL) measurement. This TDR function is useful when a vector network analyzer is not available. TDR can be used to get both differential-mode and common-mode RL by converting time domain response to its frequency S-parameter. This conversion is usually implemented by software embedded on the TDR computer. A PCIe TDR RL measurement setup is shown in Figure 5.13. The differential circuit under test—for example, a PCIe

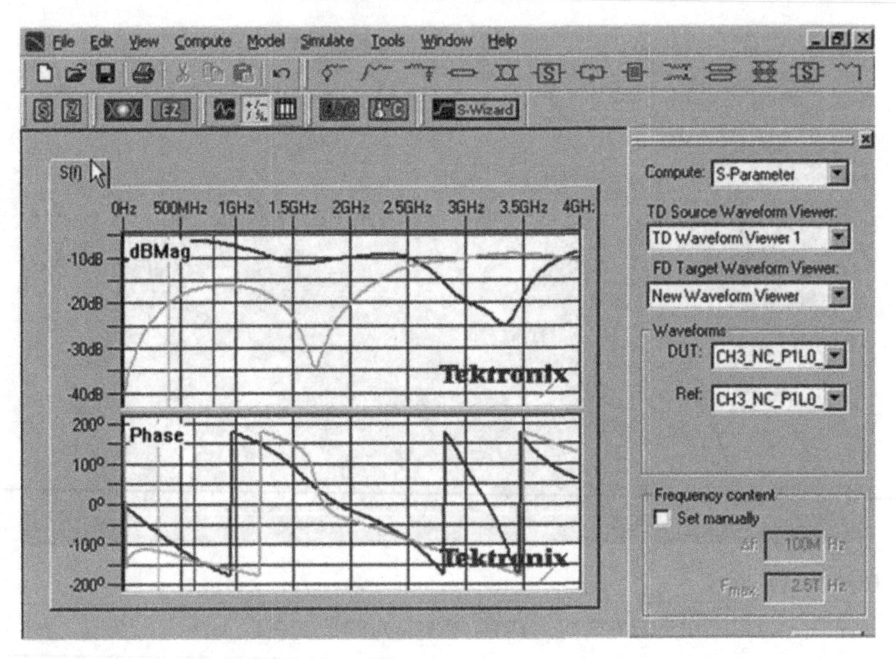

FIGURE 5.14

Time domain to frequency domain conversion software GUI.

transmitter (TX) or receiver (RX)—is routed out through a breakout channel on the test board and connected to a capacitor mimicking the AC coupling capacitor on a PCIe platform. The capacitor is connected to the TDR. RL measurements are made at the end of the respective breakout channels and require that the breakout channel's contribution to RL be de-embedded, thereby associating the RL with the PCIe TX or RX pin. The RL measurement procedures are as follows:

- Remove the CPU from the PCB socket and power down the test board.
- Measure the differential- and common-mode waveforms for de-embedding the test board breakout channel, Bias T, and TDR effects.
- Turn on the test board and measure the differential- and common-mode waveforms.
- Calculate the differential- and common-mode RLs by using the TDR time domain to S-parameter software; both reference and real measurement waveforms are required. The post-processing software GUI is shown in Figure 5.14.

Measured PCIe RX differential RL is shown in Figure 5.15. Red lines define the pass/fail mask for differential RL in the PCIe spec. The differential-mode RL covers a frequency range of 50 MHz to 4 GHz.

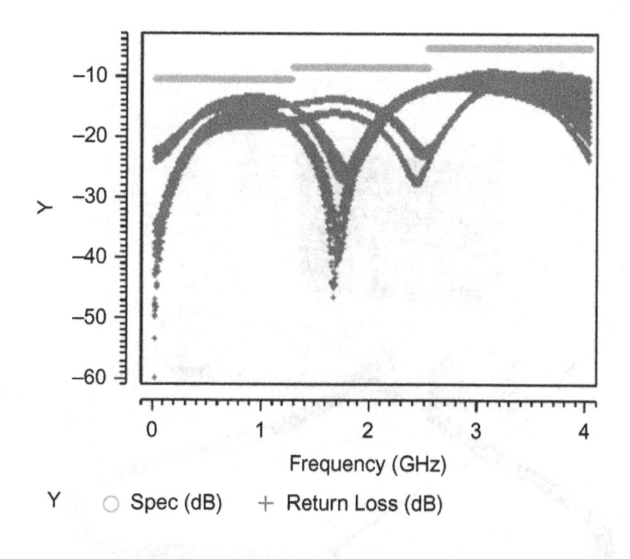

FIGURE 5.15

Results of return loss measurement using TDR.

VECTOR NETWORK ANALYZER MEASUREMENT

As mentioned in the scope calibration and filter section, platform breakout channel and test fixtures, such as cables, RF switches, and high-density connectors, have to be calibrated before starting any measurement or data acquisition. A vector network analyzer is used to measure the frequency domain response of the network consisting of the above-mentioned components. Before getting into the VNA measurement and VNA calibration procedures, it will be helpful to understand what the VNA does.

WHAT IS VNA?

A vector network analyzer is an instrument that measures the frequency response of a component or a network composed of many components, which can be both passive and active. A VNA measures the power of a high-speed signal going into and coming back from a component or a network, because power, in contrast to voltage and current, can be measured accurately at high frequencies. Both amplitude and phase of the high-frequency signal are captured at each frequency point. The built-in computer in the VNA calculates key parameters such as return loss and insertion loss of the network under test. It is also capable of visualizing the results in different formats—for example, real/imaginary, magnitude/phase, Smith chart, etc. In high-speed system tests, VNA is often used to characterize multi-port networks consisting of components such as connectors, filters, amplifiers, and transmission line/coaxial channels. VNA can be used for networks with an arbitrary number of ports—for

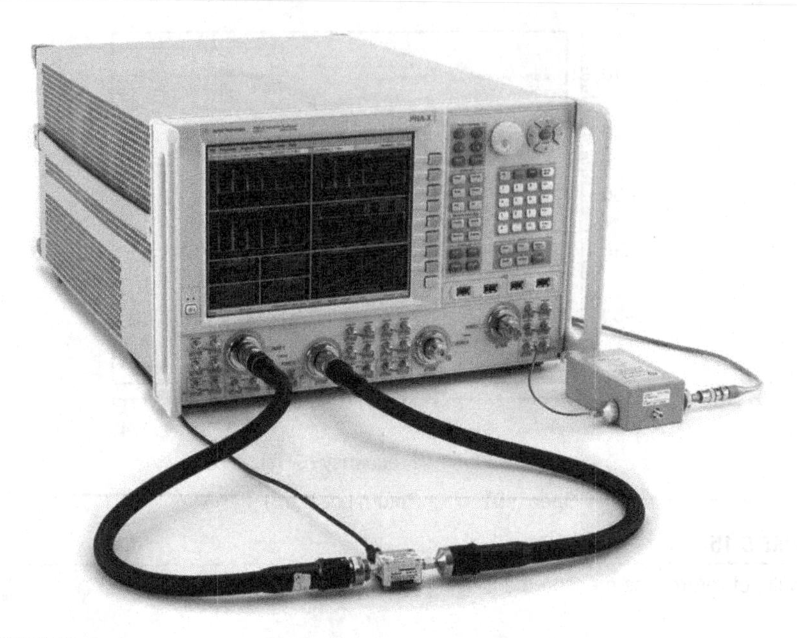

FIGURE 5.16

A VNA from Keysight Technologies, from the Keysight website.

example, the four-port differential pairs of a PCIe serial link. Figure 5.16 shows a contemporary four-port VNA with two ports connected to a device under test.

The VNA measurement frequency bandwidth and the number of frequency points across the selected frequency range are input options for a given VNA measurement. The VNA measures the high-speed signal vector response to a component or a network, one frequency at a time, by applying a continuous wave at that frequency. The magnitude of the continuous wave can also be adjusted. To faithfully characterize the measurement fixtures and generate an accurate filter for the scope to compensate for the measurement fixture losses to minimize the timing margin error, a minimum of all frequencies up to the second harmonic of the fundamental frequency of the high-speed digital/analog signal under test have to be covered in the VNA measurement bandwidth. To reduce the voltage margin error, all frequencies up to the fourth harmonic must be taken into account [5]. This translates to a minimum measurement bandwidth of 16 GHz for the PCIe Gen3 signal.

A simplified VNA architecture block diagram is shown in Figure 5.17. It involves RF sources, directional couplers (DC), dual-conversion frequency converters, reflectometers/power detectors, A/D converters, and an internal computer. The continuous wave signal used for testing is generated by the continuous wave source. The signal is swept over the specified measurement bandwidth in steps from one frequency to the next in small intervals defined by the user. A sample of the signal

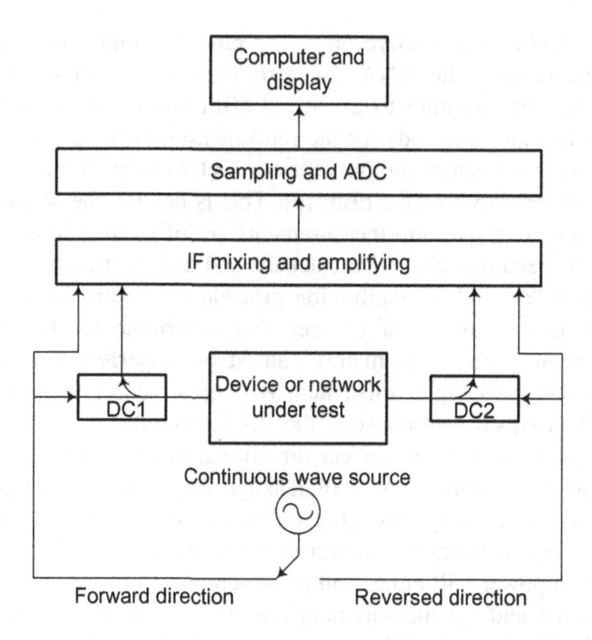

FIGURE 5.17

Basic block diagram of a VNA.

directly passes through the intermediate frequency (IF) processing unit and then to the reflectometers/power detector as a reference signal. The rest of the signal passes through a directional coupler (DC1) to the device/network under test (DUT). This directional coupler should not couple any of the incident power into its corresponding IF unit port. It is used to measure the reflected power from the DUT. The IF processing unit converts the incident and reflected signals to the IF frequency defined by the user, and the converted signals are subsequently detected by the reflectometers and converted to a digital signal. The ratio of the reverse reflected power to the incident power in digital format is then calculated by the internal computer and displayed as the mismatch at the input port of the DUT, also known as the return loss. The rest of the incident power is sent through the DUT and sampled by the second directional coupler (DC2). The ratio of the transmitted power to the incident power is the gain or loss of the DUT, also known as the insertion loss. Other imported network parameters, such as phase, standing wave ratio, group delay, and impedance, can be calculated by a VNA.

VNA ERROR SOURCES AND CALIBRATION

The error and uncertainty of a VNA fall into two major categories: the uncertainty of measuring the absolute power and the uncertainty of the ratio measurements. Unlike the relative power or ratio measurements, such as return loss and insertion loss, the VNA can also measure the absolute power. The measurement is only accurate to about ±1 dBm without calibration. However, the standard calibration methods for the ratio measurements do not help improve the absolute power measurement and a power meter calibration method is needed to reduce the uncertainty to ±0.2 dBm [6]. This is beyond the scope of this book since most of the time the ratio measurements are of interest in the characterization of the high-speed digital system channels and test fixtures.

In order to understand the calibration procedure for ratio measurements, it is worthwhile to discuss the error sources that contribute to the uncertainty of ratio measurements. Errors are mainly caused by imperfect directional coupler directivity, directional coupler mismatch, RF signal source mismatch, loss, and variations in the frequency response of the VNA system.

One example is that the imperfect directional coupler (DC1 in Figure 5.17) directivity causes a certain amount of leakage from the incident power, which is supposed to pass entirely through DC1 to the DUT, and then to the power detector. The power detector will detect both the reflected and leaked power. The unwanted leaked power will either add to or subtract from the reflected power. It cannot be ignored and the measurement inaccuracy will increase when the mismatch is small, like in most of the test cases in a high-speed compliance test system. When large reflected power occurs, the mechanics of VNA uncertainties are different from those in the above example. Detailed analysis can be found in [6]. Other than the systematic errors—i.e., the 12-term error model [6], a VNA usually requires test cables and adapters to connect from its front panel to the DUT. These cables and adapters will introduce phase shift and power attenuation, which in turn affect measurement accuracy. A VNA can only achieve high-precision measurements by correcting the intrinsic systematic errors of the instrument, test cables, adapters, and test fixtures.

The solution is to calibrate the VNA before making measurements, baseline the systematic errors, and de-embed these effects from the component and network measurement results. Different methods have been proposed: short, open, load, through (SOLT); through, reflection, load (TRL); load, reflection, match (LRM); and so on. Taking the SOLT method, for example, to perform the error correction, a set of reference standards—short, open, load, and transmission—are tested in place of the component or network. Different SOLT standards are available and used depending on whether the DUT has coaxial connectors. If the DUT has no coaxial cable access, for example, a microstrip trace on a platform, on-chip microstrip calibration standards, RF probes, and a probe station are needed to perform the SOLT calibration. A set of coaxial SOLT standards are shown in Figure 5.18. An on-chip microstrip-based SOLT standards kit on a ceramic substrate being measured by a pair of RF probes is shown in Figure 5.19. To avoid the engineers disconnecting and reconnecting the various standards, automated calibration standards such as the Agilent ECal module and software are developed to save time and reduce random errors, as shown in

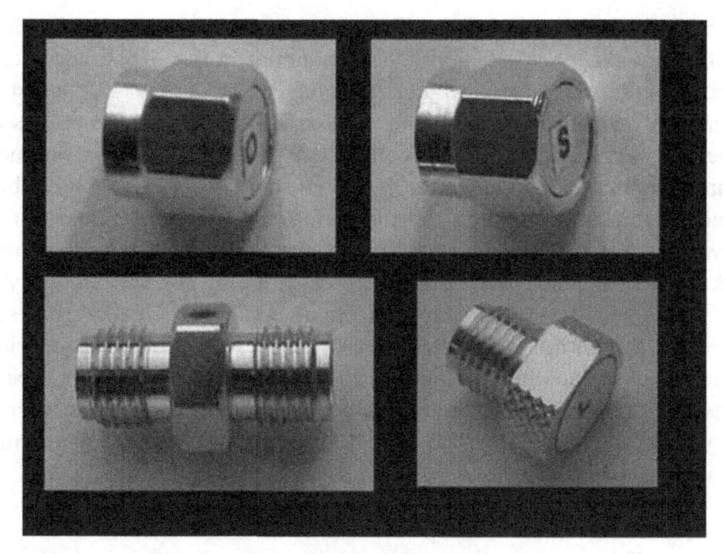

FIGURE 5.18

Coaxial SOLT calibration kit from the Clarke & Severn Electronics online store.

FIGURE 5.19

On-chip SOLT calibration kit and RF probes from CascadeMicrotech.

Figure 5.20. Before starting the calibration process, it is noteworthy that it is impossible to make a perfect short and open circuit in the GHz range because of the parasitic inductance and fringing capacitance. Modern VNAs can store the information of parasitic effects of the standard kits, and thus the effects can be de-embedded automatically in the calibration process. After measurements of phase and amplitude are made on each known standard, the differences between the standard measurements and the theoretical values are computed, stored in memory, and applied to correct each measurement of the real component and network under test. The compensation factors are only for the frequency points in the current test bandwidth setup. Keep in mind that a calibration setup is exclusive for the currently selected measurement bandwidth. Frequencies outside the range need another calibration. Calibration also needs to be repeated to compensate for test condition changes such as ambient temperature and humidity drifts over time [6]. The procedure to perform a SOLT calibration for a VNA is discussed below.

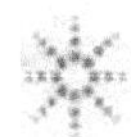

FIGURE 5.20

Automated SOLT calibration kit, ECal, from Keysight Technology.

FULL TWO-PORT SOLT CALIBRATION PROCEDURE

To carry out the full two-port reflectivity and transmission calibration, perform the following procedure:

- Check the connectors/probes for any problems such as bent pins or parts that are obviously off center [7].
- Clean the connectors/probes with compressed air at less than 60 psi. If necessary, clean the connectors with isopropyl alcohol [7].
- Torque the connectors to the specified torque. If an on-chip probe is used, the scratches on the microstrip standards from probing can be used as an indicator that good contacts have been made.
- Reduce the IF bandwidth to the smallest value that can be achieved.
- Set the test frequency range and power level.
- Connect the short, open, load, and through sequentially to the VNA on both ports. This is unnecessary if an automatic calibration kit is used.
- VNA saves the measurement data of the standards and calculates the corrections needed for the systematic errors, test cables, adapters, and so on.

EXAMPLE OF MEASUREMENT USING VNA

As discussed in previous sections and shown in Figure 5.4, test fixture effects are required to be removed. Figure 5.21 shows a PCIe base specification test fixture calibration setup. A dieless chip is positioned on a platform to introduce real package losses and provide probing pads for on-chip probing. A pair of PCIe

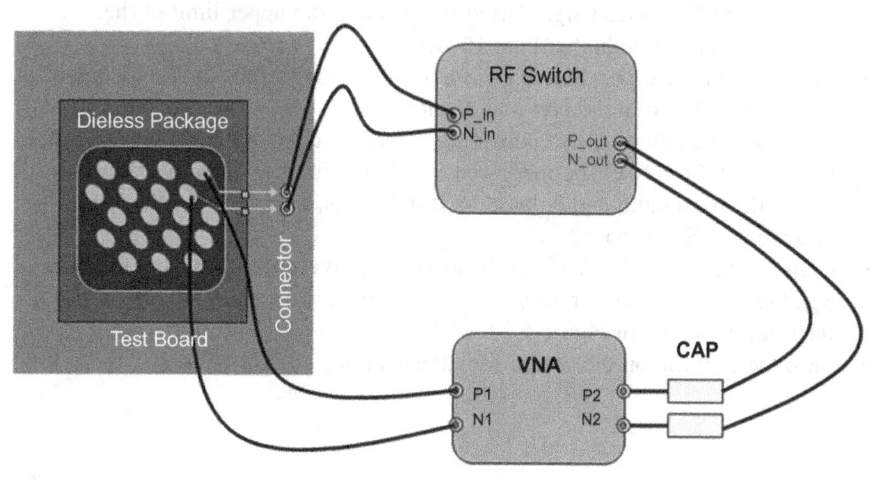

FIGURE 5.21

PCIe TX base specification test fixture removal experiment setup.

analog TX pads is connected through the transmission line breakout traces to the SubMiniature version A (SMA) cable connectors where coaxial cables are populated. To verify the transmitter is in compliance with the PCIe base specification, its signal output to the onboard connector is captured through coaxial cables, an RF switch, and a capacitor before connecting to the scope. The RF switch is used to select one PCIe lane at a time from all the PCIe lanes connected to it. The capacitor is used to represent the capacitance of a PCIe AC coupling link. To obtain the actual signal at the PCIe TX pin pads, the test setup from the pin to the front panel of the scope has to be de-embedded. As discussed in the scope section, the S-parameter of this test setup is used to generate an FIR filter for the scope to compensate this setup channel loss/distortion.

Instructions for making the test fixture measurements are given below. In this experiment the network, composed of the coaxial cables, RF switch, capacitors, and adapters, is measured with a VNA. The insertion loss of the network is measured.

VNA MEASUREMENT PROCEDURE

Steps to perform the measurements are listed as follows. The instructions are intended to be generic for most of the VNA models.

- Preset the VNA to the factory settings.
- Select S_{21} as the measurement parameter and set the vertical scale to best show the measurement curve.
- Set the measurement bandwidth from the lowest frequency of the VNA to at least the fourth harmonic of the fundamental frequency of the signal under test. For the PCIe Gen3 signal used in this case, the upper limit of the bandwidth should be higher than 16 GHz.
- Perform a full two-port SOLT calibration from the Cal menu. The detailed procedure is listed in the previous section.
- Validate the calibration by connecting a known standard (short, for example) to port 1 of the VNA. The measured S_{11} should have a magnitude of 0 dB across the bandwidth and a phase of 180 degrees. The same results should be observed for S_{22} at port 2.
- Connect the PCIe TX base specification setup to the two ports and measure S_{21}. Observe the insertion loss of the channel on the display. A typical S_{21} of the setup is shown in Figure 5.22.
- Save the S2P file on the screen for filter-making.

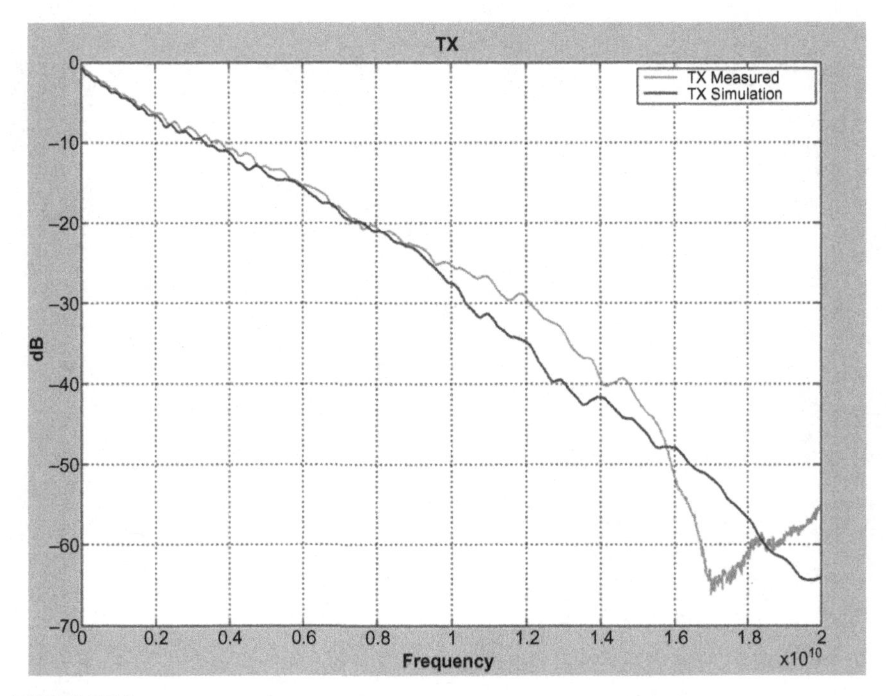

FIGURE 5.22

Example of PCIe TX base specification test setup de-embedding S_{21}.

REFERENCES

[1] Capturing the fifth harmonic: tradeoffs between sampling and real-time oscilloscopes. Agilent application note.

[2] Real-time versus equivalent-time sampling, Tektronix application note.

[3] PCIe Spec.

[4] Time Domain Reflectometry (TDR) and Time Domain Transmission (TDT) Measurement Fundamentals. Picosecond Pulse Labs application note AN15.

[5] Intel book.

[6] Scott AW, Frobenius R. RF measurements for cellular phones and wireless data system.

[7] <http://en.wikipedia.org/wiki/Network_analyzer_%28electrical%29>.

CHAPTER

Designing and validating with Intel processors

6

Design is not just what it looks like and feels like, design is how it works.

—Steve Jobs

The goal of this chapter is to provide the reader with practical full-link design and validation examples for Intel processor and chipset systems. After reading this chapter, the reader will get a peek at an intel architecture (IA)-based high-speed digital system project cycle from experienced Intel engineers. It ties together the concepts presented in previous chapters with the intent of giving the reader firsthand insights into cutting-edge Intel projects and helping readers kick-start their own IA-based products.

DESIGNING SYSTEMS WITH INTEL DEVICES

High-speed circuits are affected by all manner of noise sources on the die, in the board, or on the package. Today's high-speed circuits use adaptive circuits to compensate for the loss and remove noise from the received signal. All of these items influence the circuit performance and must be accounted for in the design to produce an accurate result. A generalized signal flow is shown in Figure 6.1. This section will provide an introduction to the equalization components and a variety of simulation methodologies used by signal integrity engineers to carry out the analysis of a full high-speed link, as shown in the figure. It integrates the models introduced in Chapter 3 and links to the equalization circuit theories of Chapter 4. After laying out how to evaluate the system design performance of a full link, readers should be able to correlate their design and simulation results to the lab validation results as discussed in the next section of this chapter.

INTERCONNECT MODEL

The interconnect models represent the board and package interconnect losses, reflections, and crosstalk. Interconnect models may come from vendors, 2D or 3D pre-layout simulations, or post-layout extractions. As the data rate increases, it

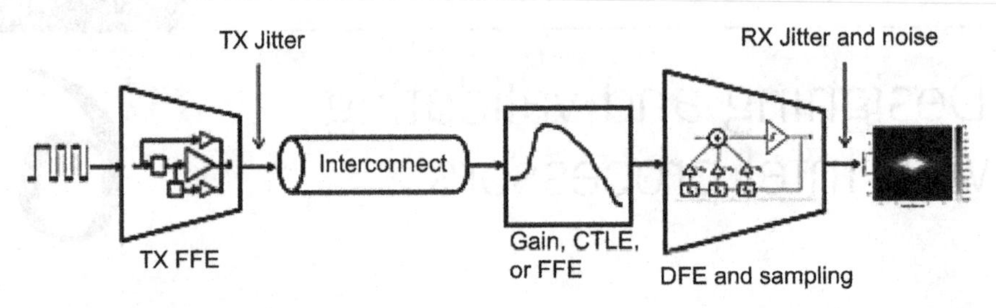

FIGURE 6.1

Simulation flow.

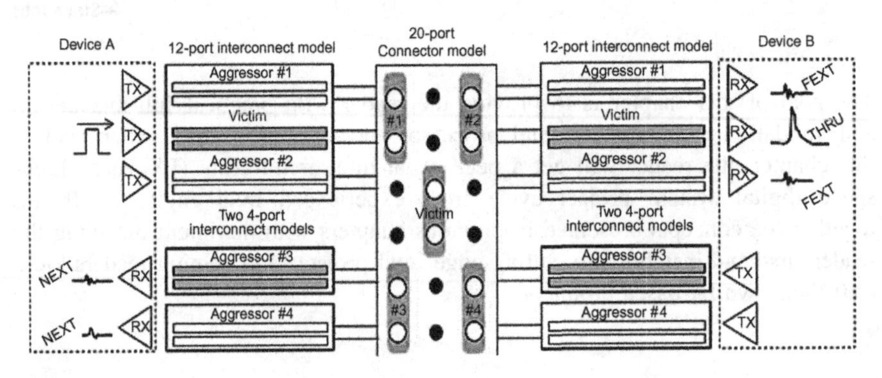

FIGURE 6.2

FEXT and NEXT interconnect models.

becomes more important to carefully account for noise sources in the system. This section discusses an interconnect example utilizing a connector with five crosstalk aggressors.

Inclusion of the relevant crosstalk sources in the system may require several models. In a transmission line, it is often sufficient to model only the next adjacent signal for coupling. The crosstalk on the second pair is often negligible. It is therefore common to see transmission line models created with 12 ports (3 pairs). However, vertical structures such as a connector or package ball grid array may contain many crosstalk aggressors on a single victim. An example of an interconnect model to capture the crosstalk in a 20-port (5-pair) connector model is shown in Figure 6.2. Since the connector is 20-port and the available interconnect model is 12-port, a combination of different models will be used. In the example, the 12-port models include all the package, via, and transmission line elements occurring before and after the connector. Additional 4-port interconnect models are used to properly account for the interconnect before and after the #3 and #4 aggressors located in the connector. The 4-port models may be a copy of any of the 12-port through-channels or newly created to represent the specific lanes.

Table 6.1 Common Equalization Types

Category	Pro	Con
TX FFE	Does not amplify noise, cancels precursor ISI	Reduces low-frequency output and power limit, back channel required for adaptation
RX FFE	No power limit like TX FFE, adapts without back channel	Amplifies noise
CTLE	Long post-cursor reach, adapts without back channel	Varies with PVT, no precursor cancellation, amplifies noise
DFE	Does not amplify noise	No precursor cancellation, complicates CDR design

These 4-port models properly attenuate the coupling observed at the connector as it travels toward an end device.

The middle pair of the 12-port model is stimulated with a transmit signal, either a step or a pulse. All remaining input ports on the left and output ports on the right must be connected to the termination impedance of the respective device. A crosstalk response may be observed at every input and output port. The response to choose for the simulation input depends on the signaling direction (transmit or receive) of the interconnect. Only the crosstalk response that is observed at a device receiver is the input to the simulation. Aggressors #1 and #2 provide far-end crosstalk (FEXT) waveforms at the output of the 12-port model. Aggressors #3 and #4 are determined to drive in the opposite direction and provide near-end crosstalk (NEXT) waveforms on the near side of the victim stimulus. The NEXT crosstalk waveform is caused exclusively by the connector model.

EQUALIZATION MODELS

High-speed circuits often contain several equalization circuits to compensate for interconnect noises and losses that must be included in the full link simulations for an accurate representation of the performance. A brief description of common equalization schemes is shown in Table 6.1. Implementation of the equalization for common interconnect simulations is often behavioral, though it varies depending on the simulation approach. These implementations are discussed throughout this section. Circuit design techniques for these equalization schemes are discussed in Chapter 4.

Transmit feed forward equalization

Transmit equalization is a popular method usually implemented with a finite impulse response (FIR) filter to remove dispersion or inter symbol interference (ISI) that is introduced by the channel. The signal is reshaped before being transmitted through the channel. Reshaping before the cursor (the bit being equalized) is pre-emphasis and after the cursor is de-emphasis. The equalization capability is limited by the transmit power. Use of pre-emphasis or de-emphasis reduces the peak output voltage swing.

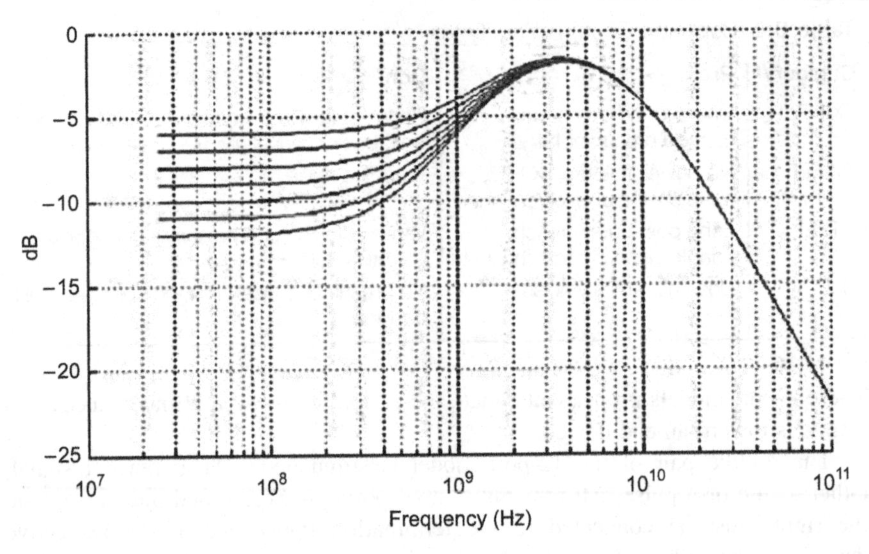

FIGURE 6.3

An 8-GT/s zero-pole CTLE.

Continuous time linear equalizer

Active designs amplify high-frequency content, while maintaining or attenuating low-frequency content. Circuit degeneration R and C values that affect the first pole and zero must be tuned to provide amplification at Nyquist frequency, while minimizing nonlinearity. A passive continuous time linear equalizer (CTLE) design attenuates low-frequency content and passes signal content at Nyquist frequency with unity gain. Generally, passive CTLE designs are linear. Depending on the design, additional gain stages may be used to attenuate or amplify the resulting waveform to a desired operating voltage. A passive CTLE design following the PCI Express (PCIe) Base 3.0 specification is shown in Figure 6.3. The design contains seven tuned zero-pole combinations that are useful to equalize various amounts of ISI. The tuned selection is typically a function of channel length, increasing the CTLE to compensate for more pulse dispersion.

Decision feedback equalizer

The decision feedback equalizer (DFE) is an effective way to remove post-cursor ISI. The DFE compensation responds to the sampled symbol (and is therefore nonlinear) and applies an FIR filter to remove the ISI. In circuits with a TX or RX FFE and a DFE at the same post-cursor position(s), there must be an optimization of the balance between FFE and DFE equalization to find the optimal margin. When there is no TX or RX FFE circuit, the DFE will fully equalize the post-cursor waveform. There are several DFE implementation techniques, including fixed position FIR, floating FIR, and infinite impulse response.

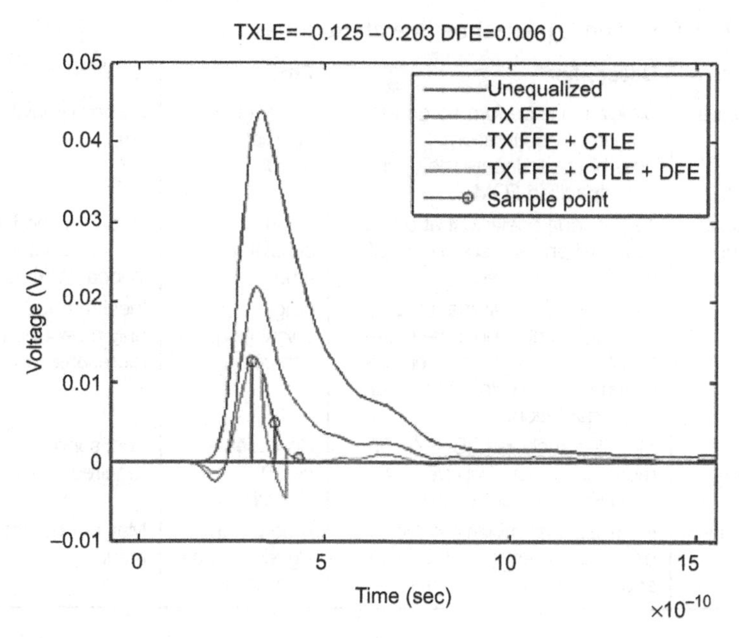

FIGURE 6.4

Pulse response reshaped by TX and RX equalization.

Equalized pulse response

The removal of ISI by each technique is well understood through observing the wave shaping that each stage performs on a single pulse response. For example, a transmitted 16-GT/s pulse with an amplitude of 1 V is dispersed through a 25-dB channel. A transmit FIR filter applies 8 dB of total boost distributed between pre-emphasis and de-emphasis. The overall amplitude of the cursor is reduced. The transmit FIR filter effectively reduces the ISI from dispersion throughout the waveform, including the precursor. The CTLE filter from Figure 6.4 with 6 dB of peaking gain is applied to the pulse response. The passive CTLE design has decreased the pulse amplitude while attenuating lower frequency noise. Finally, the receiver sample points are added to the pulse to calculate two post-cursor DFE taps. The magnitude of ISI at the sample point is equal and opposite to the DFE current, effectively eliminating the sampled ISI.

AUTOMATIC EQUALIZATION ADAPTATION

The number of ways that equalization may be configured on high-speed links is great. Complexity is increased due to equalization at both ends of the link, power consumption, gain limitations, and the number of digitized steps. Simulation is a powerful means to search through equalization possibilities to determine the

Table 6.2 Common Equalization Search Routines

Routine	Description	Pro	Con
Exhaustive	Simulation is performed on the full grid of equalization possibilities to locate the optimal user-specified FOM.	Confidence in optimal margin	Longer simulation time
Coordinate Descent	Iterates one coefficient at a time. Stops when user-specified FOM no longer improves.	Short simulation time	Requires seed value, risk of algorithm trap in local maxima
Zero Forcing	Calculates EQ that makes a targeted sample point nearest 0 mV of ISI, subject to step size. Commonly used for behavioral DFE adaptation.	Single calculation, no searching	Ideal removal of ISI, only a behavioral representation
Vendor's Module	Modules such as IBIS-AMI can represent exact adaptive behavior of vendor circuits.	Represents circuit algorithm	Tool support required
Least Mean Squares	Finds EQ that minimizes the error between the equalized waveform and a specified target value.	A common algorithm, use with any FOM	May not be optimal result

maximum achievable margin on a link. Automatic adaptation can be used to locate the best equalization setting or to mimic the adaptive behavior of today's higher speed circuits.

Some tool vendors offer multiple search options, while others may only exercise one search routine. The most common routines are shown in Table 6.2. The exhaustive and coordinate descent routines must use a figure of merits (FOM) to determine the best equalization such as eye height, eye width, sampler margin, or eye area. Different FOMs will have different margin results. The decision of which routine or FOM to use depends on the purpose of the simulation. For non-adaptive circuits, the most robust routines (such as exhaustive searches) may be used to confidently decide which equalization configuration should be used. The same FOM used for the full link pass/fail criteria should be used during the search. Use of different FOMs will vary the optimal equalization results. An additional caution is that using a different analysis method from that used for the full link pass/fail criteria can lead to suboptimal results. For example, if a peak distortion analysis (PDA) is used to find the optimal equalization (for any FOM), which is then applied to a statistical pseudo-random bit sequence (PRBS) simulation, then there can be a question as to whether the optimal equalization was truly correct. The issue is the data pattern dependency of equalized ISI and crosstalk in the channel. The cause is that equalized bits will carry different significance in the eye-opening computation between the PDA and statistical methods.

When circuits are adaptive, an effort should be made to select the search routine and FOM that most closely represent the actual circuit algorithm. Some guidance can be provided if the routine is uncertain or not yet defined. The coordinate descent and least mean squares algorithms are the most similar to common FFE and CTLE adaptation circuits. Voltage FOMs are the most likely choice for adaptive circuits, and while timing FOMs are possible, they are more complex to implement and require longer training times. Some chip manufacturers provide transmitter or receiver models in the form of Input/Output Buffer Information Specification, Algorithmic Modeling Interface (IBIS-AMI) modules. These modules contain any adaptive routines used by the circuit (complete with the FOM) in a protected intellectual property format that can be used in many electronic design automation (EDA) tool suites. When available, the IBIS-AMI module is the recommended choice.

PERFORMANCE ANALYSIS

Simulation determines the likelihood of bit errors in a system of many knowns and unknowns. In the link analysis, both the knowns and the estimations of the unknowns are combined to predict the circuit performance. Most of today's high-speed links specify the performance requirements as a bit error rate (BER). The BER is the ratio of incorrectly received bits to the total transmitted bits over an observed time. For example, the requirement of the PCIe Base specification is a BER of 1×10^{-12}. A comparison is made in Table 6.3 of various simulation methods to analyze performance.

Spice bit by bit

A spice simulator driving a transmitter circuit with a data sequence, transistor or behavioral, can pass the signal through the interconnect and several equalization stages as shown in Figure 6.5. In this simulation method, any equalization stages that require a feedback loop can only be simulated in an analog mixed-mode

Table 6.3 Comparison of Simulation Methods

Simulation Type	Achievable Number of Bits	Simulation Time	Jitter Model	Crosstalk Treatment	Simulation Linearity
Spice Bit by Bit	Up to 1×10^4 is practical	Hours to days	Budget with eye mask	Simulated phase	Nonlinear
Convolution Bit by Bit	Up to 1×10^7 is practical	Hours	Budget with eye mask	Adjustable, worst case, or specified	Linear only
PDA on Pulse	~Infinite	Seconds	Budget with eye mask	Adjustable, worst case, or specified	Linear only
Statistical on Pulse	~Infinite	Minutes	Jitter PDF	Adjustable, worst case, or specified	Linear only

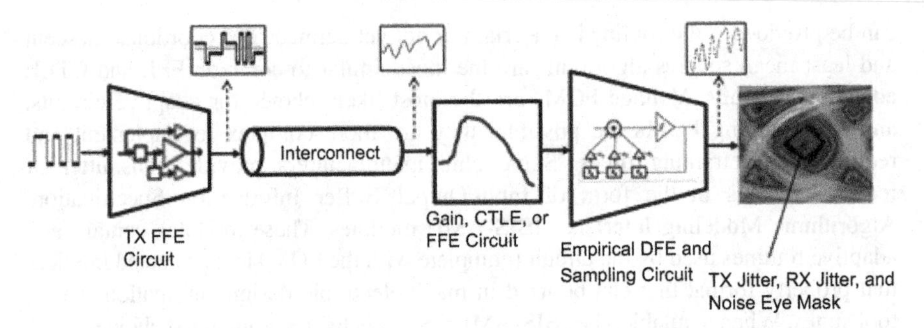

FIGURE 6.5

Bit-by-bit simulation flow.

environment. The spice bit by bit is the only method capable of modeling the nonlinear behavior of the transmitter and equalization stages. For this reason, this is the required method for circuit design. This method does have significant disadvantages. When representing the transmitter with precise transistor models, the simulation time can be extremely long to achieve a relatively short bit sequence. Even for simplified behavioral spice models that simulate faster, the reasonable number of bits achievable in simulation is lower than what is common to industry specifications. When circuits are demonstrating significant nonlinearity, the spice results may be compared to one of the other methods to understand the performance error in linear simulation methods. The spice bit-by-bit approach may model the equalization stages as actual circuit implementations or behavioral models (common for TX FFE).

Generally, spice bit-by-bit simulation does not include transmitter or receiver silicon uncertainties. The estimations for these uncertainties are represented by an eye mask applied to the received eye opening. Crosstalk may be included in bit-by-bit simulations, but the alignment of the crosstalk aggressor(s) is fixed by the phase of the simulated models. The crosstalk is added to the final victim signal, and the phase cannot be changed in post-processing.

Empirical convolution

Convolution methods require the channel impulse response to reconstruct a transient waveform at the system output. To obtain the impulse, the derivative of the step response at the end of the channel is performed. The simulation path of the step response is shown in Figure 6.6 for the empirical convolution method as well as for the next two analysis methods to be discussed. While impulse response convolution is shown with equalization impulse(s), it is possible that the step response is passed through equalization circuits if the simulation is performed in a spice environment. Convolution methods are used to obtain simulated eye-opening metrics with longer dwell times more quickly than spice bit-by-bit simulations. The convolution method may be favored to provide longer dwell time while still using detailed transistor models. The convolution technique does

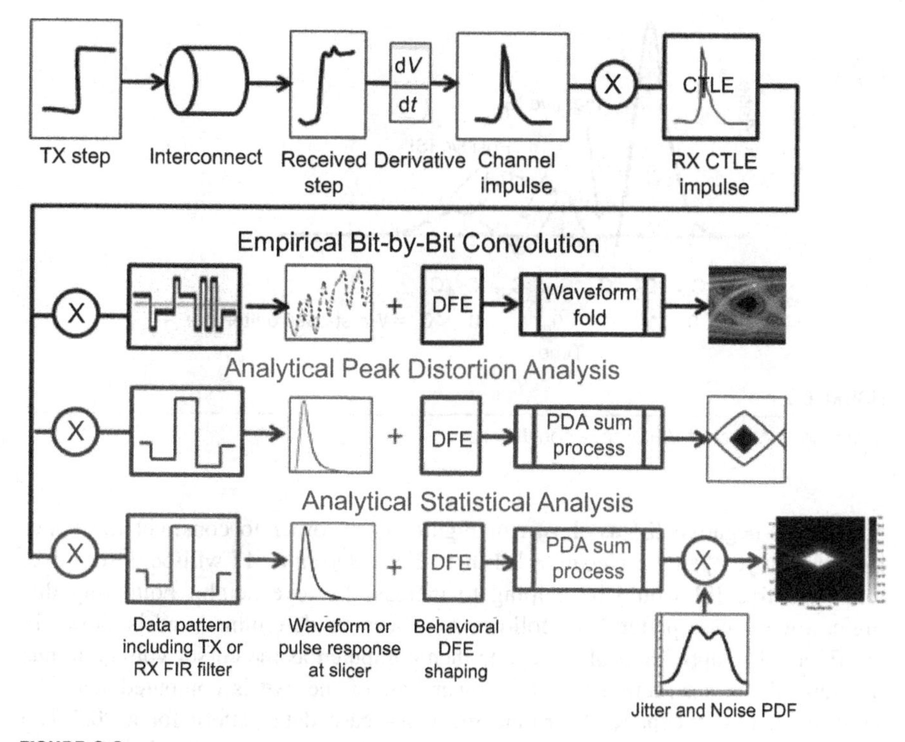

FIGURE 6.6

Signal flow for empirical, PDA, and statistical methods.

assume linearity and neglects any nonlinear behavior in the transmitter or equalization blocks. Simulation results with convolution may be compared to spice bit-by-bit results to quantify the error from nonlinearity that is absent in the convolution method. Similar to spice bit by bit, silicon uncertainties are not commonly simulated and must be accounted for through an eye mask.

In the convolution approach, TX FFE is applied by convolution of an ideal equalized pulse with the channel impulse. Likewise, receiver CTLE is applied by convolving the CTLE impulse response with the resulting TX FFE + channel impulse. The waveform shaping due to DFE is applied to the time domain response prior to constructing the eye diagram.

Peak distortion analysis

An analytical means to determine the eye opening in a channel is with the peak distortion analysis (PDA) method. The PDA method uses the pulse response at the end of a channel to construct an eye opening. The upper half ("1") and lower half ("0") of the eye opening are computed separately. Following the cursor (C_0) for a single "1" pulse, the voltage of each sample position C_{+N} is evaluated to be

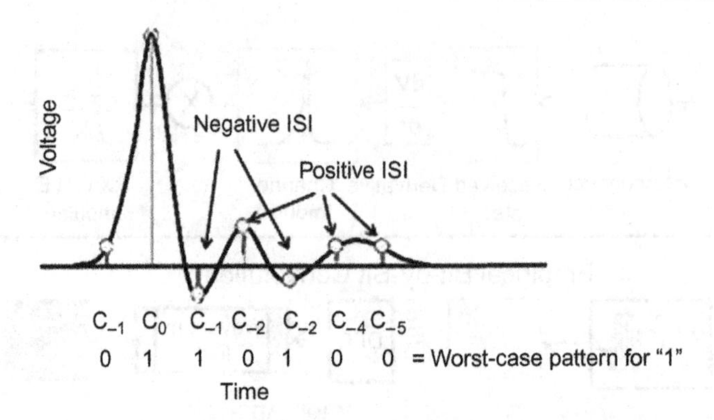

FIGURE 6.7

Selection of cursors for PDA eye construction.

positive or negative ISI as shown in Figure 6.7 in order to construct the upper half of the eye opening. Negative ISI that occurs during a "1" will be destructive, while positive ISI would be helping to increase the eye height. Following this logic, a worst-case pattern that follows a "1" can be determined and is shown in the figure. The upper half of the eye is then computed as the cursor voltage minus the sum of the destructive ISI. The lower half of the eye is computed with the same procedure except to determine the worst-case data pattern for a "0." The PDA method can be expanded to include crosstalk waveforms following the same procedure to determine a worst-case aggressing data pattern and subtracting the summation results from the cursor.

The signal path for PDA is shown in Figure 6.6. One difference from spice convolution is that the final eye-opening result is a single contour as shown in Figure 6.8. Waveform and eye density is not shown as it was in last two methods. PDA is always used with an eye mask to define the minimum eye height and width values. The eye mask is a budget to account for unsimulated silicon jitter and voltage noise. The difference between the simulated eye parameters and the eye mask is the margin. When the margin is small the channel design may be at risk of failure if any additional noise sources occur.

It may be cautioned that the correlation between PDA and measured performance begins to degrade at 5 GT/s and has significant gaps by 8 GT/s and above. The certain cause for miscorrelation is the pathological selection of the worst-case data pattern in the eye construction. In general, the probability of observing a "1" or "0" and creating destructive or constructive ISI is completely random (probability varies for encoded data). Therefore, there is a probability related to the cumulative occurrence of each bit in the worst-case pattern. Nevertheless, there are still uses for PDA in high-speed links. A few of the common uses include equalization searches and quick design tradeoff comparisons.

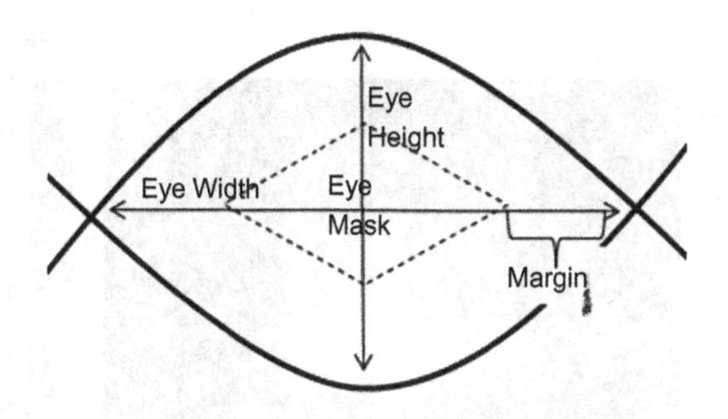

FIGURE 6.8

Eye example for PDA analysis.

Statistical analysis

As the data rates increase above 5 GT/s, the channel ISI carries a decreased likelihood of contributing to eye closure. This leads to a larger discrepancy between the probabilities associated with the worst-case analysis of PDA and the actual link performance. The need for a statistical approach is due to the dependent relationship between the probability of destructive ISI and the number of sampled ISI terms or data rate. To illustrate this, we can assume Figure 6.7 represents the fully settled pulse response of a channel sampled at 5 GT/s. We can calculate the probability of C_{+2}, C_{+4}, and C_{+5} contributing destructive ISI to a "0." If each destructive bit has a 50 percent chance of being a "0" or a "1," then the probability of the named three prior bits being "0" at the same time is the multiplication of the three probabilities: $0.5 \times 0.5 \times 0.5 = 0.125$. If the data rate is increased to 10 GT/s, the pulse width will decrease and the ISI may change, but the settling time is unchanged because it is a function of channel length. At double the data rate, there are now twice the sampled ISI voltages over the same duration that contribute to the eye opening. For simplicity, it can be estimated that the three destructive ISI samples of the positive ISI have doubled to six. The probability of six prior bits being "0" at the same time is 0.015. In this simple example, the probability of the worst-case eye opening at 5 GT/s is 12.5 percent and 10 GT/s is 1.5 percent. For a high-speed link with a long settling time, the worst-case sequence used in PDA approaches an infinite probability.

Many EDA tools provide statistical analysis, in addition to public tools available online such as StatEye and SeaSim. The methods employed by these tools are an analytical calculation to determine eye opening over time without having to calculate the waveform for each bit empirically. The simulation diagram for statistical analysis is shown in Figure 6.6.

To complete a statistical analysis, the continuous pulse response is digitized into many bins or slices. A probability of a random bit sequence is combined with the ISI bins to create an ISI probability density function (PDF). A cumulative density function is created for each bin by integration, and the family of bins

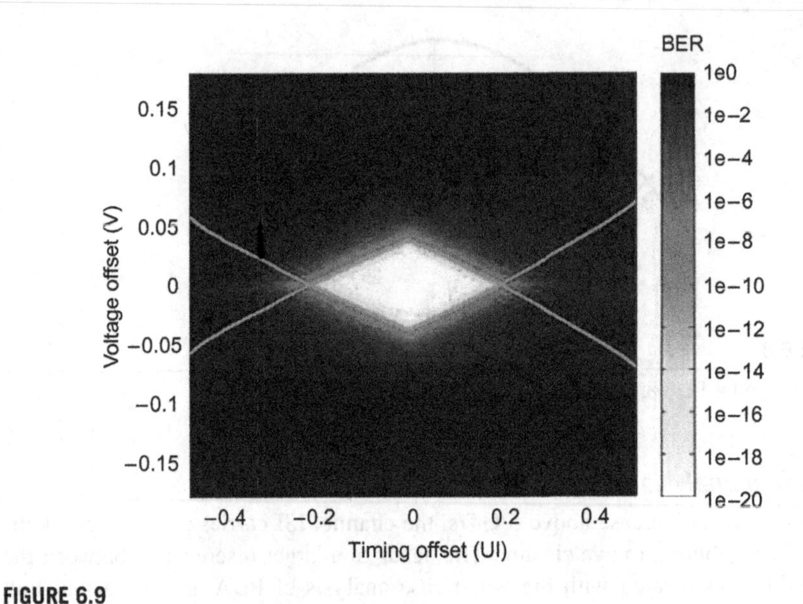

FIGURE 6.9

Example of BER versus eye contour for statistical method.

are plotted together to create a 3D eye contour as shown in Figure 6.9. A detailed explanation of the statistical method is available by Casper [1, p. 479–486].

SOLUTION FROM DESIGN OF EXPERIMENTS

The statistical behavior of a system over high-volume manufacturing can be modeled with a design of experiments (DOE). The DOE allows the development of a predictive response surface model (RSM) with fewer data points than otherwise needed to exhaustively cover a parameter space. Statistical conclusions about how signals change due to interconnect or silicon variation can be drawn from the RSM results. Empirical simulation of large distributions to understand statistical importance or defects per million is not necessary.

One such tool for creating an RSM from DOE is JMP by SAS. The prediction profiler and simulator features are shown in Figure 6.10. A predictive model for seven input variables is shown: transmitter termination, package impedances, board impedances, and PCB length. The RSM model is fit to a simulated eye height (EH). The EH was simulated for a 16-GT/s data rate on a channel with an insertion loss of approximately 25 dB at 8 GHz. The signal is equalized with TX FFE, CTLE, and DFE.

The sensitivity of each individual parameter to the output is shown. The model may be used to locate parameters with excessively high variability. Parameters with suboptimal nominal values that may require optimization can be discovered.

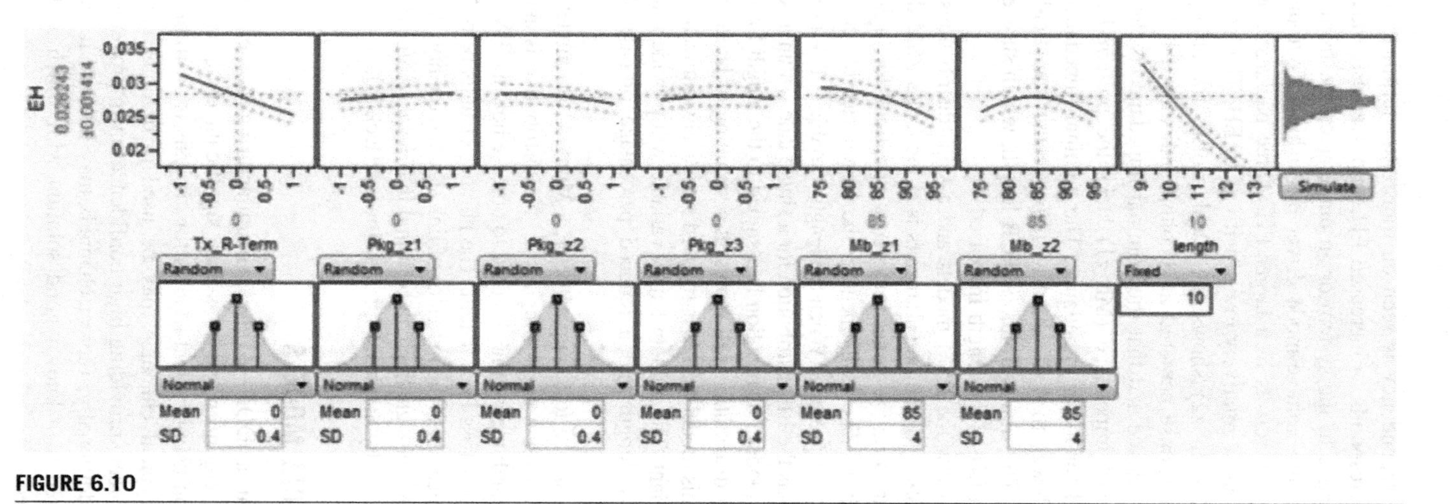

FIGURE 6.10

Prediction profiler model and simulator.

In the experiment, a high dependency is seen on transmitter termination, where lower termination (−1) corresponds to increased EH. Improved EH may be due to a better matched impedance to the system or an output increase from voltage-mode drivers. In this case, the voltage-mode driver has a higher output voltage with lower termination, which results in a larger EH. Three package impedance variables are modeled that show relatively low sensitivity to EH. The last motherboard impedance variable (Mb_z2) shows the ideal profile for an optimized impedance. The "frown" response centered at 85 ohms indicates that the EH is highly sensitive to the impedance and that the nominal impedance is the optimal result. The first motherboard impedance (Mb_z1) lacks the preferred response and indicates improved EH for lower impedance. The nominal impedance of this design should be placed lower than 85 ohms to improve the statistical performance of the interconnect. The final variable is PCB length, which showed a significant downward trend in EH as the length increased.

The simulator appears below the RSM models and assigns distributions to each variable. The standard deviation for all the parameters is at the tool's RSM default of 2.5 sigma and is not meant to define target impedance variations for high volume manufacturing (HVM) analysis. The assignment of the length variable changes the physical reality that the model represents. When the length variable is fixed to a single value, the RSM model predicts the EH distribution for a single lane within a port existing at the fixed length. If a random distribution is assigned to the length for the same sample size, then the output distribution will represent a sampling of lanes at various lengths. The exposure that this model creates is reduced statistical significance for longer lanes with lower eye height. The random length variable will provide fewer modeled cases at the maximum length and a reduced percentage of cases below a specification limit.

Once distributions are assigned to each parameter, a Monte Carlo model is simulated to predict performance over a large sample size. No distribution is assigned to the length parameter. The length metric may be used to observe how the HVM performance is expected to change across PCB length. The number of units below a specified target or industry specification is the indicator of link risk and quality. A thorough discussion on DOE and RSM for signaling can be found in *Advanced Signal Integrity for High-Speed Digital Designs* by Stephen Hall and Howard Heck [2].

SOLUTION FROM TYPICAL MODELS

Performing simulations in a DOE to cover manufacturing corners is time-consuming for the design engineer and consumes significant computing power. Running a single simulation using typical models for analysis is desirable, but are the results meaningful? Typical channels may be used to demonstrate relative tradeoff among design choices, providing high confidence on improvement without the need for DOE. Additionally, typical channels may be used to determine pass or fail when compared to a known good solution with which a statistical analysis has been performed.

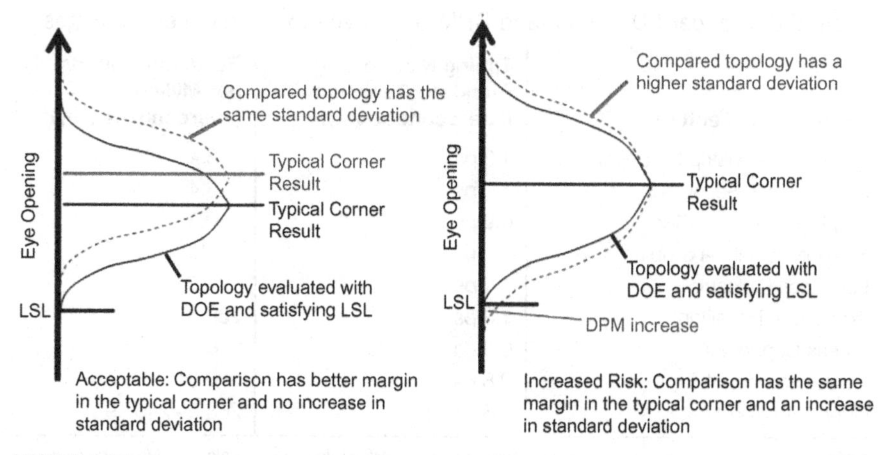

FIGURE 6.11

Distributions for acceptable and increased DPM risk.

Comparing the typical eye opening for a qualified topology for which DOE has been performed to an alternate design can offer the same level of confidence as the DOE under certain conditions. In order to perform a valid comparison, topologies must have the same statistical performance or standard deviation over volume manufacturing. At the first order, we can assume that standard deviation will be a function of the number of independent process variables. Process variables include silicon, package impedance, motherboard impedance, and mezzanine impedance. Figure 6.11 illustrates how two topologies with the same standard deviation may be compared to verify that a topology will have no more defects than a topology validated with DOE and DPM.

The graph on the left in Figure 6.11 illustrates the volume results for two topologies. For a baseline topology, a DOE and a DPM analysis are completed to determine the number of failures below a lower specification limit (LSL) that meets production requirements. A second topology with the same standard deviation is compared to the baseline by typical corner results and is determined to be acceptable. On the right side, another topology is compared to the same baseline. In this case, the standard deviation of the new topology is greater than the baseline. This topology has an increased failure count below the LSL when designed with the same typical corner result. In order to avoid this risk, it must be considered whether there is a standard deviation change in the compared topology.

What parameters really affect standard deviation and when can two topologies be compared? An experiment is performed that finds the standard deviation for several interconnect changes to understand when comparisons can be made. A baseline two-connector microstrip PCIe topology is chosen from which to make interconnect changes. The baseline topology length is chosen in order to satisfy an arbitrary requirement for this experiment. The 6-sigma standard of

Table 6.4 Standard Deviation and DPM Sensitivity to Interconnect Changes

Interconnect Feature	Timing Margin Standard Deviation (interconnect only)	Resulting Defects per Million (interconnect only)
Baseline: Microstrip 9X Separation	1.0 ps	3.4
Microstrip 5X Separation	1.0 ps	3.4
Stripline 10 mil Via Stub	0.9 ps	0.3
Stripline 80 mil Via Stub	1.1 ps	21.5
Low Loss Dielectric	1.1 ps	21.5
Add Layer Transition	1.3 ps	268.5
1 Less Connector	0.5 ps	0
Connector Mount Type	0.8 ps	0.01
Connector Form Factors	1.3–1.6 ps	268.5–2457.9

Note: *No silicon variables are used in this experiment. Standard deviation is calculated from predicted HVM results with impedance variables assigned normal distributions with three standard deviations to a stripline tolerance of 10 percent and microstrip tolerance of 15 percent. For simplicity, DPM is calculated from the standard deviation column for a standard normal distribution.*

3.4 units per million is chosen for this study. Alterations in standard deviation leading to more defections are recorded. Interconnect feature changes, standard deviation results, and DPM results are shown in Table 6.4.

The interconnect changes in the experiment demonstrated a wide variety of standard deviations. Choice of interconnect layer, signal separation, and dielectric loss are within ± 0.1 ps change, leading to a DPM increase of 18 units. New process variables like an additional routing layer (new impedance) and additional connectors are seen to have more significant impact in DPM. Connector mount types of through-hole mount (baseline) and press fit (experiment) were considered when comparing mount types and demonstrated a moderate change in standard deviation. Connector form factors were changed to alternative and nonstandard PCIe connectors with very different performance profiles. These connectors had a significant impact on standard deviation, greatly increasing the risk when comparing channels.

It should be noted that performance distributions are not perfectly normal and have tails, as shown in the prior section. When exchanging interconnect features, margins and DPM will significantly increase or decrease, preventing an ideal comparison of DPM without fine-tuning length in order to align typical corner margins. For simplicity, standard deviation provides a good enough merit to assess changes in the distribution.

Silicon variation plays a role not present in this study that should not be overlooked. The random variation of the transmitter and receiver will RSS with interconnect variation and may decrease the sensitivities shown in this study. If silicon variability is higher than the interconnect variability in Table 3.13, the DPM increases will be significantly attenuated.

Simulation results have also shown a moderate correlation between channel ISI and standard deviation. It is recommended to complete a similar study to assess standard deviation changes when silicon's interaction with the interconnect changes, including different interfaces, data rates, or equalization schemes. When these terms have changed, DPM analysis can be completed on a selection of different interconnect features to assess any significant standard deviation changes. The study reviewed here can summarize the following allowable changes for comparing typical corners:

Low-Risk Changes

- Stripline and microstrip layer choices
- Pair-to-pair separation
- Dielectric loss
- Connector-mounting types

Increased-Risk Changes

- Adding via or layer transition
- Adding or removing connectors
- Connector form factor
- Data rate, silicon, and equalization

The benefit of using the typical corner will accelerate layout optimization. With low-risk changes, including common layout tradeoffs like PCB loss, board layer, and separation, effective design decisions can be made in less time.

SYSTEM VALIDATION WITH INTEL DEVICES

Perhaps the most exciting time in a project is when your design is ready for the first power-on. Rarely does everything work just right as soon as you flip the power switch and attempt to boot. But with careful preparation, you can make this process as smooth as possible. In this section, we will discuss some of the steps you can take before, during, and after power-on to get your product to market with quality and efficiency.

POWER-ON PREPARATIONS

Proper planning can make the difference between a smooth power-on experience and a hectic period of chaos. It is strongly recommended that considerable thought and planning be put into the power-on long before the power-on actually occurs. Following is a checklist of some common items that should be considered:

1. Test and probing capabilities—During the design phase, consideration should be given to provide access for probing critical signals or providing workaround options for risky circuits.

2. Access to rework equipment and spare components—If there are any design or manufacturing errors, you will want to quickly work around the issues. You will need ready access to soldering equipment and spare components.
3. Access to measurement equipment—Equipment for troubleshooting power-on issues is essential. This includes multimeters, oscilloscopes, logic analyzers, and a ITP-XDP (JTAG) debugger.
4. Access to experts—You will want key design, software, and validation team members available to assist you with any issues that might arise.
5. Power-on BIOS—Depending on the design, there may be several I/O-specific deliverables to the BIOS. For buses like PCIe, you may need to specify design-dependent equalization settings (TX equalization settings, CTLE peaking settings, etc.). Be sure to review component specifications for guidance on what settings will need to be adjusted for your design.
6. Power-on software—A careful review of all required software, drivers, utilities, and test programs should be done prior to power-on.
7. Access to sufficient quantities of boards and components—The validation methodologies discussed in this section refer to the need to test certain minimum quantities to evaluate design health for high-volume manufacturing. Ordering or reserving the appropriate number of boards is best accomplished in the pre-test preparation stage.

TYPES OF I/O DESIGN VALIDATION

Validation of your high-speed I/O design can be categorized by three main types of tests:

1. Functional testing—This is the most basic type of test. Can the I/O reliably operate without errors across variations in boards, silicon, end-point devices, data patterns, and environmental conditions?
2. Specification compliance testing—This is the traditional approach to validation of an I/O, but is not necessarily the most efficient or cost-effective. The specific details are documented in the relevant industry specifications, such as the PCI-SIG PCIe Card Electromechanical Specification (CEM). It is often required for certification, but is generally done in a limited fashion. For example, certification often requires demonstrating that a single board meets the required specification. While an important part of the certification process, this alone is not sufficient to provide confidence that the product will meet quality metrics over high-volume production.
3. I/O system margin testing—Using on-die features, I/O system margining provides an efficient evaluation of the I/O performance covering the entire link. This is accomplished through silicon features that allow voltage and timing offsets at the receiver to measure the width and height of the eye during actual system bus operation. Various bit patterns are sent on the

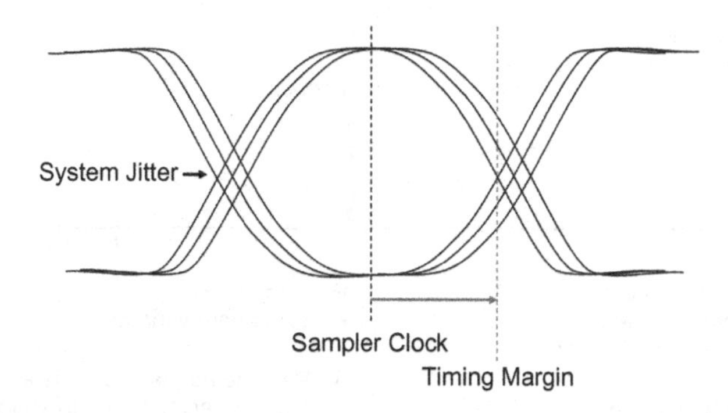

FIGURE 6.12

Timing margin at the receiver.

interface, and through different mechanisms, checked to verify that the pattern was received without error. If no error is found, the voltage or timing offset is incremented for that direction, and the process is repeated. When a miscomparison of the pattern is found, the offset being applied is recorded as the margin boundary. Figure 6.12 shows a conceptual diagram of how the timing margin can be captured.

While all three methods are important in an overall validation plan for a high-speed I/O design, the remainder of this chapter will focus on I/O system margin testing and the typical capabilities and methods provided by Intel for validation of their devices. Memory buses, like DDR3 or DDR4, have unique challenges compared to high-speed serial I/O buses like PCIe. For this reason, memory buses and high-speed serial I/O buses will be discussed separately. However, we'll start with a basic overview of system margin validation as it pertains to both memory buses and serial I/O buses.

SYSTEM MARGINING VALIDATION OVERVIEW

A system margin provides a holistic view of I/O health because it includes the full effects of the complete interconnect and is a measurement of the eye as seen at the receiver. This is accomplished in different ways depending on the circuit design but results in the ability to determine the eye height and eye width and, in many cases, can even create a composite eye diagram.

The challenge of any validation exercise is that it must rely on a limited snapshot of the design behavior to provide confidence in the lifetime behavior of the design. Additionally, most validation coverage schemes will only evaluate a limited set of pre-production parts, yet attempt to project the behavior of the total system across HVM.

Evaluation of the system margin result requires establishing how big the eye must be to ensure that the I/O bus will meet the specified BER requirements

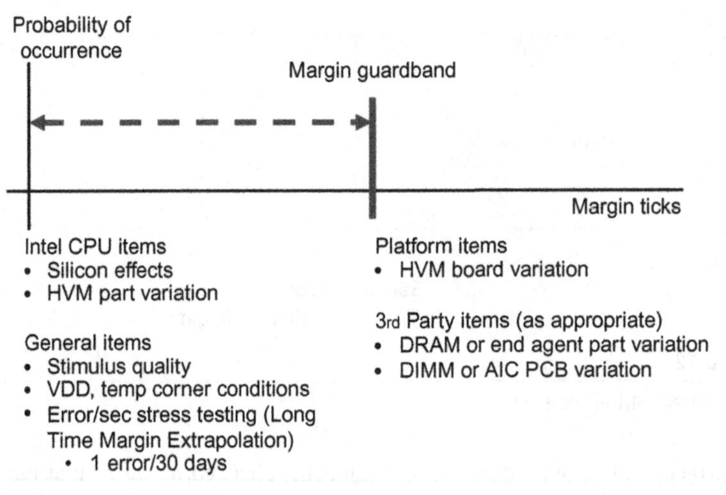

FIGURE 6.13

Margin guardband.

across HVM. This "margin requirement" or "margin guardband" is determined by Intel and includes several items as shown in Figure 6.13.

Margin is not a constant; it is a distribution. For example, there is some variation from part to part, or even from lane to lane. Other factors, such as voltage and temperature, can also impact margin. Margin can even change from boot to boot. The guardband is provided to account for these factors, which may not be fully expressed in the small number of test runs conducted on a particular design. Intel typically recommends collecting margin measurements on a minimum of 5 systems with a minimum of 5 repeats on each system. This is often referred to as a 5 × 5, or 5 systems × 5 repeats. This provides the basis for evaluating the margin results as a distribution instead of a single value. Intel margin guidelines take into account this minimal sample size along with other key factors, such as the length of the margin test and other factors that are not part of the test (board variation, silicon variation, etc.). The idea is to predict, over high volume, the quality of the system interconnect. With a distribution, it is possible to extrapolate to determine how many systems would be at risk of dropping below a minimum margin guardband where excessive BER or user-visible errors may occur (Figure 6.14).

The evaluation of the margin results collected in a 5 × 5 requires engineering judgment and assumptions as to the minimum margin that is sufficient to build confidence that the design will work in the real world. As shown previously, Figure 6.13 provides an illustration of what components may be part of a margin guardband requirement. If corner conditions for temperature and VDDQ (the I/O power rail) can be applied in the margin data collection, the guardband can be lower than if margining is done only with nominal temperature and VDDQ conditions. Or, if every board manufactured can be tested, the guardband can be

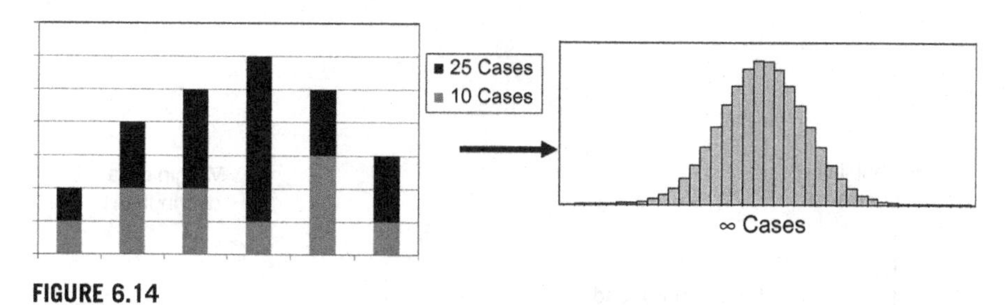

FIGURE 6.14

Sample quantity illustration: Gaussian distribution expectation.

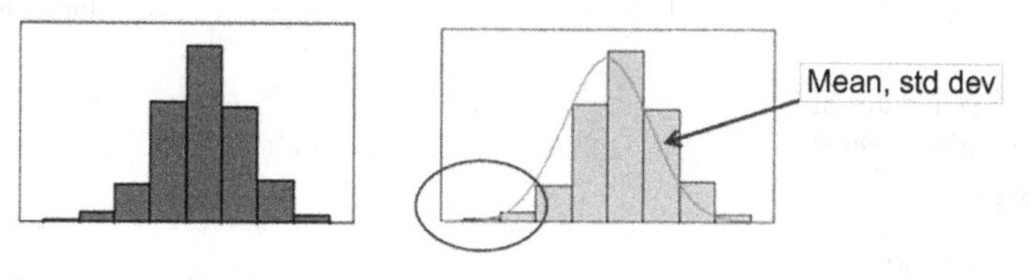

FIGURE 6.15

Sample quantity illustration: Gaussian distribution expectation.

lower than if margining is done only with 5 pre-production boards. However, a 5×5 is generally recommended as the minimum validation quantity in order to control the amount of total testing that must be done.

Statistical principles indicate that any margin result is possible, given enough samples. If a 5×5 collection (25 samples) shows 10 steps of margin as the worst result, it is possible that 50 samples may have indicated 9 steps as the worst result. It is statistically possible that, with an even greater number of samples, 8, 7, or 6 steps, or even worse, would be revealed. However, to stay efficient, we can use the distribution profile of the margin behavior found from the 5×5 to extrapolate for the behavior of a larger sample size.

Figure 6.14 provides an example of forming a Gaussian distribution with a greater sample count. In the example, 10 samples capture a certain collection of margin behavior. With 25 samples, more results are included, and generally form a recognizable Gaussian distribution shape. As the sample count goes to infinity, it is generally expected that the distribution shape will continue to fill in.

Figure 6.15 shows how we can use a 25-sample set to fit a distribution curve to the data, using statistical software to perform a Gaussian distribution fit on the 5×5 margin results. This fitting exercise identifies a projected mean and standard deviation for the distribution behavior. This same distribution fit is then applied in Figure 6.16. The distribution fit provides a probability function for predicting the tail-end behavior via extrapolation; that is, the probability that 8 steps, 7 steps, 6 steps, 5 steps, etc., will occur. This is compared to the margin guardband established for the 5×5 condition. The area under the tail to the left

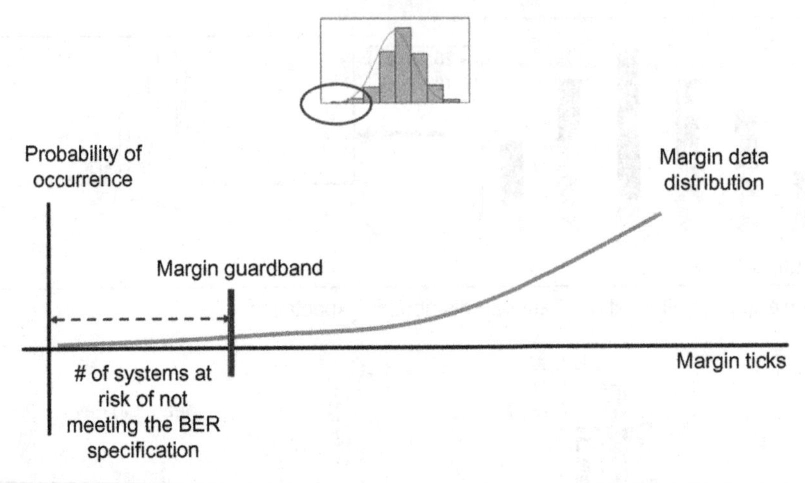

FIGURE 6.16

Margin guardband.

of the margin guardband line is described as the units per million (UPM) quantity. This UPM number represents the number of units out of a million that would be at risk of exceeding the expected BER, given the guardband assumptions.

The UPM approach is a more realistic way of judging risk than a static eye mask. Meeting a static eye mask requirement is a common way to achieve a specification or logo requirement but may not provide a full picture of design health. If the eye mask's guardband assumptions are too conservative (larger than necessary), the system will not likely produce user-visible error, but there will inherently be extra cost to the system designer due to overdesign. A static eye mask also ignores statistical principles, which indicates that given enough samples and a long enough run time, there will be an eye transition that violates the eye mask. The key is whether the violation matters or not.

UPM is a way to quantify if the inevitable occurrence of a negative margin is something that a system designer must care about. Clearly, if 100,000 out of 1 million manufactured systems have negative margin issues, this will be a significant problem as many systems sold to end customers would have to be returned and replaced by new copies, which would be a costly endeavor. The concern would be much lower if it were only 5 systems out of 1 million, for example, even though it is a non-zero figure. That figure, whether it is 5, 10, 100, 500, etc., systems out of 1 million, is ultimately up to the system designer to identify, based on quality and reliability goals. It should also be kept in mind that margining typically considers only bit miscomparisons, which may not necessarily translate to user-visible correctable or uncorrectable errors (i.e., some buses have error detection and correcting or retry schemes).

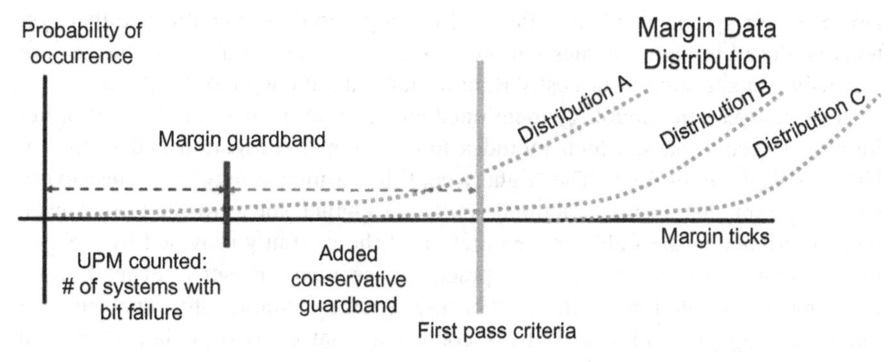

FIGURE 6.17

The first-pass methodology.

The calculation of UPM can be relatively quick to run once the guardband factors are identified and statistical scripts are written to calculate the Gaussian fit of the 5 × 5 data and extract the area under the curve for UPM. However, it is not an instant process, as the data must be collected and then analyzed to produce this answer. Intel often provides a margin criteria, which is intended as a quick first-pass identification of risk, in order to screen out those cases that would easily clear a UPM calculation. For those cases, meeting the first-pass criteria means that a full UPM calculation is unnecessary.

In Figure 6.17, three theoretical distributions are shown, representing 5 × 5 data collections. The margin guardband used for the UPM calculation is shown, along with the first pass criteria that have additional guardband added. This additional guardband is simply to establish a higher bar that, if cleared, indicates that a UPM analysis is expected to show low risk and does not necessarily need to be run.

- Distribution A has some data points from the 5 × 5 below the first-pass criteria. A UPM analysis is needed to determine the risk. In this illustration, there is some area under the tail below the margin guardband, so the UPM is a non-zero number.
- Distribution B has some data points from the 5 × 5 below the first-pass criteria. A UPM analysis is needed to determine the risk. In this illustration, there is no noticeable area under the tail below the margin guardband, so the UPM may be a very small number, perhaps approaching zero.
- Distribution C has no data points from the 5 × 5 that are below the first-pass criteria. A UPM analysis is not necessary, as it is expected that the calculation will always show a very small UPM result.

It is important to note that the concept of a hard "pass/fail" criteria has not been mentioned yet. For component validation according to an industry specification, this kind of criterion may be expressed as a requirement to achieve qualification of an industry logo or certification for marketing and compatibility purposes.

However, for system validation, the goal is simply to ensure confidence that systems produced in high volumes will not experience issues in the field that require expensive on-site support or costly Return Material Authorizations (RMAs).

The first-pass methodology mentioned above is not a "pass/fail" criterion but implies a second pass, which includes low-margin investigation and statistical UPM analysis as options. The highest confidence metric may be to test every single system that is manufactured under the actual software loads that each will experience in the field for the duration of the warranty provided by the system vendor. This is obviously impractical for many reasons. Instead, Intel recommends validation methods that use smaller samples to represent the manufacturing population and stress conditions that can extrapolate long-term margin behavior.

For this reason, the 5 × 5 methodology is employed. It is not intended to capture the worst margins that the system will ever produce but to provide a sample that can be used for statistical extrapolation. It should again be stressed that a quantity of 5 is considered a small sample size for making statistical assumptions, particularly about the variance, and especially when attempting to model the far tail of the distribution. The guardband attempts to account for this to some extent. The guardband and methodology make some simplifying assumptions about the distribution in an attempt to balance risk and validation effort and cost. Typically, the distribution is assumed to be Gaussian with a similar variance as seen by the higher volume testing done at Intel. If there are known concerns or other indications of issues, it is strongly recommended to increase the sample size.

DDR SYSTEM MARGINING VALIDATION

A unique challenge of system margin validation for the DDR interface is the fact that a single board layout topology may be required to support different Dual Inline Memory Module (DIMM) populations per channel at the highest speeds possible. A typical Intel design may include support for one, two, or three DIMMs populated per channel (DPC). For Intel server designs, two or three slot-per-channel (SPC) designs are most common. The ability to populate more DIMMs overall in a given system can be a capacity differentiator in the marketplace. Likewise, the ability to run those DIMM populations at higher speeds can be an important performance differentiator. Accordingly, the validation of a particular system design must be comprehensive enough to build confidence that all supported configurations and speeds have robust margins.

As mentioned previously, system validation attempts to build confidence for HVM by evaluating a limited set of pre-production systems. For DDR, the following components must be considered:

- Memory controller (silicon, package)
- System board (sockets, connectors, PCB routing, stackup, etc.)
- DIMM memory module (DIMM PCB routing, stackup, and DRAM)

In a typical 5 × 5 coverage scheme, 5 system boards will be populated with CPUs (acting as the memory controller), as well as with DIMMs, according to the configuration under study. The purpose of using 5 boards is to capture a representation of manufacturing variation from the PCB vendor. Accordingly, boards of the same design, but from different PCB vendors, will require separate 5 × 5 coverage for each PCB vendor in order to separately capture the manufacturing sample variation for each vendor. In the same way, the use of multiple DIMMs is intended to represent the manufacturing variation of the DIMMs and the DRAM or buffer silicon on the DIMM. A separate 5 × 5 will be necessary when testing with DIMMs from memory supplier "A" versus memory supplier "B." Additionally, since the silicon design and behavior are typically different for different DRAM part types, a separate 5 × 5 is necessary when testing with 2Rx4 DIMMs versus 2Rx8 DIMMs versus 1Rx4 DIMMs, and so on. Finally, other factors. such as memory interface speed, will each require a separate 5 × 5 data collection.

The very large scope of collecting margin results from 5 boards with 5 boots each, representing different PCB vendors, memory suppliers, DIMM part types, loading configurations, and operational speeds, calls for a margining tool that can execute very quickly. For DDR, Intel typically provides a utility, the Rank Margining Tool, which executes as part of the system initialization process (BIOS). Since the tool itself runs locally to the target, it can run with little overhead and can complete a margin assessment of all populated DIMMs in a matter of minutes.

Figure 6.18 is an example of the Rank Margining Tool output, which is expressed in terms of a number of steps, using the margining approach shown in Figure 6.1. The result shown indicates the number of steps from the nominal position that the strobe (RxDqLeft, RxDqRight, TxDqLeft, TxDqRight) or Vref (RxVLow, RxVHigh, TxVLow, TxVHigh) was shifted before a bit error occurred. This information is provided for each installed rank in each channel for each CPU. Nx indicates the node or CPU. Cx indicates the channel. Dx indicates the DIMM. Rx indicates the rank.

A classical method of margining and stress is to apply a stressor in the operating system, preferably one that can excite thorough stress patterns on all high-speed interfaces, as well as exercise the power delivery networks so as to generate power

	RxDqLeft	RxDqRight	RxVLow	RxVHigh	TxDqLeft	TxDqRight	TxVLow	TxVHigh
N0. C0. D0. R0:	−18	26	−27	31	−26	25	−31	30
N0. C0. D1. R0:	−11	28	−19	22	−26	25	−31	31
N0. C1. D0. R0:	−18	26	−27	31	−26	25	−31	30
N0. C1. D1. R0:	−11	28	−19	22	−26	25	−31	31

FIGURE 6.18

Rank Margining Tool example.

noise that could affect signal integrity quality. Intel's primary memory margining approach of using the Rank Margining Tool does not focus on producing these conditions, as these effects are limited through the use of a data scrambler, or otherwise through reserving guardband in the margin criteria formation. However, it can be useful to complement Rank Margining Tool testing with functional stress, to build confidence in the system design, or in cases where the system design may be expected to have greater than normal exposure to cross-interface or power delivery effects. For DDR in particular, it is always a good idea to perform some sanity checks by margining with real functional stress. It can also be useful for high-speed serial I/O, although the risk is much lower due to the differential signaling and typical random-like encoding of the bus traffic.

The quality of the stress involved requires its own investigation, especially as different designs may generate different responses to different stress programs. To identify an appropriate stress program to be used, it is recommended to try different software programs and evaluate how margins are affected. Intel does not make specific recommendations for software that will produce stress, although benchmarks and memory stress software exist commercially.

The duration of stress, or dwell time, is another factor that must be investigated. Worse margins are expected to be revealed when a stress is run for longer periods of time, due to the assumption of unbounded added white Gaussian noise. The Intel tools and associated guardbands take this into account when using the Rank Margining Tool. The appropriate dwell time for functional stress testing can be influenced by the reliability goals of the design, as well as practical considerations for testing.

In order to conduct functional stress testing, a fundamental capability that is usually needed is the ability to adjust the timing and Vref parameters in the operating system environment. This enables the margining approach to be used by identifying the boundaries at which errors occur. System designers are encouraged to contact Intel for tools that can enable this function. In this model, a determination of minimum margin may also be established in order to evaluate whether the margin found with functional stress is acceptable for design confidence.

HIGH-SPEED SERIAL I/O MARGINING VALIDATION

As with DDR, system margin testing on high-speed I/O interfaces provides the best method for determining the electrical quality of a given interface topology over HVM. Even when margin testing is performed on a small sample of systems (such as 5×5 testing), models can be applied to the measured margin data to provide a prediction of the number of systems over high volume that will exceed a certain quality criteria. For high-speed serial I/O, this quality criterion is the specified I/O bit error rate target for the given interface. For example, system margin testing can be used to estimate the number of systems over volume for which the BER on a PCIe link for a given design will exceed the specified 1e-12–BER limit. As a result, system margin testing has become an increasingly important method for validating high-speed I/O performance for both Intel product release qualification and system design qualification.

Intel provides various tools to assist with system qualification. The Intel® Electrical Margin Tool (Intel® EMT) is an example of a system margin validation tool that Intel provides to customers to characterize the high-speed I/O electrical quality of their platform interconnects. Intel EMT runs on the target system itself under an operating system and it electrically stresses the platform interconnects in functional mode until the detection of correctable link errors. To stress the receiver path (end point driving to CPU/PCH receiver) the tool will typically offset the receiver sample point in the voltage and time axis from the nominal (non-margining) position. To stress the transmit path (CPU/PCH driving to an end point) the tool will typically reduce the transmitted voltage swing and, if the interface supports it, add jitter to the transmitted output to stress the link timings. With this capability the Intel® EMT can be used to identify board routing issues, optimize equalization settings, and qualify platforms for high-volume production readiness.

For silicon product testing, Intel electrical validation teams will perform margin testing across a range of silicon process, voltage, and temperature conditions. While additional assurance of system product performance may be obtained by testing across these conditions, it is not required due to the added cost and burden of doing so. It is assumed that the silicon temperature and board voltage conditions under which the margin testing is performed reflect the typical use conditions the product would experience in the field when used by an end consumer. It is recommended that for interfaces that lead to an external connector (i.e., USB) various cable/device combinations should be tested to ensure robust electrical performance across a range of device configurations. If testing resources are limited, Intel recommends that priority be placed on covering topologies that result in the longest and shortest interface lengths. Additionally, when multiple ports exist on a platform, all ports should be tested to ensure coverage across ports. Target-based margining tools, such as the Intel® EMT, allow the possibility to integrate margining testing into the production test flow for high-volume manufacturing data collection and analysis. This high-volume data can provide extremely valuable data, especially if the design is pushing the solution space limits.

Typical data collection for assessing the electrical robustness of a high-speed I/O (HSIO) link consists of executing the recommended 5×5 testing method as mentioned previously. It is important to note that the 5×5 margin results are specific to the platform and configuration that was tested. Therefore, any modification to the electrical characteristics of the platform, including changes to interconnect topology, end-point device, or equalization settings, would require re-testing of the platform.

A typical margin analysis result is shown in Figure 6.19. Shown is a "margin diamond" plot for a single port where the average system margins for the 5×5 runs are shown on the left side and the minimum margins observed across the 5×5 are shown to the right. The corners of the diamond represent the high, low, right, and left margins in steps from the nominal sampler position.

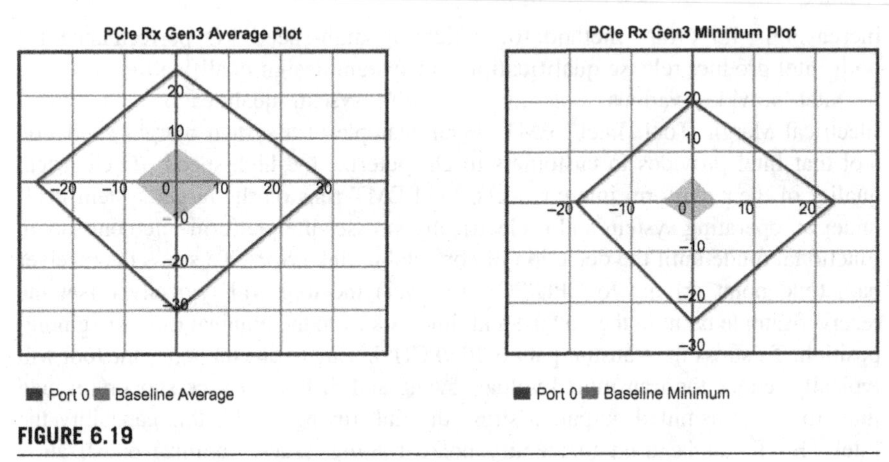

FIGURE 6.19

Typical high-speed I/O margin analysis result.

To determine the margin quality of a link, Intel provides average baseline reference values, minimum baseline values, and average customer reference board (CRB) margin values for each interface and data rate, where available. The average and minimum baseline values are used for pass/fail determination, while the average CRB values are used for reference only and represent the general average margins observed during internal testing on Intel's CRBs. When comparing margins to the reference values, two requirements must be met in order to obtain a pass indication. First, the average margins calculated for a given bus and data rate must meet or exceed the corresponding average baseline reference value. Second, all measured margin values must meet or exceed the appropriate minimum baseline reference values. Not meeting either of these conditions will result in a fail indication.

Several pass/fail examples are shown in Figure 6.20. The three sections represent the measured margins of different systems. Each data point in the plot represents the margin value for a particular run. The solid horizontal line indicates the average margin for each data grouping. The leftmost figure shows an average margin that is below the average baseline limit, but all margin values exceed the minimum baseline threshold. Since the average margin of the data group is below the average baseline limit, a fail result is given. The margins in the center figure also produce a fail indication because the average margin is below the average baseline threshold and at least one margin run is below the minimum baseline threshold. Finally, margins in the right-most figure produce a pass indication because the average margin exceeds the average baseline and no single margin value was below the minimum baseline.

There is no normative specification for the amount of system margin required on an interface. A pass indication denotes a lower risk of signal integrity issues for the given interface and data rate. It does not, however, guarantee flawless operation across HVM. Similarly, a fail indication does not necessarily mean a system will experience interface failures. It does, however, indicate an elevated risk of signal integrity issues across HVM. If a fail indication is reported, additional testing will likely be required to find the root cause issues and determine

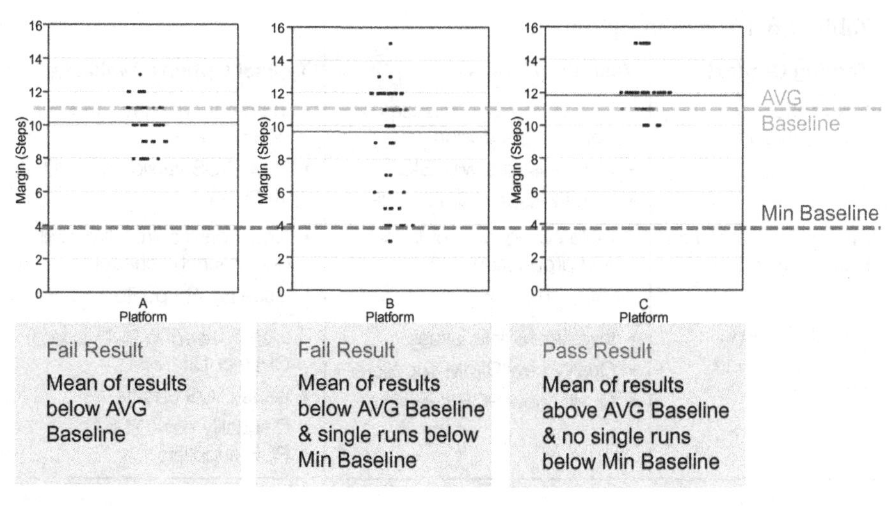

FIGURE 6.20

Example of high-speed I/O margin results.

the risk level. Additionally, Intel application engineers (AEs) can review the data and provide additional support. The pass/fail indicators from 5 × 5 testing can help to identify areas on which to focus design and validation efforts.

LOW-MARGIN DEBUG GUIDANCE

In the event that low margins are found during testing, it becomes important to identify the cause or causes of the low-margin behavior. Is the design faulty? Was the low-margin result due to the memory or I/O controller, the board, the DIMMs, or the end devices that were populated? Or are there other causes? In accordance with the scientific method, repeatability of low-margin behavior is important in order to prove that margin issues exist and to identify the cause(s) of the low-margin behavior (Table 6.5).

Table 6.5 Possible Root Cause When Low Margin is Observed

Low-Margin Observation	Possible Root Cause
Repeatable only on one board copy, not others.	Low-margin board may represent a corner condition or may indicate a faulty copy.
Repeatable on multiple copies of the board design.	May indicate improvement needed with the board layout.
Repeatable with one DIMM part number or end device but not others.	The affected DIMM part number or end device may need improvement.
Repeatable on different board designs.	Points to the BIOS or CPU as the likely root cause or may indicate a common design aspect that needs to be improved.
Low margin seen once but cannot be reproduced.	May be due to transient effect, such as poor seating of DIMMs or add-in card.

Table 6.6 Mitigation Option

Testing Context	Validation Goals	Typical Options Available
Pre-Production Board Version	• Find and fix board issues or optimize the design • Identify issues with BIOS • Identify issues with the CPU	• Intercept changes into new board spin • New BIOS version or CPU stepping
Production Candidate Board Version	• Build confidence for launch and production	• New board spin, which will delay launch schedule • Issue BIOS update
Final Production Board Version/In Market	• Investigate field issues • Qualify new DIMM suppliers • Qualify new PCB suppliers	• Identify need to RMA board, CPU, or DIMMs • Issue BIOS update • Disqualify new DIMM or PCB suppliers

It is also useful to understand the context of when low margins are found. Low margins may be expected with pre-production boards, for example, but not with boards that are already in the field. The context of the low margin can help to determine the next steps and their goals (Table 6.6).

SUMMARY

This chapter has provided some practical guidelines and considerations for designing and validating a system based on Intel silicon devices. The goal of validation is to determine the risk of shipping the product in volume. For I/O, this means confirming that the design meets the expected BER across the supported operating conditions and manufacturing variations. Intel provides a methodology and set of tools to accomplish this in an efficient manner. The bus margining methodology provides a holistic view of the health of the I/O bus. The use of statistical methods provides a balance between validation cost and product risk.

REFERENCES

[1] Casper BK, Balamurugan G, Jaussi JE, Kennedy J, Mansuri M. Future microprocessor interfaces: analysis, design and optimization. In: Proceedings of the IEEE custom integrated circuits conference. Sept. 2007. p. 479–86.

[2] Hall S, Heck H. Advanced signal integrity for high-speed digital designs. New Jersey: John Wiley & Sons; 2009.

Index

Note: Page numbers followed by "*f*" and "*t*" refer to figures and tables, respectively.

U

V

W